Using Social and Information Technologies for Disaster and Crisis Management

Murray E. Jennex
San Diego State University, USA

Information Science
REFERENCE

Managing Director:	Lindsay Johnston
Editorial Director:	Joel Gamon
Book Production Manager:	Jennifer Yoder
Publishing Systems Analyst:	Adrienne Freeland
Assistant Acquisitions Editor:	Kayla Wolfe
Typesetter:	Christy Fic
Cover Design:	Jason Mull

Published in the United States of America by
Information Science Reference (an imprint of IGI Global)
701 E. Chocolate Avenue
Hershey PA 17033
Tel: 717-533-8845
Fax: 717-533-8661
E-mail: cust@igi-global.com
Web site: http://www.igi-global.com

Library of Congress Cataloging-in-Publication Data

Using social and information technologies for disaster and crisis management / Murray E. Jennex, editor.
    p. cm.
 Includes bibliographical references and index.
 Summary: "This book highlights examples of disaster situations in recent years in which social and information technologies were useful in distributing and receiving information updates"--Provided by publisher.
  ISBN 978-1-4666-2788-8 (hbk.) -- ISBN 978-1-4666-2789-5 (ebook) -- ISBN 978-1-4666-2790-1 (print & perpetual access) 1. Emergency management--Information technology. 2. Emergency management--Information services. 3. Emergency communication systems. I. Jennex, Murray E., 1956-
 HV551.2.U85 2013
 363.34'8--dc23
                                        2012032561

British Cataloguing in Publication Data
A Cataloguing in Publication record for this book is available from the British Library.

The views expressed in this book are those of the authors, but not necessarily of the publisher.

Nigel Snoad, *Microsoft, USA*

Suha Ulgen, *United Nations Secretariat, USA*

Firoz Verjee, *NOAA, USA*

Peter Wooders, *Department of Internal Affairs, New Zealand*

Sisi Zlatanova, *Delft University of Technology, The Netherlands*

Table of Contents

Detailed Table of Contents

Chapter 1

John W. Barbrey, Longwood University, USA

In 2009, the U.S. Department of Education published an Action Guide for Emergency Management at Institutions of Higher Education (U.S. Department of Education, 2009). In 2006, the Virginia State Crime Commission issued a prescient "Final Report: Study on Campus Safety (HJR 122)" regarding Virginia's colleges and universities (Virginia State Crime Commission, 2006). Gray (2009) provided results from a "Columbine 10-Year Anniversary Survey", which reviewed recent campus safety improvements of 435 K-12 and university respondents. From the three documents, prescribed campus safety activities were identified that could be consistently found in the stated programs and policies on university websites. Of these activities, 18 separate criteria upon which a university's online emergency preparedness/safety/ security messages could be evaluated through content analysis were conceptualized (coding: 1= school has criterion, 0= does not), to estimate the quality of the overall preparedness message of each institution in the small sample (n = 99) of universities, representing all 50 states in 2010.

Chapter 2

Albert Y. Chen, University of Illinois at Urbana-Champaign, USA
Feniosky Peña-Mora, Columbia University, USA
Saumil J. Mehta, University of Illinois at Urbana-Champaign, USA
Stuart Foltz, Construction Engineering Research Laboratory, USA
Albert P. Plans, Universitat Politècnica de Catalunya, Spain
Brian R. Brauer, Illinois Fire Service Institute, USA
Scott Nacheman, Thornton Tomasetti, USA

The efficiency of Urban Search and Rescue operations depends on the supply of appropriate equipment and resources, and an efficient damage assessment facilitates deployment of these resources. This paper presents an Information Technology (IT) supported system for on-site data collection to communicate structural condition, track search and rescue status, and request and allocate appropriate resources. The system provides a unified interface for efficient gathering, storing, and sharing of building assessment information. Visualization and access of such information enable rescuers to respond to the disaster more efficiently with better situational awareness. The IT system also provides an interface for electronic resource requests to a geospatial resource repository service that enables a spatial disaster management environment for resource allocation. Request and deployment of critical resources through this system enables lifesaving efforts, with the appropriate equipment, operator, and materials, to become more efficient and effective. System development at the Illinois Fire Service Institute has shown promising results.

Tiziana Catarci, Sapienza Università di Roma, Italy
Massimiliano de Leoni, Technical University of Eindhoven, The Netherlands
Andrea Marrella, Sapienza Università di Roma, Italy
Massimo Mecella, Sapienza Università di Roma, Italy
Alessandro Russo, Sapienza Università di Roma, Italy
Renate Steinmann, Salzburg Research Forschungsgesellschaft mbH, Austria
Manfred Bortenschlager, Samsung Electronics Research Institute, UK

In complex emergency/disaster scenarios, persons from teams from various emergency-response organizations collaborate to achieve a common goal. In these scenarios, the use of smart mobile devices and applications can improve the collaboration dynamically. The lack of basic interaction principles can be dangerous, as it could increase the level of disaster or can make the efforts ineffective. This paper examines the main results of the project WORKPAD finished in December 2009. WORKPAD worked on a two-level architecture to support rescue operators during emergency management. The use of a user-centered design methodology during the entire development cycle has guaranteed that the architecture and resulting system meet end-user requirements. The feasibility of its use in real emergencies is also proven by a demonstration showcased with real operators. The paper includes qualitative and quantitative results and presents guidelines that can be useful in developing emergency-management systems.

Brianna Terese Hertzler, San Diego State University, USA
Eric Frost, San Diego State University, USA
George H. Bressler, San Diego State University, USA
Charles Goehring, San Diego State University, USA

The events of September 11, 2001, the Indian Ocean tsunami in 2004, and Hurricane Katrina in 2005 awakened American policymakers to the importance of the need for emergency management. This paper explains how a cloud computing environment can support social networks and logistical coordination on a global scale during crises. Basic cloud computing functionality is covered to show how social networks can connect seamlessly to work together with profound interoperability. Lastly, the benefits of a cloud computing solution is presented as the most cost-effective, efficient, and secure method of communication during a disaster response, with the unique capability of being able to support a global community through its massive scalability.

Dan Harnesk, Luleå University of Technology, Sweden
Heidi Hartikainen, Luleå University of Technology, Sweden

This paper draws on the socio-technical research tradition in information systems to re-conceptualize the information security in emergency response. A conceptual basis encompassing the three layers—technical, cognitive, and organizational—is developed by synthesizing Actor Network Theory and Theory of Organizational Routines. This paper makes the assumption that the emergency response context is built on the relationship between association and connectivity, which continuously shapes the emergency action network and its routines. Empirically, the analysis is based on a single case study conducted across three emergency departments. The data thus collected on information security, emergency department routines, and emergency actions is used to theorize specifically on the association/connectivity relationship. The resultant findings point to the fact that information security layers have a meaning in emergency response that is different from mainstream definitions of information security.

Chapter 6

Andrea Kavanaugh, Virginia Tech, USA

Steven Sheetz, Virginia Tech, USA

Francis Quek, Virginia Tech, USA

B. Joon Kim, Indiana University-Purdue University, USA

Many proposed technological solutions to emergency response during disasters involve the use of cellular telephone technology. However, cell phone networks quickly become saturated during and/or immediately after a disaster and remain saturated for critical periods. This study investigated cell phone use by Virginia Tech students, faculty and staff during the shootings on April 16, 2007 to identify patterns of communication with social network ties. An online survey was administered to a random sample pool to capture communications behavior with social ties during the day of these tragic events. The results show that cell phones were the most heavily used communication technology by a majority of respondents (both voice and text messaging). While text messaging makes more efficient use of bandwidth than voice, most communication on 4/16 was with parents, since the majority of the sample is students, who are less likely to use text messaging. These findings should help in understanding how cell phone technologies may be utilized or modified for emergency situations in similar communities.

Chapter 7

Suradej Intagorn, USC Information Sciences Institute, USA

Kristina Lerman, USC Information Sciences Institute, USA

Up-to-date geospatial information can help crisis management community to coordinate its response. In addition to data that is created and curated by experts, there is an abundance of user-generated, user-curated data on Social Web sites such as Flickr, Twitter, and Google Earth. User-generated data and metadata can be used to harvest knowledge, including geospatial knowledge that will help solve real-world problems including information discovery, geospatial information integration and data management. This paper proposes a method for acquiring geospatial knowledge in the form of places and relations between them from the user-generated data and metadata on the Social Web. The key to acquiring geospatial knowledge from social metadata is the ability to accurately represent places. The authors describe a simple, efficient algorithm for finding a non-convex boundary of a region from a sample of points from that region. Used within a procedure that learns part-of relations between places from real-world data extracted from the social photo-sharing site Flickr, the proposed algorithm leads to more precise relations than the earlier method and helps uncover knowledge not contained in expert-curated geospatial knowledge bases.

Chapter 8

Teresa Onorati, Universidad Carlos III de Madrid, Spain

Alessio Malizia, Universidad Carlos III de Madrid, Spain

Paloma Díaz, Universidad Carlos III de Madrid, Spain

Ignacio Aedo, Universidad Carlos III de Madrid, Spain

The interaction design for web emergency management information systems (WEMIS) is an important aspect to keep in mind due to the criticality of the domain: decision making, updating available resources, defining a task list, and trusting in proposed information. A common interaction design strategy for WEMIS seems to be needed, but currently there are few references in literature. The aim of this study is to contribute to this lack with a set of interactive principles for WEMIS. From the emergency point of view, existing WEMIS have been analyzed to extract common features and to design interactive principles for emergency. Furthermore, the authors studied design principles extracted from a well-known (DERMIS) model relating them to emergency phases and features. The result proposed here is a set of design principles for supporting interactive properties for WEMIS. Finally, two case studies have been considered as applications of proposed design principles.

To aid emergency response teams in training and planning for potential community-wide emergency crises, two coordinated research teams centered in King County, Washington have developed software-based tools to provide cognitive aids for improved planning and training for emergency response scenarios. After reporting the results previously of using the tools in pilot studies of increasing complexity, the implementation teams have been searching out community-wide emergency response teams working on emergency response plans that might benefit from use of the tools. In this paper, the authors describe the tools, the application of them to a countywide hospital evacuation scenario, and the evaluation of their value to emergency responders for improving situation awareness and insight generation.

As part of its expanding role, particularly as an agent of peace building, the United Nations (UN) actively participates in the implementation of measures to prevent and manage crisis/disaster situations. The purpose of such an approach is to empower the victims, protect the environment, rebuild communities, and create employment. However, real world crisis management situations are complex given the multiple interrelated interests, actors, relations, and objectives. Recent studies in healthcare contexts, which also have dynamic and complex operations, have shown the merit and benefits of employing various tools and techniques from the domain of knowledge management (KM). Hence, this paper investigates three distinct natural crisis situations (the 2010 Haiti Earthquake, the 2004 Boxing Day Asian Tsunami, and the 2001 Gujarat Earthquake) with which the United Nations and international aid agencies have been and are currently involved, to identify recurring issues which continue to provide knowledge-based impediments. Major findings from each case study are analyzed according to the estimated impact of identified impediments. The severity of the enumerated knowledge-based issues is quantified and compared by means of an assigned qualitative to identify the most significant attribute.

Chapter 11

In this paper, the authors present a business rules-based decision support system for the allocation of traumatized patients. The assignment of patients to vehicles and hospitals is a task that requires detailed up-to-date information. At the same time, it has to be carried out quickly. The authors propose supporting medical staff with an IT system. The proposed system could be used in cases of mass incidents, as it is problematic, but essential, to provide all injured with adequate healthcare as fast as possible. The contribution is a system based on business rules, which is a novel approach in this context. Its feasibility is proven by prototypic implementation. In this paper, the authors describe the development project's background as well as the system's requirements and implementation details. The authors present an exemplary scenario to show the strengths of the proposed approach.

Having the right information at the right time is crucial to make decisions during emergency response. To fulfill this requirement, emergency management systems must provide emergency managers with knowledge management and visualization tools. The goal is twofold: on one hand, to organize knowledge coming from different sources, mainly the emergency response plans (the formal knowledge) and the information extracted from the emergency development (the contextual knowledge), and on the other hand, to enable effective access to information. Formal and contextual knowledge sets are mostly disjoint; however, there are cases in which a formal knowledge piece may be updated with some contextual information, constituting composite knowledge. In this paper, the authors extend a knowledge framework with the notion of composite knowledge, and use spatial hypertext to visualize this type of knowledge. The authors illustrate the proposal with a case study on accessing to information during an emergency response in an underground transportation system.

Chapter 13

This paper considers how emergency response organizations utilize available social media technologies to communicate with the public in emergencies and to potentially collect valuable information using the public as sources of information on the ground. The authors discuss the use of public social media tools from the emergency management professional's viewpoint with a particular focus on the use of Twitter. Limited research has investigated Twitter usage in crisis situations from an organizational perspective. This paper contributes to the understanding of organizational innovation, risk communication, and technology adoption by emergency management. An in-depth longitudinal case study of Public Information Officers (PIO) of the Los Angeles Fire Department highlights the importance of the information evangelist within emergency management organizations and details the challenges those organizations face engaging with social media and Twitter. This article provides insights into practices and challenges of new media implementation for crisis and risk management organizations.

Chapter 14

Tina Comes, Karlsruhe Institute of Technology (KIT), Germany

Niek Wijngaards, Thales Research and Technology & D-CIS Lab, The Netherlands

Michael Hiete, Karlsruhe Institute of Technology (KIT), Germany

Claudine Conrado, Thales Research and Technology & D-CIS Lab, The Netherlands

Frank Schultmann, Karlsruhe Institute of Technology (KIT), Germany

Decision-making in emergency management is a challenging task as the consequences of decisions are considerable, the threatened systems are complex and information is often uncertain. This paper presents a distributed system facilitating better-informed decision-making in strategic emergency management. The construction of scenarios provides a rationale for collecting, organising, and processing information. The set of scenarios captures the uncertainty of the situation and its developments. The relevance of scenarios is ensured by gearing the scenario construction to assessing alternatives, thus avoiding time-consuming processing of irrelevant information. The scenarios are constructed in a distributed setting allowing for a flexible adaptation of reasoning (principles and processes) to both the problem at hand and the information available. This approach ensures that each decision can be founded on a coherent set of scenarios. The theoretical framework is demonstrated in a distributed decision support system by orchestrating experts into workflows tailored to each specific decision.

Chapter 15

Austin W. Howe, San Diego State University, USA

Murray E. Jennex, San Diego State University, USA

George H. Bressler, San Diego State University, USA

Eric G. Frost, San Diego State University, USA

Can populations self organize a crisis response? This is a field report on the first two efforts in a continuing series of exercises termed "Exercise24 or x24." The first Exercise 24 focused on Southern California, while the second (24 Europe) focused on the Balkan area of Eastern Europe. These exercises attempted to demonstrate that self-organizing groups can form and respond to a crisis using low-cost social media and other emerging web technologies. Over 10,000 people participated in X24 while X24 Europe had over 49,000 participants. X24 involved people from 79 nations while X24 Europe officially included participants from at least 92 countries. Exercise 24 was organized by a team of workers centered at the SDSU Viz Center including significant support from the US Navy as well as other military and Federal organizations. Dr. George Bressler, Adjunct Faculty member at the Viz Center led both efforts. Major efforts from senior professionals EUCOM and NORTHCOM contributed significantly to the preparation for and success of both X24 and especially X24 Europe. This paper presents lessons learned and other experiences gained through the coordination and performance of Exercise24.

Chapter 16

Alexander Smirnov, St. Petersburg Institute for Informatics & Automation of the Russian Academy of Sciences, Russia

Tatiana Levashova, St. Petersburg Institute for Informatics & Automation of the Russian Academy of Sciences, Russia

Nikolay Shilov, St. Petersburg Institute for Informatics & Automation of the Russian Academy of Sciences, Russia

Alexey Kashevnik, St. Petersburg Institute for Informatics & Automation of the Russian Academy of Sciences, Russia

Ubiquitous computing opens new possibilities in various aspects of human activities. The paper proposes an approach to emergency situation response that benefits from the ubiquitous computing. The approach is based on utilizing profiles to facilitate the coordination of the activities of the emergency response operation members. The major approach underlying idea is to represent the operation members jointly with information sources as a network of services that can be configured via negotiation of participating parties. Such elements as profile structure, role-based emergency response, negotiation scenarios, and negotiation protocols are described in detail.

Preface

Are there more disasters and crises or are we just more aware of those that occur? An interesting question but one whose answer, while important, is not the purpose of this preface. What is important is that more crises or not, we are more aware of them and we expect those with authority to respond well to these crises. 2011 and 2012 saw hurricanes, wild fires, war, mass shootings, floods, draughts, etc. Businesses had to deal with crises such as bad products, bad public relation decisions, communication gaffes, industrial accidents, service outages, etc. Individuals, schools, organizations of all kinds all had to deal with various sorts of crises. So while we suspect there are more crises now than in the past, we definitely are more aware of them; and because we see them play out in real time on television, the Internet, or social media; we are more demanding in and critical of the observed crisis response. 2012 saw the Italian courts convict scientists for giving bad emergency advice showing that we are expecting our experts to give advice that is reasonable and that society will hold the experts accountable. A precedent that may have far reaching consequences as experts become reluctant to advise in uncertain situations. 2012 also saw a public skeptical of the ability of FEMA (Federal Emergency Management Agency) after the debacle of the Hurricane Katrina response; do a better job responding to Hurricane Sandy. Crisis response and our perception of how our leaders responded contributed to the election of the President of the United States when Barack Obama was perceived to be presidential and a good leader in a crisis and continue to influence public perception of government officials.

More access to information is bringing the world's population closer together. The tsunami in Japan and the crisis at Fukushima were witnessed by viewers around the world. We are also seeing the advent of crowdsourcing crisis response. A crisis occurs, we see the response, and we get involved. Citizens could text in a donation to the Red Cross for Hurricane Sandy victims (as well as other crises) with the result that micro donations are changing the way crisis response is being funded. Technology is enabling this real time observation and participation. Drones with digital cameras bring events to our screens and live data to crisis managers and first responders. More bandwidth is enabling the transmission of more data, information, and knowledge to crisis response managers and even the first responders in the field. Better sensors, models, displays, maps, and communications are being incorporated into crisis response information systems. These improved crisis response information systems are improving our ability to respond and to manage the response. These crisis response information systems are also impacting our ability to make decisions. Managers are starting to display decision paralysis. Decision paralysis occurs when decision makers need more and more data before they can make a decision. The fear is that they will miss something that could cause injury or death to a first responder or a victim. I recently read a criticism of commanders in Afghanistan as it is taking them longer to evacuate personnel than it did in Vietnam. The article blamed lack of leadership and initiative but I believe it is decision paralysis. Commanders in Vietnam had limited data availability, especially when compared to commanders in Afghanistan. I am a former United States Navy officer, trained in the late 1970s. At that time we were trained to take into account all available data and information and then take action. When data and information are limited you can assimilate it quicker and make quicker decisions. However, as the flood of data, information, and knowledge grows it takes longer to process it and determine appropriate courses of action. Also, managers and commanders are being held liable for the decisions they make using the perception of "what they should have known." This is where decision paralysis becomes dangerous, the need to always have more or even perfect data, information, and knowledge before a decision can be made. The result is longer decision making times. Additionally, there is great concern that the L'Aquila, Italy convictions of the Italian government's seismic experts will further reduce the willingness of crisis managers and commanders to make decisions without having all available data, information, and knowledge and leading many of them to wait for more data, information, and knowledge rather than acting on what they have.

We are also becoming so reliant on these crisis response information systems and have embedded them as an imperative aspect of successful decision making practices that when they fail, the repercussions can have devastating effects. Preventative measures cannot fully protect organizations from such mishaps, which is why effective crisis management protocols are critical for businesses' survival. This is leading to a research agenda with respect to information systems for crisis response that explores questions such as:

- How do we validate that the real time data, information, and knowledge being collected and provided to crisis managers are valid?
- How do we process the large amounts of real time data, information, and knowledge being collected into actionable intelligence?
- How do we prevent decision paralysis and information overload in crisis response?
- What role do social media have in crisis response?
- What technologies can be applied to assist decision making in crisis response?
- What policies and planning should be done to prepare organizations to respond to crises?
- What systems do we need for situational awareness and early warning of crises?
- How do we analyze and use communications data to improve crisis response?
- How do we quickly capture lessons learned and generate best practices from crises, and then disseminate them to the crisis response community?
- How do we prepare individuals, organizations, and communities to be ready and self-sustaining during crises?
- What decision models do we need to use?
- What processes, procedures, and techniques work in crisis response?
- What organizational structures work best for crisis response organizations?

This book incorporates research that addresses much of the above research agenda. The following chapters provide readers with an inside view of the current discussion of trends in crisis response and disaster recovery methods in a wide variety of situations and industries, including education, healthcare, corporate, and environmental sectors.

John W. Barbrey begins by "Evaluating Campus Safety Messages at 99 Public Universities in 2010." In 2009, the U.S. Department of Education published an Action Guide for Emergency Management at Institutions of Higher Education (U.S. Department of Education, 2009). In 2006, the Virginia State Crime Commission issued a prescient "Final Report: Study on Campus Safety (HJR 122)" regarding Virginia's colleges and universities (Virginia State Crime Commission, 2006). Gray (2009) provided results from a "Columbine 10-Year Anniversary Survey," which reviewed recent campus safety improvements of 435 K-12 and university respondents. From the three documents, prescribed campus safety activities were identified that could be consistently found in the stated programs and policies on university websites. Of these activities, 18 separate criteria upon which a university's online emergency preparedness/ safety/ security messages could be evaluated through content analysis were conceptualized (coding: 1= school has criterion, 0= does not), to estimate the quality of the overall preparedness message of each institution in the small sample (n = 99) of universities, representing all 50 states in 2010.

"Equipment Distribution for Structural Stabilization and Civilian Rescue" by Albert Y. Chen et al. investigates how the efficiency of Urban Search and Rescue operations depends on the supply of appropriate equipment and resources, and an efficient damage assessment facilitates deployment of these resources. This chapter presents an Information Technology (IT) supported system for on-site data collection to communicate structural condition, track search and rescue status, and request and allocate appropriate resources. The system provides a unified interface for efficient gathering, storing, and sharing of building assessment information. Visualization and access of such information enable rescuers to respond to the disaster more efficiently with better situational awareness. The IT system also provides an interface for electronic resource requests to a geospatial resource repository service that enables a spatial disaster management environment for resource allocation. Request and deployment of critical resources through this system enables lifesaving efforts, with the appropriate equipment, operator, and materials, to become more efficient and effective. A system development at the Illinois Fire Service Institute has shown promising results.

Next, Tiziana Catarci et al. describe "WORKPAD: Process Management and Geo-Collaboration Help Disaster Response." In complex emergency/disaster scenarios, persons from teams from various emergency-response organizations collaborate to achieve a common goal. In these scenarios, the use of smart mobile devices and applications can improve the collaboration dynamically. The lack of basic interaction principles can be dangerous, as it could increase the level of disaster or can make the efforts ineffective. This chapter examines the main results of the project WORKPAD, finished in December 2009. WORKPAD worked on a two-level architecture to support rescue operators during emergency management. The use of a user-centered design methodology during the entire development cycle has guaranteed that the architecture and resulting system meet end-user requirements. The feasibility of its use in real emergencies is also proven by a demonstration showcased with real operators. The chapter includes qualitative and quantitative results and presents guidelines that can be useful in developing emergency-management systems.

Chapter 4

The events of September 11, 2001, the Indian Ocean tsunami in 2004, and Hurricane Katrina in 2005 awakened American policymakers to the importance of the need for emergency management. "Experience Report: Using a Cloud Computing Environment during Haiti and Exercise24" by Brianna Terese Hertzler et al. explains how a cloud computing environment can support social networks and logistical coordination on a global scale during crises. Basic cloud computing functionality is covered to show how social networks can connect seamlessly to work together with profound interoperability. Lastly, the benefits of a cloud computing solution is presented as the most cost-effective, efficient, and secure method of communication during a disaster response, with the unique capability of being able to support a global community through its massive scalability.

Chapter 5, "Multi-Layers of Information Security in Emergency Response" by Dan Harnesk and Heidi Hartikainen, draws on the socio-technical research tradition in information systems to re-conceptualize the information security in emergency response. A conceptual basis encompassing the three layers—technical, cognitive, and organizational—is developed by synthesizing Actor Network Theory and Theory of Organizational Routines. This chapter makes the assumption that the emergency response context is built on the relationship between association and connectivity, which continuously shapes the emergency action network and its routines. Empirically, the analysis is based on a single case study conducted across three emergency departments. The data thus collected on information security, emergency department routines, and emergency actions is used to theorize specifically on the association/connectivity relationship. The resultant findings point to the fact that information security layers have a meaning in emergency response that is different from mainstream definitions of information security.

Chapter 6

Many proposed technological solutions to emergency response during disasters involve the use of cellular telephone technology. However, cell phone networks quickly become saturated during and/or immediately after a disaster and remain saturated for critical periods. In "Cell Phone Use with Social Ties During Crises: The Case of the Virginia Tech Tragedy," Andrea Kavanaugh et al. investigated cell phone use by Virginia Tech students, faculty, and staff during the shootings on April 16, 2007 to identify patterns of communication with social network ties. An online survey was administered to a random sample pool to capture communications behavior with social ties during the day of these tragic events. The results show that cell phones were the most heavily used communication technology by a majority of respondents (both voice and text messaging). While text messaging makes more efficient use of bandwidth than voice, most communication on 4/16 was with parents, since the majority of the sample is students, who are less likely to use text messaging. These findings should help in understanding how cell phone technologies may be utilized or modified for emergency situations in similar communities.

Next, Suradej Intagorn and Kristina Lerman explore "Mining Geospatial Knowledge on the Social Web." Up-to-date geospatial information can help the crisis management community to coordinate its response. In addition to data that is created and curated by experts, there is an abundance of user-generated, user-curated data on Social Web sites such as Flickr, Twitter, and Google Earth. User-generated data and metadata can be used to harvest knowledge, including geospatial knowledge that will help solve real-world problems including information discovery, geospatial information integration, and data management. This chapter proposes a method for acquiring geospatial knowledge in the form of places and relations between them from the user-generated data and metadata on the Social Web. The key to acquiring geospatial knowledge from social metadata is the ability to accurately represent places. The authors describe a simple, efficient algorithm for finding a non-convex boundary of a region from a sample of points from that region. Used within a procedure that learns part-of relations between places from real-world data extracted from the social photo-sharing site Flickr, the proposed algorithm leads to more precise relations than the earlier method and helps uncover knowledge not contained in expert-curated geospatial knowledge bases.

Chapter 8

The interaction design for web emergency management information systems (WEMIS) is an important aspect to keep in mind due to the criticality of the domain: decision making, updating available resources, defining a task list, and trusting in proposed information. A common interaction design strategy for WEMIS seems to be needed, but currently there are few references in literature. The aim of the chapter by Teresa Onorati et al., "Interaction Design Principles for Web Emergency Management Information Systems," is to contribute to this lack with a set of interactive principles for WEMIS. From the emergency point of view, existing WEMIS have been analyzed to extract common features and to design interactive principles for emergency. Furthermore, the authors studied design principles extracted from a well-known (DERMIS) model relating them to emergency phases and features. The result proposed here is a set of design principles for supporting interactive properties for WEMIS. Finally, two case studies have been considered as applications of proposed design principles.

In Chapter 9, "RimSim Response Hospital Evacuation: Improving Situation Awareness and Insight through Serious Games Play and Analysis," Bruce Campbell and Chris Weaver describe how, to aid emergency response teams in training and planning for potential community-wide emergency crises, two coordinated research teams centered in King County, Washington have developed software-based tools to provide cognitive aids for improved planning and training for emergency response scenarios. After reporting the results previously of using the tools in pilot studies of increasing complexity, the implementation teams have been searching out community-wide emergency response teams working on emergency response plans that might benefit from use of the tools. In this chapter, the authors describe the tools, the application of them to a countywide hospital evacuation scenario, and the evaluation of their value to emergency responders for improving situation awareness and insight generation.

"Knowledge-Based Issues for Aid Agencies in Crisis Scenarios: Evolving from Impediments to Trust" by Rajeev K. Bali investigates crisis management from an international perspective. As part of its expanding role, particularly as an agent of peace building, the United Nations (UN) actively participates in the implementation of measures to prevent and manage crisis/disaster situations. The purpose of such an approach is to empower the victims, protect the environment, rebuild communities, and create employment. However, real world crisis management situations are complex given the multiple interrelated interests, actors, relations, and objectives. Recent studies in healthcare contexts, which also have dynamic and complex operations, have shown the merit and benefits of employing various tools and techniques from the domain of knowledge management (KM). Hence, this chapter investigates three distinct natural crisis situations (the 2010 Haiti Earthquake, the 2004 Boxing Day Asian Tsunami, and the 2001 Gujarat Earthquake) with which the United Nations and international aid agencies have been and are currently involved, to identify recurring issues which continue to provide knowledge-based impediments. Major findings from each case study are analyzed according to the estimated impact of identified impediments. The severity of the enumerated knowledge-based issues is quantified and compared by means of an assigned qualitative to identify the most significant attribute.

In "Supporting the Allocation of Traumatized Patients with a Decision Support System," Tim A. Majchrzak et al. present a business rules-based decision support system for the allocation of traumatized patients. The assignment of patients to vehicles and hospitals is a task that requires detailed up-to-date information. At the same time, it has to be carried out quickly. The authors propose supporting medical staff with an IT system. The proposed system could be used in cases of mass incidents, as it is problematic, but essential, to provide all injured with adequate healthcare as fast as possible. The contribution is a system based on business rules, which is a novel approach in this context. Its feasibility is proven by prototypic implementation. In this chapter, the authors describe the development project's background as well as the system's requirements and implementation details. The authors present an exemplary scenario to show the strengths of the proposed approach.

Having the right information at the right time is crucial to make decisions during emergency response. To fulfill this requirement, emergency management systems must provide emergency managers with knowledge management and visualization tools. In "Visualizing Composite Knowledge in Emergency Responses using Spatial Hypertext," José H. Canós describe this goal as twofold: on one hand, to organize knowledge coming from different sources, mainly the emergency response plans (the formal knowledge) and the information extracted from the emergency development (the contextual knowledge), and on the other hand, to enable effective access to information. Formal and contextual knowledge sets are mostly disjoint; however, there are cases in which a formal knowledge piece may be updated with some contextual information, constituting composite knowledge. In this chapter, the authors extend a knowledge framework with the notion of composite knowledge, and use spatial hypertext to visualize this type of knowledge. The authors illustrate the proposal with a case study on accessing to information during an emergency response in an underground transportation system.

"Emergency Management, Twitter, and Social Media Evangelism" by Mark Latonero and Irina Shklovski considers how emergency response organizations utilize available social media technologies to communicate with the public in emergencies and to potentially collect valuable information using the public as sources of information on the ground. The authors discuss the use of public social media tools from the emergency management professional's viewpoint with a particular focus on the use of Twitter. Limited research has investigated Twitter usage in crisis situations from an organizational perspective. This chapter contributes to the understanding of organizational innovation, risk communication, and technology adoption by emergency management. An in-depth longitudinal case study of Public Information Officers (PIO) of the Los Angeles Fire Department highlights the importance of the information evangelist within emergency management organizations and details the challenges those organizations face engaging with social media and Twitter. This chapter provides insights into practices and challenges of new media implementation for crisis and risk management organizations.

Decision-making in emergency management is a challenging task as the consequences of decisions are considerable, the threatened systems are complex, and information is often uncertain. "A Distributed Scenario-Based Decision Support System for Robust Decision-Making in Complex Situations" by Tina Comes presents a distributed system facilitating better-informed decision-making in strategic emergency management. The construction of scenarios provides a rationale for collecting, organizing, and processing information. The set of scenarios captures the uncertainty of the situation and its developments. The relevance of scenarios is ensured by gearing the scenario construction to assessing alternatives, thus avoiding time-consuming processing of irrelevant information. The scenarios are constructed in a distributed setting, allowing for a flexible adaptation of reasoning (principles and processes) to both the problem at hand and the information available. This approach ensures that each decision can be founded on a coherent set of scenarios. The theoretical framework is demonstrated in a distributed decision support system by orchestrating experts into workflows tailored to each specific decision.

In "Exercise24: Using Social Media for Crisis Response," Austin W. Howe et al. ask, "Can populations self organize a crisis response?" This chapter is a field report on the first two efforts in a continuing series of exercises termed "Exercise24 or X24." The first Exercise24 focused on Southern California, while the second (24 Europe) focused on the Balkan area of Eastern Europe. These exercises attempted to demonstrate that self-organizing groups can form and respond to a crisis using low-cost social media and other emerging web technologies. Over 10,000 people participated in X24, while X24 Europe had over 49,000 participants. X24 involved people from 79 nations while X24 Europe officially included participants from at least 92 countries. Exercise24 was organized by a team of workers centered at the SDSU Viz Center, including significant support from the US Navy as well as other military and Federal organizations. Dr. George Bressler, Adjunct Faculty member at the Viz Center led both efforts. Major efforts from senior professionals EUCOM and NORTHCOM contributed significantly to the preparation for and success of both X24 and especially X24 Europe. This chapter presents lessons learned and other experiences gained through the coordination and performance of Exercise24.

Ubiquitous computing opens new possibilities in various aspects of human activities. The last chapter, "Ubiquitous Computing for Personalized Decision Support in Emergency" by Alexander Smirnov, proposes an approach to emergency situation response that benefits from the ubiquitous computing. The approach is based on utilizing profiles to facilitate the coordination of the activities of the emergency response operation members. The major approach underlying idea is to represent the operation members jointly with information sources as a network of services that can be configured via negotiation of participating parties. Such elements as profile structure, role-based emergency response, negotiation scenarios, and negotiation protocols are described in detail.

Chapter 1
Evaluating Campus Safety Messages at 99 Public Universities in 2010

John W. Barbrey
Longwood University, USA

ABSTRACT

In 2009, the U.S. Department of Education published an Action Guide for Emergency Management at Institutions of Higher Education (U.S. Department of Education, 2009). In 2006, the Virginia State Crime Commission issued a prescient "Final Report: Study on Campus Safety (HJR 122)" regarding Virginia's colleges and universities (Virginia State Crime Commission, 2006). Gray (2009) provided results from a "Columbine 10-Year Anniversary Survey", which reviewed recent campus safety improvements of 435 K-12 and university respondents. From the three documents, prescribed campus safety activities were identified that could be consistently found in the stated programs and policies on university websites. Of these activities, 18 separate criteria upon which a university's online emergency preparedness/safety/security messages could be evaluated through content analysis were conceptualized (coding: 1= school has criterion, 0= does not), to estimate the quality of the overall preparedness message of each institution in the small sample (n = 99) of universities, representing all 50 states in 2010.

INTRODUCTION

This academic arrived at Longwood University in Farmville, Virginia in 2008, in the Criminal Justice Studies program. The attention given within the Commonwealth of Virginia to campus safety and security issues, given the shootings at

Virginia Polytechnic Institute and State University (Virginia Tech) in April 2007, quickly became palpable due to Longwood's status as an institution in the same state educational system as Virginia Tech. Consequently, this research was the product of personal academic and professional curiosity, beginning with the salient question: are universities in the United States doing everything they should in preparation for their next emergency?

DOI: 10.4018/978-1-4666-2788-8.ch001

Quarantelli (1997) and Perry and Lindell (2003) recommended similar guidelines for a general disaster planning process, while Templeton et al. (2009) and Fields (2009) suggested steps for creating college campus emergency plans. These recommendations, when combined with recent National Incident Management System (NIMS) or Incident Command System (ICS) protocols fostered by the Federal Emergency Management Agency (FEMA), suggested an emergency planning process and written plan for a college campus that incorporates the following: data (e.g., threats, supplies, infrastructure); formal step-by-step procedures for each organizational sub-unit; relevant laws, regulations, and mandates; inter- and intra-organizational cooperation using both committees and a quasi-military chain-of-command for communication purposes; use of new technology; a focus on good decision-making in a politically charged environment during different scenarios based on past precedent; ongoing training and revision; and a willingness to be flexible and ignore past precedent. With all these potential variables, how is any leader at any college supposed to develop a single emergency plan, revamp the planning process, or know that she/he is doing everything possible to protect her/his campus? One possible solution is the same as it was decades ago: find the appropriate federal manual, state policy or program, or practitioner journal, and use the documents as templates for one's own organizational planning and programmatic needs.

For example, in 2009, the U.S. Department of Education (ED) published an *Action Guide for Emergency Management at Institutions of Higher Education*. In 2006, the Virginia State Crime Commission (the Commission), at the request of the Commonwealth's General Assembly two years earlier, issued a prescient "Final Report: Study on Campus Safety" regarding Virginia's colleges and universities. Within the report was a list of twenty-seven "best practices" used by schools studied by the Commission. Gray (2009), in *Campus Safety*

Magazine, provided results from a "Columbine 10-Year Anniversary Survey", in which the publication evaluated the campus safety improvements of 435 K-12 and university respondents.

The premise behind this study's research methods was to create a nationwide snapshot of college campus safety practices that could provide valid comparisons to this researcher's institution. Therefore, the small sample (n=99) used in this study includes only publicly-funded, four-year universities or colleges. To create a national sample, and to allow for comparisons between large and relatively small schools (Longwood has a total enrollment of approximately 4800), each state's "flagship" university and its smallest university (based upon undergraduate enrollment) are used in the sample.

Using the ED (2009) manual, the Commission's (2006) report, and Gray's (2009) survey, a list was created of prescribed campus safety/security/emergency preparedness activities that could (in this researcher's estimation) possibly be found in the form of stated programs and policies on university websites in 2010. After a preliminary review of a handful of university web sites, using the list of prescribed safety activities as a guide, 18 distinct concepts were identified that commonly appeared in the form of online written safety/security/preparedness statements. Using the 18 concepts, or *criteria* (listed in Table 2), an index score was created to evaluate the overall stated emergency preparedness/safety/security messages of each institution in the sample in the summer of 2010.

GOVERNMENT EMERGENCY PLANS BY THE BOOK

In many ways, the public preparedness/response/management message in the United States, from post World War II to post-9/11, has not changed. Although U.S. federal entities are now prescribing an *all-hazards* approach to their emergency preparedness activities, this multi-threat approach/ paradigm/model is not new, particularly when one looks at examples of the preparatory recommendations in federal manuals and other publications from the 1950s to present. Despite constants that exist in preparedness messages to the U.S. public over time (e.g., have a medical kit and appropriate shelter), emergency manuals and plans do evolve over time. They are re-written depending upon the known possible cataclysmic dangers posed from internal sources within territorial borders and from external or foreign sources, and depending upon lessons learned from ongoing evaluations of written statutes, policies, and procedures when plans are put into practice.

The current list of threats to the American public includes all possible natural (e.g., flood, tornado, forest fire, and earthquake) and man-made (e.g., acts of terrorism, biohazards, chemical spills) dangers the government (at all levels, both elected and appointed officials) is expected to attempt to either prevent or mitigate. Perry and Lindell (2003) write, in a review of the emergency planning process literature, that "community emergency preparedness [consists of] planning, training, and written plans…." (p. 337); and "[p]reparedness is a state of readiness to respond to environmental threats" (p. 338). However, there is no such thing as one all-encompassing emergency plan that covers every possible situation. They concede "…there may not be a direct translation of results from [natural and technological] disasters to terrorist events" (p. 337) due to the possible variation in size and scope of terrorist attacks.

Forms of threats can be conceptualized in multiple ways, complicating organizational planning, policy, and practice. For example, a *terrorist act* can be a pipe bomb that injures a few people, a dirty bomb that irradiates thousands, or a cyber attack on a corporation. Application of the USA PATRIOT Act (Uniting and Strengthening America by Providing Appropriate Tools Required to Intercept and Obstruct Terrorism Act of 2001) can hypothetically sentence a successful computer

virus creator or hacker as if he or she were plotting to blow something up with dynamite.

Worse yet for policymakers is the reality that best efforts *pre-event* are no guarantee of agency or individual effectiveness or efficiency *during* an event. Hurricane Katrina demonstrated in 2005 that it might not be possible for governments to fully prepare for large-scale emergencies, especially if the public is not receptive to emergency preparedness messages, and if all levels of government are ill-prepared for the scope of a disaster. No government entity can fully predict or prevent the impact of a determined, mentally disturbed, lone gunman on a mission. College campuses have porous perimeters, and it is often impossible to prevent pedestrians from entering, who may be criminals in search of targets.

Public emergency preparedness since the 1940s in the United States is a good example of post-WWII federalism. The national dialogue about public preparedness over almost seven decades has been led by federal agencies, presidential administrations, and national public-service organizations such as the American Red Cross. Due to concerns about national security and international threats, the federal agencies have been the primary authors of the ongoing emergency preparedness public messages and concomitant manuals. State and local governments still focus much of their preparations upon regional natural disasters and weather-related emergencies (Redlener & Berman, 2006). Despite the logical separation of roles, given the possible scale of emergencies (the beginning of WWIII versus a tornado that destroys a single neighborhood), the overall preparedness message at lower levels of government (as in the policy areas of education, transportation, and healthcare) are directly (through agency policy mandates and statutes) and indirectly (through grants and other expenditures) influenced by decisions at the federal level.

As any U.S. citizen old enough to remember knows, during the height of the Cold War, especially during the 1950s with the Korean War, and the '60s with the Cuban Missile Crisis and Vietnam, the national preparedness message was largely driven by the threat of nuclear war. In 1950, the National Security Resources Board (NSRB), within Truman's Executive Office of the President, published sobering information on the potential impacts of a nuclear explosion, based upon lessons learned from the bombing of Nagasaki and Hiroshima (NSRB, 1950a). The NSRB also created a 154-page manual titled *United States Civil Defense* (NSRB, 1950b). The manual demonstrated a top-down, hierarchical policy approach to the nuclear threat and an emphasis on community planning at the state and local levels, "on whom rest the primary responsibility for civil defense" (p. III). The local communities were supposed to be led by *volunteer wardens*, and local responses were supposed to be bolstered by federal planning, advice, and supplies during a disaster (NSRB, 1950b).

Similar to FEMA and the ED (2009) guidance today, *Civil Defense* (NSRB, 1950b) discussed command and control (i.e., current NIMS and ICS), in addition to the support role of the military (warning, advice, technical assistance, and personnel and equipment if necessary), public affairs programs (education, information, and relations), and finance and legislative issues (primarily at the state and local levels). Like the ED (2009) and Commission (2006) documents used in this study, *Civil Defense* suggested local organizational charts, as well as issues directly related to agency responses (e.g., warning systems, evacuation, mutual aid, power plant protection, communication, and supply services). Training was supposed to focus on individual preparedness, such as first-aid and fire prevention (NSRB, 1950b).

The *all-hazards* term originated with the inception of FEMA under the Carter Administration. The federal government's message historically was to prepare for "low probability, high consequence events" (Redlener & Berman, 2006, p. 4). If one can find paper copies of FEMA documents from as late as the Reagan era, one will be struck by FEMA's apparent remaining desire to instruct families in building their own emergency shelters. The 1980s shelter plans were intended as protection against multiple hazards, including nuclear fallout radiation, hurricanes, tornadoes, and earthquakes, and to provide "limited protection from the blast and fire effects of a nuclear explosion" (FEMA, 1980, p. 2).

An air-raid shelter alone was not sufficient. The Cold War manuals always stressed the necessity of having food, water, and other basic supplies for sustaining a family for up to *two weeks* (FEMA, 1983, 1985). Other 1980s documents are all-inclusive as to threats. *In Time of Emergency…A Citizen's Handbook,* FEMA (1983) mentions all forms of natural disasters, nuclear power plant accidents (perhaps a nod to Three Mile Island and environmental group pressure), fire, and nuclear attack.

In a three-ring binder titled, *State and Local Mitigation Planning, How-to Guide: Understanding Your Risks, Identifying Hazards and Estimating Losses*, published by FEMA in August 2001, the possible hazards only include forms of *natural* disasters. In the document's Introduction, FEMA (2001) states, "While this guide does not provide specific direction for all hazards, the basic procedures explained here could be adapted for any natural hazard with variations that respond to the peculiar nature of each hazard" (pp. v-vi).

Perry and Lindell (2003) comment about a shift toward *dual use* manuals (p. 345), or the movement away from a Cold War split between nuclear (federal) and natural disasters (local), to the current (post Cold-War) integrated, or comprehensive management of all-hazards approach. Redlener and Berman (2006) argue that the newly created Department of Homeland Security (DHS) and the Red Cross gave more attention to the all-hazards approach, as part of the national preparedness message following 9/11, because the two organizations wanted to make disaster preparedness efforts applicable to *highly probable natural disasters*.

WRITTEN DOCUMENTS DON'T ALWAYS LEAD TO GOOD PRACTICE

Clearly, the 1950s civil defense model of public preparedness does not exist commonly in the United States today. The residents and workers in individual cities, businesses, buildings, or other locations that are likely potential locations for a terrorist attack or predictable natural disaster (e.g., Florida regularly sees hurricanes) may be more prepared than others, but most Americans could probably not identify by name a single community leader to call or turn to in times of disaster, particularly at a level lower than municipal government.

In practice, mayors and chiefs of police and/ or fire will typically take the lead in public/media relations during large-scale emergencies. Under NIMS or ICS emergency communication protocols promulgated today at the federal level by multiple agencies, the highest-ranking uniformed officer in a Law Enforcement or Fire Protective Service agency will always take the lead in matters of operations (use of communication and other emergency equipment) and use of governmental authority. When there is chaos in the streets and citizens are imperiled, elected officials and most non-operational administrators are forced to temporarily take a backseat to the decision-making of emergency service agency personnel. This arrangement does nothing to ensure that individuals or communities, however one defines them (e.g., campus, dorm, the student body), are prepared for an emergency, or know exactly what to do in case of one. Myths and misconceptions about disaster victims persist, which can cause decision-makers to make crucial mistakes when time is at a premium.

Victims typically do not panic and often decide for themselves whether to evacuate, ignoring the warnings of authorities. Survivors are typically the first to search for others needing help, they are the first to try to protect property from additional damage, and they turn to friends and relatives for assistance before asking for help from government

agencies or the Red Cross. Looting and a widespread increase in crime after a disaster are rare. Because government actors incorrectly expect citizens to panic and/or loot, they may place too much emphasis on restricting information and providing security immediately after a disaster, thereby causing citizens to be more confused and less cooperative with orders from authorities (Perry & Lindell, 2003).

After 9/11, the National Center for Disaster Preparedness (NCDP) at Columbia University began conducting an annual national survey, which included questions related to personal and community preparedness, and confidence in and perceptions of governmental responses. By comparing the annual NCDP survey results to similar surveys conducted by the Gallup Organization between 1940 and 2000, Redlener and Berman (2006) attempted to gauge the American public's readiness over time. The 2005 NCDP report revealed that most Americans did not know what *prepared* means, the U.S. public was not listening to the national preparedness dialogue, and Americans did not understand who would be in charge during a disaster. Families were not prepared for a disaster, and more than 1/3 said they would not evacuate if they lacked confidence in who issued the order. The general lack of preparedness and awareness of national preparedness messages (e.g., stocking food and water, creating family emergency plans, or buying duct tape to seal windows against a chemical attack) was not related to any particular community size or region of the country. Redlener and Berman (2006) also found that for decades, the American public has believed it is not prepared, regardless of the source of the potential threat (international terrorism, nuclear war, or natural disaster) or source of the national preparedness message.

Another 2008 study by the NCDP tried to determine what parents of minor children would do if ordered to immediately evacuate. Despite consistent federal and state preparedness messages to the contrary, the researchers concluded that parents would likely place a reckless priority on retrieving their children before evacuating (Redlener et al., 2008).

The *mental noise model* of information processing according to Ferrante (2010) would be applicable to panicked parents. In the model, as perceived risk increases, the ability to receive information and act upon it decreases. According to R.M. Sandman's *Risk = Hazard + Outrage* theory of crisis communication, the public must feel there is a hazard, be afraid of it, and know exactly what to do before they will perceive a threat and take appropriate action. Consequently, public officials must calmly, rationally, and empathetically explain dangers to the public, or the public will ignore instructions from the ineffective leaders who choose to do otherwise (Ferrante, 2010).

This researcher attended two meetings in 2009. The first was in Washington, DC, with the American Association for the Advancement of Science; the second was in Monterrey, CA, at the Naval Post-graduate School. At both, a small collection of academics from around the country and officials from either a state-level Centers for Disease Control and/or the DHS were present. The general consensus was that no public sector agency seemed to be as prepared as one would hope they would be.

There is literature to belabor the above grim point. Lurie et al. (2004), in a 2003 RAND study, evaluated public health preparedness through tabletop exercises in seven jurisdictions in California, representing more than 1/3 of the state population. Dausey et al. (2007) coordinated 31 tabletop exercises around the U.S. over three years, working with local and state governments. Both studies found considerable variation in the levels of readiness among their samples. Both studies found problems in how organizations disseminated information to the public and the media, a general lack of knowledge regarding federal policies and responses to local events, and a general need for more training and policy revision.

From the late 1970s to late 1990s, universities in the U.S. were forced by court cases and the eventual passage of the Crime Awareness and Campus Security Act of 1990 (the Clery Act, named after murdered Lehigh University freshman Jeanne Ann Clery) to redress their minimal security efforts. The *in loco parentis* doctrine had dominated campus administrators' decisions regarding the safety of students during most of the 1960s and 1970s, thereby allowing administrators to have more discretion. State and federal courts began eroding the doctrine during the 1970s and '80s, when universities were held liable for crimes on campus involving student victims (Fisher, 1995).

The courts have used the *doctrine of foreseeability* [sic, emphasis added] as the standard for establishing liability in lawsuits filed by victimized students….[However] the courts have not always agreed on the interpretation of foreseeability or set consistent standards for adequate security protection…. (Fisher, 1995, p. 90)

The ED wanted to use the Clery Act as a panacea to force institutions to distribute campus safety policies and crime statistics to inform parent and student consumers of university services and potential dangers, so individuals could make better safety decisions. Despite the Clery Act's purpose of shedding light on levels of campus crime, studies show that students, like other victims, often do no report crimes. Clery reports focus on raw counts not crime rates (e.g., offenses/1000 students), possibly creating a false sense of security at some schools; individual offender and victim characteristics, and thefts are not included in the Clery statistics (Fisher, 1995). As anyone who has spent time on a college campus knows, theft is a common occurrence. The 2008 Higher Education Act amended the Clery Act, but didn't remove all the deficiencies. Colleges and universities must now provide the ED with fire safety reports, annually test their emergency procedures, create missing persons policies, and report hate crimes and drug violations (McCarter, 2008).

ONLINE CRISIS COMMUNICATION

Today, online networks are an important part of emergency or crisis communication. Less than 24 hours after the 2007 Virginia Tech shootings, students using Facebook, with little official information from the university, had identified all 32 victims prior to the school's official public release of the information. The website of San Diego's National Public Radio station, Google maps, and online blogs were used by residents of southern California to track wildfires in the fall of 2007. Online public bulletin boards were used to share information after the 2008 earthquakes in China (Winerman, 2009). As this paper is being written, social networks are being credited for regime change in Egypt.

Patricelli et al. (2009) wrote about the high probability of a complete failure of terrestrial telecommunications systems during a large-scale disaster, either caused by the event itself or as phone systems are deluged with calls for service. Huang et al. (2010), in a study of the emergency response to damage caused to Taiwan in 2009 by typhoon Morakot, argued the importance of internet-based two-way communications. Internet users in Taiwan, who were familiar with popular social networks such as Twitter, created unofficial websites to gather real-time situation reports, which were later linked to local government websites, when voice reporting systems were overloaded and the government's response during the first days of flooding was slow. Huang et al. (2010) conceded that internet-based communication is limited in poor countries. Nsiah-Kumi (2008) believed the particular concerns and characteristics of special populations (e.g., the poor, elderly, or illiterate) should be taken into consideration when public officials must communicate with them during a crisis.

One could argue that many first-world college students are a unique population; one which is very comfortable using the Internet. At Longwood University, all incoming freshmen are required to possess a laptop computer, and new students receive a Longwood email account and access to user support services. Longwood University is on Twitter and Facebook. A study conducted by Wireless and Mobile News (2010) found that undergraduates at the Claremont Colleges in California send an average 26.6 text messages per day; students text while in class, while working out at the gym, when using the bathroom, and even while bicycling.

DATA AND METHODS

Because Longwood University is neither the Commonwealth of Virginia's "flagship" (it is the University of Virginia), nor the smallest (Univ. of VA College at Wise) four-year, public, non-specialty (e.g., maritime, art, military) university or college in the state, it was not included in this study's sample. The 50 flagship public universities were primarily included in the sample for size comparison to smaller public universities from each state.

In addition, one might expect flagship institutions, which are representatives of their states' governments, with considerable public funding and large student bodies, to have greater levels of advertised safety/security/preparedness measures among universities in the United States. They should, compared to any other form of higher education institution, be more professional and bureaucratic, typically knowing how to follow and write formal policies, and probably using marketing, public relations, and IT services, which would likely maximize their opportunities for using web-based forms of communication. There were 99 schools in the sample instead of 100, because Wyoming has one four-year public school, its flagship, the University of Wyoming.

The collegeboard.com website was used in May 2010 to locate university websites, confirm current enrollments to determine which public four-year university was smallest in each state, and to col-

7

lect secondary current* (year 2009) enrollment numbers for the 99 schools in the sample. Identifying which school was the perceived flagship university from each state was sometimes open to a bit of interpretation, after reading university web pages. For example, a New Yorker had to be consulted to definitively determine that Stony Brook was the NY flagship.

The ED (2009) manual consulted for this study seemed more appropriate than what seemed to be the most relevant FEMA (2009) manual (titled *Developing and Maintaining State, Territorial, Tribal, and Local Government Emergency Plans*). Two vitally important criteria for campus safety were easy to conceptualize from the many standards established by the ED (2009) manual: 1) a publicly *available emergency plan document* and 2) some representation of *all-hazards* or multiple dangers/threats within the emergency plan or other safety tips.

The Commission's (2006) report stated twenty-seven specific safety and security "best practices" used by Commonwealth schools, which yielded a dozen specific criteria for this study:

- Any statement of having a *safety committee*, comprised of vital members from throughout the university organization.
- An explicit statement of adhering to *NIMS or ICS* protocols.
- *Crime Prevention Through Environmental Design (CPTED)* concepts are mentioned (for sampled schools, CPTED concepts usually appeared within general safety tips (e.g., paying attention to one's physical surroundings)).
- A *self-defense or Rape Aggression Defense (RAD)* course (service offered by Police Dept. or other campus unit, not an academic course (e.g., Recreation)).
- Mention of a student or citizen police *academy*, or training for a student non-sworn police/security force.

- Campus Police Dept. (PD) or other unit provides free rides by car or shuttle after dark, or a walking *security escort upon request*, using students or full-time staff.
- Use of a survey or email link, asking for comments or *feedback* from users, linked from a webpage.
- The PD possesses or seeks *accreditation* (from the Commission on Accreditation for Law Enforcement Agencies (CALEA) or from another entity).
- Any statement of membership(s)/participation in any *professional organization* (not including accreditation organizations), by any top safety/security/emergency personnel.
- Explicit statement of having *concurrent jurisdiction* over an area/town/city external to the campus facilities.
- A victims' rights policy, or procedural statement or page specifically mentioning some range of *victim services*/programs/agencies.
- A PD *crime investigation policy*, or any statement of who/when/how investigations are conducted by the PD.

The Commission's (2006) report recommended two best practices (among the original 27) that were initially considered for inclusion in this study. They were ultimately not included in this research, because every school in the sample did them: a) adopt a list of prohibited/illegal student behaviors, and b) develop a list of possible sanctions for the student behaviors. Both items were always in an online Student Handbook or Code of Conduct for the sampled schools.

Gray's (2009) *Campus Safety Magazine* article produced the last four criteria used in this study:

- A separate crime reporting system from main phone or 911, usually separate phone *tip line* or anonymous "silent witness" email form shown on PD webpage.

- The university has a *mass notification system* (text messages, emails, sirens).
- *Security cameras* or closed-circuit television are used on campus (as indicated by any statement about their use).
- Any mention of *new* tactical *equipment*, technology, or vehicles (the magazine survey asked respondents about the acquisition of a variety of new equipment and technology).

The content analysis portion of this study, as this researcher systematically read university web pages and associated documents, looking for any verbal indication of any of the 18 criteria, was conducted during June-August, 2010. Much of the information needed to determine whether a school performed/provided a criterion could often be found on the web pages of University/Campus Police or Security departments, Emergency Management or Health & Safety offices, or Student Affairs offices. The total points received for each of the 18 criteria for each school (coding: 1= has criterion, 0= does not) were combined to create each school's Index Score (for a possible score range of 0 – 18).

If it appeared as though a school had very few of the 18 criteria, and would have a low Index Score, extra time (approximately 30 minutes – 1 hour) was spent using online search tools, embedded within university online homepages, looking for key words (e.g., security, safety, emergency, investigation, etc.) to ensure no relevant web pages or other digital documents had been missed. The logic of the Index Score is simple: the more criteria a school demonstrated on its web pages in the summer of 2010, the better a school seemed to represent the recommendations from the above documents.

Some of this study's limitations are glaring. Possibilities for human error are everywhere, particularly in the construction of the list of 18 criteria, and the content analysis of university web pages. Other public or private manuals, reports, or texts could have been consulted to generate a different or larger list of criteria upon which the safety/security messages of schools in the sample were evaluated. Perhaps FEMA or other federal agencies had better guidance than the ED, or other states had commission reports that applied to campus safety/security. The design is cross-sectional, the sample could be much larger among only four-year public schools, and findings cannot be generalized to any other form of higher educational institution (e.g., junior and private colleges).

The biggest caveat for this research is the high probability that multiple schools in the sample perform more security/safety-related functions in reality than they reveal on their websites. Schools which seem to offer few of the studied criteria could simply have poor website content, or schools may choose to not advertise some activities, such as video surveillance of their students. The Index Score may simply be a measure of a school's ability to advertise its safety/security/preparedness policies and programs online, at one point in time.

It is also likely that a few of the criteria are inherently problematic (discussed below), which is why so few schools seem to perform those activities. And, should all the schools in the sample get credit for having formal student codes of conduct and sanction policies (excluded from the Index Score)? Despite these methodological problems, the low level of adoption of the 18 study criteria among the 99 schools, even among criteria one would expect most schools to have, should cause the reader concern. The 18 criteria can be grouped into four natural categories: Emergency Plan Components, PD Programs & Services, Communication, and measures of Professionalism.

FINDINGS

Although not included in the sample, Longwood University (for full disclosure) had the same Index Score as the other Virginia schools in this study. A complete list of the schools in the sample, along with their index scores and 2009 undergraduate enrollments can be seen in the Appendix (see Table A1). As shown in Table 1, the highest Index Score in the sample was 15 (scored by the University of Maryland and University of Minnesota). Six universities had zero (0) Index Scores: Harris-Stowe State University, Nevada State College, Mayville State University, Eastern

Oregon University, Dakota State University, and Lyndon State College. The 2009 undergraduate enrollment (labeled: undergrad09) range between the largest and smallest schools, and the enrollment standard deviation of approximately 10,800 students, indicates likely monstrous student-generated resource differences (funding generated by tuitions, faculty and staff to support more classes and classrooms, etc.) between schools in the sample. The statistical mean of 7.94 and a mode of 10 for the Index Score, demonstrates that a lot of schools in the sample are not advertising many of the 18 study criteria on their websites.

Using SPSS, the bivariate correlation between the enrollment size of universities in the sample (undergrad09 in Table 1) and the Index Score compiled for each university is statistically significant at the .01 level (sig. = .000), and indicates a fairly strong statistical relationship between the two criteria (Pearson correlation = .649). Finding that the size of the school is related to the amount of stated security measures is not surprising, given that larger schools would likely have more resources for every organizational function, including the creation and dissemination of information about campus safety and security.

Table 2 shows each of the 18 criteria (simplified from the Data & Methods section into a

short phrase) used in this study, comparing frequencies for the criteria that appear in the total (99) sampled schools, to the subsets of Flagships Only and Smaller Schools Only. Proportions of criterion adoption, based on the total sample and subsets are also shown. What is most striking about Table 2 is that not only have all the schools in the sample not adopted any single criterion, but there is also a large difference in the safety messages at the flagships compared to the smaller schools. Only 59% of the sample has a publicly available emergency plan, which the ED (2009) and FEMA (2009) documents discussed above mention in great detail. In the total sample, as well as in each subset, universities mentioned mass notification systems the most, compared to any other criteria (81% of 99, 96% of flagships, 65% of smaller). Mentioning all-hazards is next (68 of 99 total, 42 of 50 flagships, 26 of the 49 smaller schools), followed by a night escort (26 of the smaller schools had a nighttime escort program).

Making statements about memberships in professional organizations seems to be something very few schools in the sample do, as well as advertising new equipment. Not mentioning professional organization memberships makes some sense, because such detailed information is the stuff of professional resumes or *curricula vitae*, not public sector agency web pages. Nevertheless, many of the larger police departments at sampled schools have "messages from the chief", or even photos of each staff member, which would be logical places to mention the information. Delaware State University, for example, only mentions that officers can participate in professional associations.

Schools in the sample may not always mention new equipment because they are state institutions

Table 1. Descriptive statistics for undergraduate enrollment (2009) and index score

	N	Range	Mode	Minimum	Maximum	Mean	Std. Deviation
undergrad09	99	39978	-----	200	40178	11700.94	10829.322
Index	99	15	10	0	15	7.94	4.028

Table 2. 18 criteria frequency: total sample, flagships only, and smaller schools only

Criteria Categories	The 18 Criteria	Sample	Flagships	Smaller
		n=99	n=50	n=49
		Frequency (proportion)		
Emergency Plan Components	Emergency Plan	58 (0.59)	36 (0.72)	22 (0.45)
	All Hazards	**68 (0.69)**	**42 (0.84)**	**26 (0.53)**
	Safety Committee	64 (0.65)	41 (0.82)	23 (0.47)
	NIMS or ICS	50 (0.51)	32 (0.64)	18 (0.37)
Police Dept. Programs & Services	CPTED	59 (0.60)	40 (0.80)	19 (0.39)
	Self-defense or RAD	45 (0.45)	32 (0.64)	13 (0.27)
	Academy	32 (0.32)	26 (0.52)	6 (0.12)
	Night Escort	**67 (0.68)**	**41 (0.82)**	**26 (0.53)**
Communication	Crime Tip/Hotline	58 (0.59)	37 (0.74)	21 (0.43)
	Mass Notify System	**80 (0.81)**	**48 (0.96)**	**32 (0.65)**
	Feedback device	32 (0.32)	23 (0.46)	9 (0.18)
	Security Cameras	19 (0.19)	14 (0.28)	5 (0.10)
Professionalism	PD Accreditation	20 (0.20)	17 (0.34)	3 (0.06)
	Prof org membership	**3 (0.03)**	**1 (0.02)**	**2 (0.04)**
	New equipment	**13 (0.13)**	**10 (0.20)**	**3 (0.06)**
	Concurrent Jurisdiction	20 (0.20)	14 (0.28)	6 (0.12)
	Victim Rights Policy	**56 (0.57)**	**34 (0.68)**	**22 (0.45)**
	Investigation Policy	42 (0.42)	30 (0.60)	12 (0.24)

in tight financial circumstances like other state institutions; therefore, they don't have anything to mention. Still, campus police and security offices in the sample typically show pictures of their vehicles, K-9 units, and bicycle patrol units if they have them, so perhaps they are showing new equipment online, but are just not stating it explicitly. Any new sworn police officer would likely get some new equipment.

The flagships alone had frequencies in the single digits for only one of the criteria: professional organization membership (3). By contrast, the smaller schools subset (n=49) had frequencies in the single digits for *seven criteria*: academy programs (6), feedback devices (9), security cameras (5), PD accreditation (3), professional org. membership (2), new equipment (3), and concurrent jurisdiction (6).

A feedback device could be as simple as an email link for comments, yet smaller schools with fewer overall safety messages may not see it as a priority, compared to basic security and 911 services. Concurrent jurisdiction arrangements were somewhat uncommon in the total sample (20%), and were usually the product of campus police being sworn in more than one jurisdiction. Perhaps the most unusual concurrent jurisdiction in the sample can be seen at Framingham State University, where its University Police officers are "Special State Police and Deputy Sheriffs," which includes the Town of Framingham.

Academy programs, security cameras, PD accreditation, and memberships in associations may simply be unaffordable or unattainable by smaller schools with smaller budgets and fewer personnel. However, many law enforcement officials belong to professional organizations, and saying so on a university web page may be seen as unusual.

Of the four categories of criteria, in the total sample those related to the emergency plan were mentioned the most, with at least 50% of the schools in the sample stating them (ranging from 50-68%). The professionalism category was the most problematic for the total sample, with four criteria being adopted by 1/5 of the sample or less. However, the two policy-related criteria in the professionalism category, a victims' rights policy (57% of the total sample) and an investigation policy (42% of the total sample), were possibly mentioned more because writing a brief policy statement is relatively quick and inexpensive, compared to the other criteria in the category.

CONCLUSION AND FUTURE CONSIDERATIONS

Some schools in the sample need better online messages, some may need more programs and policies that meet the criteria, and some of the very low scoring schools in the sample must need both. Some of the Flagships did not score as high as one might expect, given the enrollment of any flagship. Perhaps the lack of safety/security information on university web pages stems solely from poor website design, or not enough personnel with the skills required to update web pages. Schools may intentionally try to hide all mention of crime or safety on their web pages in attempts to downplay anything that may scare away scarce student tuition dollars.

Federal vs. state, and emergency management vs. law enforcement, traditional agency roles for providing protective services have been constant particularly when terrorists, or any sort of external man-made events, are the source of a given threat. Consequently, the Federal Bureau of Investigation resides in the U.S. Department of Justice, and not within DHS, which houses FEMA. In the sample, campus "safety" or "security" is typically the domain of university police departments, and separate Emergency Management/Health & Safety departments are typically tasked with maintaining universities' emergency plans and occupational safety programs. Many of the 18 criteria would fall under the responsibility of a campus security or police agency, which should not be surprising, since 12 of them were recommended by the Virginia State Crime Commission (2006).

Despite these traditional institutional divisions, the massive re-organization of the federal government in the years following 9/11 has in some ways resulted in a blurring of institutional roles. For example, introductory criminal justice texts (Dempsey & Forst, 2010; Gaines & Miller, 2009) argue that Joint Terrorism Task Force models are the norm for relationships between federal and local law enforcement in locales with potential targets in their jurisdictions. Similarly under the concept of Homeland Security, municipal police are now expected to know about biological weapons (e.g., white powder that makes your eyes instantly water), and have a broader knowledge of first aid. At a typical larger university in the sample, which has an online Emergency Plan, the Director of Campus Safety/Police Chief would likely be the primary decision-maker during *any* large-scale emergency on campus, following NIMS or ICS protocols.

Most campus safety and security issues are handled by law enforcement entities, because campus police units often have more round-the-clock personnel, allowing more division of labor, compared to campus emergency management or occupational safety units, which may consist of only a handful of administrative staff, who work regular office hours. Even the feedback emails and security camera systems mentioned by schools in the sample were typically managed by campus police. Smaller schools, with very small and sometimes part-time security staff, who do not have sworn officers or 24/7 capabilities, may not have the resources to implement any form of emergency preparedness improvement.

Perry and Lindell (2003) noted that during a disaster, too much emphasis may be placed on providing security and restricting information, instead of communication, which may later lead to resistance from the public to orders. Campus police or security units provide most of the safety services on a campus, including some of the communication lines vital during a crime or disaster, and are most likely to take command during a disaster, if only temporarily. There is a potential danger of schools intentionally withholding information, or downplaying safety problems, before and/or during bad events out of a greater concern about damaging a school's public image.

There are some differences between the language of the Cold War and 2010. Foreign nuclear attack, once the focus of all worst-case scenario preparations from the 1950s–'80s, is seldom mentioned in the current federal government message to the public, although large-scale "dirty" radiological threats, biological weapons, or chemical contamination are often included in manuals detailing the all-hazards paradigm.

After comparing hard copy Cold War and current online publication guidance, one could argue that *communities* used to be prepared for emergencies under the rubric of civil defense; yet today, *individuals* or *single households* are being encouraged to fend for themselves for 72 hours, instead of the two weeks recommended in the 1950s. If a college campus was deemed a *community*, then many of them are likely not prepared to be self-sufficient for 24 hours, much less the two weeks recommended in the 1950s. On university websites in the sample, "Shelter In Place" instructions seem to now be common in university emergency plans or safety tips, in recognition that some emergency events (e.g., an active shooter on campus or a tornado) require campus community members to not evacuate. Future research could evaluate the prevalence of individual threats within all-hazards messages, or the content of campus Emergency Plans.

To date, no data has been collected for dollars spent on the universities' public safety messages (i.e., marketing, public relations, IT resources), because merely finding consistent enrollment data for the schools in the sample proved so difficult. University budget numbers may be very hard to obtain. A larger university sample, perhaps including other types of higher educational institutions, utilizing a survey instrument, might yield a more complete picture of actual campus safety/preparedness/security efforts, or the content of the messages that appear online at schools within the United States. Additional criteria about the organizational structures (e.g., number of personnel in relevant units, personnel roles, chains-of-command, sworn v. civilian staff), or a better conceptualization of the level of professionalization or bureaucracy at each school could be sought. Criteria regarding Student Affairs could be included (e.g., existence of Greek student organizations or alcohol on campus, residential v. commuter student populations, contents of Student Handbooks or Code of Conducts), as well as ecological factors (i.e., surrounding town or region).

According to the NCDP research discussed above, even if schools had excellent safety messages, by any measure, there may never be a guarantee that the campus community receives it, or obeys instructions before or during an event and its immediate aftermath. So one must constantly revise, and plan, and adjust to the ever-changing, unpredictable threats.

ACKNOWLEDGMENT

Collegeboard.com secondary enrollment data for "degree-seeking undergrads" was used for multiple reasons. It quickly became apparent that schools in the U.S. don't have similar enrollment reporting schemes when one looks at their websites. Some institutions do not separate degree from non-degree seeking students, some only gave enrollment total (grad + undergrad), and some do not have data from 2009 on their websites. Also, the degree-seeking numbers from collegeboard.com, when compared to numbers reported on school websites, tended to be smaller by a few dozen for smaller schools, to a few hundred for larger schools (although for the schools still reporting 2007 or 2008 data, the collegeboard.com number could be higher than what the school reported). Ultimately, the choice was made to use the 2009 secondary data from a single source, instead of 99 sources using multiple reporting schemes. Enrollment data for 2010 was not available from most universities or collegeboard.com at the time of this research.

REFERENCES

College Board. (2010). *College Search*. Retrieved from http://collegesearch.collegeboard.com/search

Dausey, D. J., Buehler, J. W., & Lurie, N. (2007). Designing and Conducting Tabletop Exercises to Assess Public Health Preparedness for Manmade and Naturally Occurring Biological Threats. *BioMed Central Public Health, 7*, 92. Retrieved May 15, 2010, from http://search.ebscohost.com.proxy.longwood.edu

Dempsey, J. S., & Forst, L. S. (2010). *An Introduction to Policing* (5th ed.). Clifton Park, NY: Cengage.

Federal Emergency Management Agency (FEMA). (1980). *Aboveground Home Shelter*. Washington, DC: Government Printing Office.

Federal Emergency Management Agency (FEMA). (1983). *In Time of Emergency: A Citizen's Handbook*. Washington, DC: Government Printing Office.

Federal Emergency Management Agency (FEMA). (1985). *Protection in the Nuclear Age*. Washington, DC: Government Printing Office.

Federal Emergency Management Agency (FEMA). (2001). *State and Local Mitigation Planning, How-to Guide: Understanding Your Risks, Identifying Hazards and Estimating Losses* (Tech. Rep. No. 386-2). Washington, DC: Government Printing Office.

Federal Emergency Management Agency (FEMA). (2009). *Developing and Maintaining State, Territorial, Tribal, and Local Government Emergency Plan: Comprehensive Preparedness Guide 101*. Retrieved February 2, 2010, from http://www.fema.gov

Ferrante, P. (2010). Risk & Crisis Communication. *Professional Safety, 55*(6). Retrieved January 20, 2011, from http://search.ebscohost.com.proxy.longwood.edu

Fields, J. W. (2009). 10 Steps to Creating a Campus Security Master Plan. *Campus Safety Magazine*. Retrieved January 19, 2010, from http://www.campussafetymagazine.com/Channel/Security-Technology/Articles/2009/03/10-Steps-to-Creating-A-Campus-Security-Master-Plan.aspx

Fisher, B. S. (1995). Crime and Fear on Campus. *Annals of the American Academy of Political and Social Science, 539*, 85-101. Retrieved February 2, 2010, from http://www.jstor.org/stable/1048398

Gaines, L. H., & Miller, R. L. (2009). *Criminal Justice in Action* (5th ed.). Belmont, CA: Thomson-Wadsworth.

Gray, R. (2009). Columbine 10 Years Later: The State of School Safety Today. *Campus Safety Magazine*. Retrieved January 8, 2010, from http://www.campussafetymagazine.com/Channel/School-Safety/Articles/2009/03/Columbine-10-Years-Later-The-State-of-School-Safety-Today.aspx

Huang, C.-M., Chan, E., & Hyder, A. A. (2010). Web 2.0 and Internet Social Networking: A New tool for Disaster Management? -- Lessons from Taiwan. (from http://search.ebscohost.com.proxy. longwood.edu). *BMC Medical Informatics and Decision Making, 10*, Retrieved January 20, 2011. doi:10.1186/1472-6947-10-57

Lurie, N., Wasserman, J., Soto, M., Myers, S., Namkung, P., Fielding, J., & Valdez, R. B. (2004). Local Variation in Public Health Preparedness: Lessons from California. (from http://search. ebscohost.com.proxy.longwood.edu/). *Health Affairs, 23*, w341–w353. Retrieved May 15, 2010.

McCarter, M. (2008). A New Standard for Campus Security. *Homeland Security Today*. Retrieved February 25, 2010, from http://hstoday.us

National Security Resources Board (NSRB). (1950a). *Medical Aspects of Atomic Weapons*. Washington, DC: U.S. Government Printing Office.

National Security Resources Board (NSRB). (1950b). *United States Civil Defense*. Washington, DC: U.S. Government Printing Office.

Nsiah-Kumi, P. A. (2008). Communicating Effectively With Vulnerable Populations During Water Contamination Events. (from http://search. ebscohost.com.proxy.longwood.edu). *Journal of Water and Health, 6*, 63–75. Retrieved January 20, 2011. doi:10.2166/wh.2008.041

Patricelli, F., Beakley, J. E., Carnevale, A., Tarabochia, M., & von Lubitz, D. (2009). Disaster Management and Mitigation: The Telecommunications Infrastructure. [from http://search.ebscohost.com. proxy.longwood.edu]. *Disasters, 33*(1), 23–37. Retrieved January 20, 2011. doi:10.1111/j.1467-7717.2008.01060.x

Perry, R. W., & Lindell, M. K. (2003). Preparedness for Emergency Response: Guidelines for the Emergency Planning Process. [from http://search. ebscohost.com.proxy.longwood.edu]. *Disasters, 27*(4), 336–350. Retrieved February 21, 2010. doi:10.1111/j.0361-3666.2003.00237.x

Quarantelli, E. L. (1997). Ten Criteria for Evaluating the Management of Community Disasters. [from http://search.ebscohost.com.proxy.longwood.edu]. *Disasters, 21*(1), 39–56. Retrieved February 21, 2010. doi:10.1111/1467-7717.00043

Redlener, I., & Berman, D. A. (2006). National Preparedness Planning: The Historical Context and Current State of the U.S. Public's Readiness, 1940-2005. [from http://search.ebscohost.com. proxy.longwood.edu]. *Journal of International Affairs, 59*(2), 81–103. Retrieved February 12, 2010.

Redlener, I., Grant, R., Abramson, D., & Johnson, D. (2008). *Annual Survey of the American Public by the National Center for Disaster Preparedness*. New York, NY: National Center for Disaster Preparedness, Columbia University Mailman School of Public Health and The Children's Health Fund. Retrieved May 23, 2010, from http://www.ncdp. mailman.columbia.edu/files/white_paper_9_08. pdf

Templeton, D. E., Ellerman, G., & Branscome, T. (2009). Case Study: Radford University Overcomes Emergency Management Hurdles. *Campus Safety Magazine*. Retrieved February 21, 2010, from http://www.campussafetymagazine.com

U.S. Department of Education. (2009). *Action Guide for Emergency Management at Institutions of Higher Education*. Retrieved February 2, 2010, from http://www.ed.gov/emergencyplan

Virginia State Crime Commission. (2006). *HJR 122 Final Report: Study on Campus Safety*. Retrieved December 15, 2009, from http://leg2. state.va.us/DLS

Winerman, L. (2009). Social Networking: Crisis communication. [from http://search.ebscohost.com.proxy.longwood.edu]. *Nature, 457,* 376–378. Retrieved January 20, 2011. doi:10.1038/457376a

Wireless and Mobile News. (2010). *WMN Exclusive: College Students Text Often & During All Sorts of Activities.* Retrieved January 20, 2011, from http://www.wirelessandmobilenews.com/2010/12/text-use-in-college-students.html

APPENDIX

Table A1. Index score and undergrad enrollment of sampled schools (in order of score)

State	School Name	Index Score	Undergrad Enrollment
MARYLAND	University of Maryland	15	25898
MINNESOTA	University of Minnesota	15	29921
FLORIDA	University of Florida	14	33038
IOWA	University of Northern Iowa	14	11061
KENTUCKY	University of Kentucky	14	18806
OREGON	University of Oregon	14	18213
ARKANSAS	University of Arkansas	13	14861
CALIFORNIA	University of California, Berkeley	13	25530
DELAWARE	University of Delaware	13	15407
MISSOURI	University of Missouri	13	23529
NEVADA	University of Nevada, Reno	13	12878
NORTH CAROLINA	University of North Carolina at Chapel Hill	13	17565
TEXAS	University of Texas at Austin	13	37464
ALABAMA	University of Alabama	12	22046
ARIZONA	University of Arizona	12	29340
GEORGIA	University of Georgia	12	25882
MICHIGAN	University of Michigan	12	26033
NEW YORK	State University of New York at Stony Brook	12	16034
TEXAS	Texas A&M University, Texarkana	12	1030
ARIZONA	Northern Arizona University	11	18044
FLORIDA	New College of Florida	11	825
ILLINOIS	University of Illinois at Urbana-Champaign	11	30400
INDIANA	Indiana University, Bloomington	11	31087
IOWA	University of Iowa	11	20079
NEW JERSEY	Rutgers, The State University of NJ: New Brunswick	11	28817
NORTH DAKOTA	University of North Dakota	11	10440
OHIO	Ohio State University	11	40178
PENNSYLVANIA	Pennsylvania State University	11	37855
CALIFORNIA	University of California, Merced	10	3190
COLORADO	University of Colorado at Boulder	10	26405
INDIANA	Indiana University East	10	2081
MAINE	University of Maine	10	8868
NEW HAMPSHIRE	University of New Hampshire	10	12226
NEW YORK	State University of New York College at Potsdam	10	3691
RHODE ISLAND	University of Rhode Island	10	12520

continued on following page

Table A1. Continued

State	School Name	Index Score	Undergrad Enrollment
SOUTH CAROLINA	University of South Carolina	10	20135
TENNESSEE	University of Tennessee, Martin	10	6476
VIRGINIA	University of Virginia	10	14225
VIRGINIA	University of VA College at Wise	10	1612
WASHINGTON	University of Washington	10	30554
WISCONSIN	University of Wisconsin, Superior	10	2515
CONNECTICUT	University of Connecticut	9	16459
KANSAS	University of Kansas	9	20240
MARYLAND	University of Maryland, Baltimore	9	840
MASSACHUSETTS	University of Massachusetts	9	20287
MASSACHUSETTS	Framingham State College	9	3485
MISSISSIPPI	University of Mississippi	9	13024
MONTANA	University of Montana, Missoula	9	12196
TENNESSEE	University of Tennessee, Knoxville	9	20849
VERMONT	University of Vermont	9	10371
WEST VIRGINIA	West Virginia University	9	21720
DELAWARE	Delaware State University	8	3222
HAWAII	University of Hawaii, Manoa	8	13583
IDAHO	Lewis-Clark State College	8	3239
NEBRASKA	University of Nebraska, Lincoln	8	18955
NEW HAMPSHIRE	University of New Hampshire at Manchester	8	844
NEW JERSEY	Rutgers, The State University of NJ: Camden	8	4067
NEW MEXICO	Western New Mexico University	8	1753
UTAH	University of Utah	8	20983
CONNECTICUT	Eastern Conn State University	7	4896
KANSAS	Emporia State University	7	4063
LOUISIANA	University of Louisiana, Lafayette	7	14240
MISSISSIPPI	Mississippi Valley State University	7	2357
NEW MEXICO	University of New Mexico	7	19613
OHIO	Ohio University: Eastern Campus	7	774
OKLAHOMA	University of Oklahoma	7	20685
WISCONSIN	University of Wisconsin, Madison	7	28690
WYOMING	University of Wyoming	7	9586
ALASKA	University of Alaska, Fairbanks	6	5273
ALASKA	University of Alaska Southeast	6	1145
NORTH CAROLINA	Elizabeth City State University	6	3045
SOUTH DAKOTA	University of South Dakota	6	6043
IDAHO	University of Idaho	5	8757
LOUISIANA	Southern University at New Orleans	5	2603
SOUTH CAROLINA	University of South Carolina at Beaufort	5	1319

continued on following page

Table A1. Continued

State	School Name	Index Score	Undergrad Enrollment
WASHINGTON	University of Washington, Bothell	5	2329
ARKANSAS	Southern Arkansas University	4	2647
COLORADO	Adams State College	4	2347
MONTANA	University of Montana, Western	4	1207
NEBRASKA	Peru State College	4	1997
PENNSYLVANIA	Penn State Wilkes-Barre	4	594
RHODE ISLAND	Rhode Island College	4	7500
ALABAMA	University of West Alabama	3	1867
GEORGIA	Georgia Southwestern State University	3	2594
HAWAII	University of Hawaii, West Oahu	3	1253
ILLINOIS	Governors State University	3	2791
MAINE	University of Maine at Machias	3	565
MICHIGAN	Lake Superior State University	2	2528
MINNESOTA	University of Minnesota, Rochester	2	200
KENTUCKY	Kentucky State University	1	2326
OKLAHOMA	Oklahoma Panhandle State University	1	1177
UTAH	Dixie State College of Utah	1	6484
WEST VIRGINIA	Glenville State College	1	1191
MISSOURI	Harris-Stowe State University	0	1844
NEVADA	Nevada State College	0	1990
NORTH DAKOTA	Mayville State University	0	887
OREGON	Eastern Oregon University	0	3160
SOUTH DAKOTA	Dakota State University	0	1450
VERMONT	Lyndon State College	0	1500

This work was previously published in the International Journal of Information Systems for Crisis Response and Management, Volume 3, Issue 1, edited by Murray E. Jennex and Bartel A. Van de Walle, pp. 1-18, copyright 2011 by IGI Publishing (an imprint of IGI Global).

Chapter 2
Equipment Distribution for Structural Stabilization and Civilian Rescue

Albert Y. Chen
University of Illinois at Urbana-Champaign, USA

Feniosky Peña-Mora
Columbia University, USA

Saumil J. Mehta
University of Illinois at Urbana-Champaign, USA

Stuart Foltz
Construction Engineering Research Laboratory, USA

Albert P. Plans
Universitat Politècnica de Catalunya, Spain

Brian R. Brauer
Illinois Fire Service Institute, USA

Scott Nacheman
Thornton Tomasetti, USA

ABSTRACT

The efficiency of Urban Search and Rescue operations depends on the supply of appropriate equipment and resources, and an efficient damage assessment facilitates deployment of these resources. This paper presents an Information Technology (IT) supported system for on-site data collection to communicate structural condition, track search and rescue status, and request and allocate appropriate resources. The system provides a unified interface for efficient gathering, storing, and sharing of building assessment information. Visualization and access of such information enable rescuers to respond to the disaster more efficiently with better situational awareness. The IT system also provides an interface for electronic resource requests to a geospatial resource repository service that enables a spatial disaster management environment for resource allocation. Request and deployment of critical resources through this system enables lifesaving efforts, with the appropriate equipment, operator, and materials, to become more efficient and effective. System development at the Illinois Fire Service Institute has shown promising results.

DOI: 10.4018/978-1-4666-2788-8.ch002

INTRODUCTION

Critical resources such as heavy construction equipment are required in conditions when human power is not sufficient to perform Urban Search and Rescue (US&R) operations. The performance of search and rescue depends on the delivery of these critical resources. As US&R operations involve the location, rescue, and initial medical stabilization of victims trapped in confined spaces, inefficient equipment delivery could delay civilian rescue. At the same time, safety of the rescuers is one of the most important responsibilities of US&R and structural stability of damaged infrastructures is a key component of rescuers' safety. FEMA US&R Structural triage is the process of evaluating structurally compromised buildings to determine operational priority (US Army Corps of Engineers, 2008). The priority is set based on factors such as occupancy, known victims, probability of live victims, collapse mechanism and structural condition. To keep track of search and rescue information at structurally compromised buildings, building marking systems (BMS) are used in the current practice. As technical rescue operations for major disasters tend to be in the order of hours/days, these marking systems are imperative for effective communication and allocation of rescue forces. In a lifesaving scenario, standardized information for building identification, conditions assessment, hazards and victim status is of great importance. However, challenges in the current practice have been identified. This paper presents an Information Technology (IT) supported system that addresses the challenges for on-site data collection to communicate structural condition, status of US&R operations, and to request resources for stabilization of those structures for search and rescue within those structures.

GAP IDENTIFICATION

From lessons learned in recent disasters, information gathering for critical decision making has been recognized as one of the greatest challenges in disaster response. Response efforts cannot reach their full potential without the information needed to make critical decisions. For example, after the 9/11 terrorist attacks, the authorities were not fully aware of available resources and did not have complete access to available information. As a result, resources were deployed inefficiently, which compromised the effectiveness of response operations (National Commission on Terrorist Attacks Upon the United States, 2004).

Distribution of resources during disaster response operations has been characterized by various shortcomings that inhibit efficient and effective decision making. Setting priorities for allocation of limited resources is one of the challenges (National Commission on Terrorist Attacks Upon the United States, 2004). Efficient information gathering and decision making for distribution of resources is critical to support disaster response efforts.

A large number of engineering parameters such as the type of structure, patterns of collapse, and shoring alternatives play important roles for decision making. These factors contribute to decision making for prioritization of rescue activities and in some cases are vital in ensuring the safety of the rescuers. For example, in-structure route selection is critical to quickly and safely access victims trapped under a partial collapsed building. As such, structural triage and BMS has been one of the key features carried out by the engineering workforce on US&R operations (McGuigan, 2002). The information gathered (such as the structural triage) is then disseminated to the stakeholders for decision making (setting up operational priority for buildings). In other words, critical information needs to be communicated to or retrievable by numerous levels of command at different times and stages of disaster response. However, the information is usually transferred through paper copies, which cannot be effectively distributed.

For example, once a triage or BMS is complete, the primary communication method in the current practice is in paper format. It can easily take 24 hours for the paper copy to reach the Incident Command. Since triage is a base of information decision makers use to set priorities of response efforts, the delay in information dissemination compromises the timing for decision making. An alternative would be through verbal radio transmission but this can be both incomplete and unreliable. If other actors need access to the information, locating and retrieving the information in paper format or radio transmission is also difficult. For example, engineers develop structural stability information and stabilization plans for building structures. These plans include critical information—such as needed equipment, material and manpower—required to carry out the plans. Poor communication of this information compromises

the effectiveness of underlying response efforts. As a result, an effective communication mechanism for such critical information is of great importance.

Search and rescue information—such as stability of a structure, victim location within the structure, and search assessment—are communicated through BMS. Different types of building assessment methods (see Figure 1) and their building markings may be deployed at the same time in the disaster area each with its respective purpose.

Though a prime objective of each building assessment method is the safety of the rescuers, each method has a different focus and scope. However, when response teams do not conform to the standard BMS, information cannot be properly communicated. This leads to confusion, which may require re-assessment of the structure. In the current practice, the standard way of communicating BMS information in the field is through marking with the international orange spray paint on buildings. Re-assessment leads to re-marking and updating of the orange spray paint, which is

Figure 1. Four categories of BMS defined in the national US&R response system

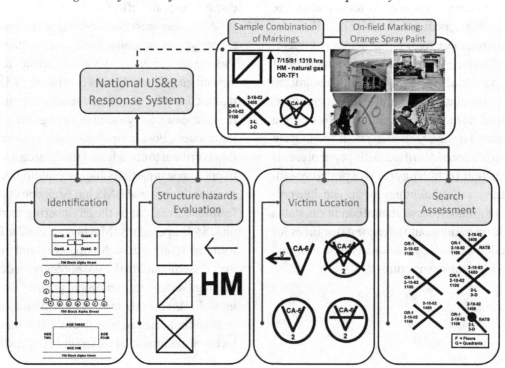

also hard to maintain and organize due to its physical limitations. In addition, it is difficult to resolve, or even to detect, contradicting assessment information. There is no unified interface to allow access to and collaboration between various types of building assessment methods and BMS from various organizations (see Table 1). This could potentially lead to ineffective communication that adversely affects rescuers' situational awareness and safety. Additionally, the orange spray paint not only limits the amount of information that could be communicated to the rescuers, but also hinders their access to and situational awareness of the structure's condition. The building markings may not be visible to rescuers from time to time due to the weather conditions, dust, darkness, or debris. As a result, a unified interface that communicates more and better information more quickly and reliably improves decision making and the underlying disaster response efforts. Table 1 depicts BMSs among different organizations and Figure 1 shows building markings of 4 BMS categories established by the FEMA National US&R Response System to be sprayed on building structures.

From past deployment of BMS, issues in the following have been highlighted: information flow, coordination of data and its integration,

Table 1. Different types of BMS

BMS	Organization/s	Reference
National US&R Response System	• FEMA • US Army Corps of Engineers • Local/Regional responders	(FEMA, 2003)
INSARAG	• International Search and Rescue Advisory Group (UN) • International US&R teams	(INSAR, 2005)
ATC Applied Technology Council	• Applied Technology Council (ATC) • FEMA • US Army Corps of Engineers	(ATC, 2001)

paper-based and error-prone forms, communication channels and information update from the structure's location to the Incident Command (Dawes et al., 2004). Table 2 summarizes these problems.

PROBLEM STATEMENT

Because time is a critical factor in terms of lifesaving during disasters, delay of response efforts could lead to unnecessary casualties (Gill, 2007). The probability of rescuing victims trapped under a collapsed structure greatly decreases after the first 24 hours (Mituno, 1999). In addition, how efficiently required resources are provided to the US&R teams is of great importance. As a result, this paper focuses on efficient information gathering at the disaster zone to enable a better resource distribution decision making.

Methodology

In order to resolve the communication and information retrieval issues highlighted, the use of Radio Frequency Identification (RFID) tags with computer networks for gathering, storing and sharing information is proposed. The onsite information gathering will be achieved through a mobile application utilizing RFID tags as storage for the assessment information (Chen et al., 2010). First responders will attach RFID tags on to buildings during disaster response, equivalent to spraying orange paint. Furthermore, with the digital assessment information, efficient decision making for resource deployment is enabled through a geospatial resource management service, which could facilitate improved efficiency and effectiveness of emergency response. The geospatial resource management service will compile the information gathered from the mobile application and process critical decisions. Figure 2 shows a use case for the system. First responders, through this system, will be able to create structural triage

Table 2. Problems faced in deployment of BMS

Problem	Origin
Marking Systems overlaps	Local/State/Federal Response
Remarking/Rework	Multiple Marking Systems
Communication Channels	Human work-cycle (TF/ICC)
Information Flow	Information not automatically updated
Building Marks Hidden	Smoke, debris, dynamic scenario

for a building structure, create BMS assessment, access and edit information for assessment information on the map, and request resources. The geospatial resource management service, ARMS (Automated Resource Management System), will receive the resource request for further decision making (Chen et al., in press). The challenges aforementioned are expected to be addressed by these following IT components.

Structural Assessment

Building Assessment System (BAS) is a mobile application developed to support building assessment procedures on the disaster site. BAS incorporates standard structural triage and BMS procedures from the FEMA National US&R Response System (Aziz et al., 2009; Peña-Mora et al., 2008). Figure 3 shows the structural/hazard evaluation markings embedded in BAS. In order to accommodate the situation where no communication infrastructure is operational, BAS operates on a mobile ad hoc network (Peña-Mora et al., 2010). BAS has been developed to run on laptops, tablet PCs and Personal Digital Assistants (PDA). The programming language BAS was implemented with is Visual C#. A user interface consisting of a map control through which the disaster site can be seen (see Figure 4) is embedded. The map control enables the visualization of the assessed buildings on the disaster site. When evaluating a building, the user places a RFID tag on the structure. As soon as the building is evaluated, a reduced format of

the evaluation is broadcast to other devices on the network. The complete document is stored on the RFID tag attached to the building and on the local device through which the building is assessed. The complete and reduced format is a design choice to avoid overloading the network. Currently, the location and safety status have been chosen for propagation to others on the network. On the map, a RFID tag is shown as a square marker with color-coding of green, yellow and red corresponding to safe, restricted, or dangerous of the condition of the building (see Figure 3 and Figure 4). Anyone with sufficient authentication who is connected within the network, as long as connected and, can view the map and access building assessment information for any building, by simply clicking on the particular marker they are interested in. The network protocol seeks for the complete document on the network, either from a local copy on a device or the RFID tag. Figure 3 shows the building structural collapse hazard marking system and its embedment in BAS.

Resource Deployment

Once onsite information is gathered, appropriate resource responses are required. Critical decisions need to be made to mitigate the damage resulting

Figure 2. Use case

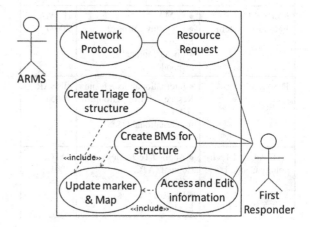

Figure 3. Structural/hazard evaluation marking and its embedding in BAS

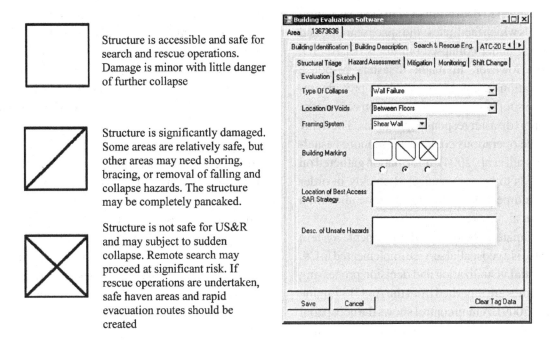

Structure is accessible and safe for search and rescue operations. Damage is minor with little danger of further collapse

Structure is significantly damaged. Some areas are relatively safe, but other areas may need shoring, bracing, or removal of falling and collapse hazards. The structure may be completely pancaked.

Structure is not safe for US&R and may subject to sudden collapse. Remote search may proceed at significant risk. If rescue operations are undertaken, safe haven areas and rapid evacuation routes should be created

Figure 4. Graphical user interface of BAS with a map and structural triage embedded

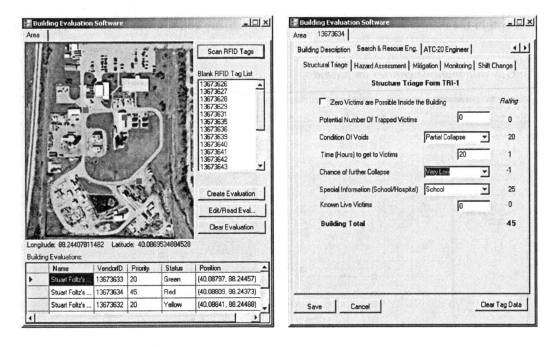

from the disaster. As mentioned before, access to heavy equipment is critical to disaster response operations (Gentes, 2006; SBC, 2006; Kevany, 2005; Bissell et al., 2004). Heavy equipment is a necessity during operations such as (1) rapid debris clearance of the transportation network so that first response teams can reach blocked hazard zones, (2) careful lifting of damaged structural elements in conditions when human power is not sufficient, and (3) selected debris removal to clear structural materials which facilitates void space searches and tunneling under collapsed buildings (ELANSO, 2009). However, in major disasters response organizations are not always able to provide immediate supply of heavy construction equipment to support disaster response operations. The delay in rescue operations could result in more casualties (Bissell et al., 2004). Information gathered on site needs to be communicated quickly in order to make timely decisions for resource allocation (Chen et al., in press).

Automated Resource Management System (ARMS) is a geospatial service implemented in C#. Geospatial visualization and decision processing capabilities are inherited from the ArcGIS Engine (ESRI, 2009). A map control shows the geospatial information gathered onsite and resource information in the geospatial database maintained prior the incident. Data is updated during the disaster response operations. Figure 5 shows critical information within the database such as available resources, hospital and schools. Figure 6 depicts a spatial query for backhoe loaders with capacity type I, which follows the definition of the National Incident Management System (NIMS). The green (light) circle marker on the map (see Figure 6) represents the locations that have the backhoe while the red (dark) marker represents other locations.

By introducing proper standard NIMS typing from the outset, when the resource is requested from the field, the right resource can be secured from the right vendor at the right time, taking much of the guesswork out of the equipment request process. This also enables resource requests to flow smoothly into the planning section (responsible for identifying the specific resources needed), the logistics section (responsible for the acquisition of resource), and the finance section (responsible for making arrangements for payment and then actual payment of a given resource).

ARMS is equipped with automated tools such as geocoding of incident location with address/coordinate and shortest routes with partial road-

Figure 5. GIS with multiple entities on the map

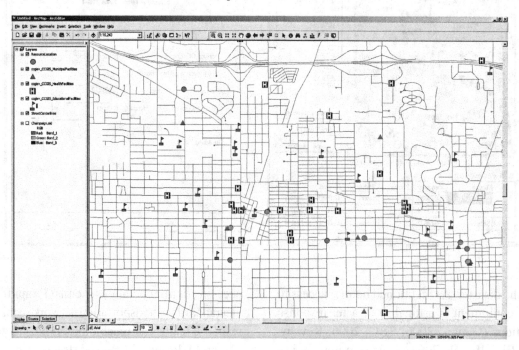

blocks for resource deployment. These tools are wrapped around with user-friendly interfaces and extension of existing tools. The graphical user interface (GUI) of the ARMS service is shown in Figure 7. The service listens to incoming connections and reads in the gathered data from BAS.

The incoming data is sent through XML. Figure 7 shows a resource request from BAS and a test run of the route generation from resource depots that has the requested resources to the target location, which is the Illinois Fire Service Institute, on the map.

Figure 6. Spatial query of desired resource type; highlighted are the resource matched

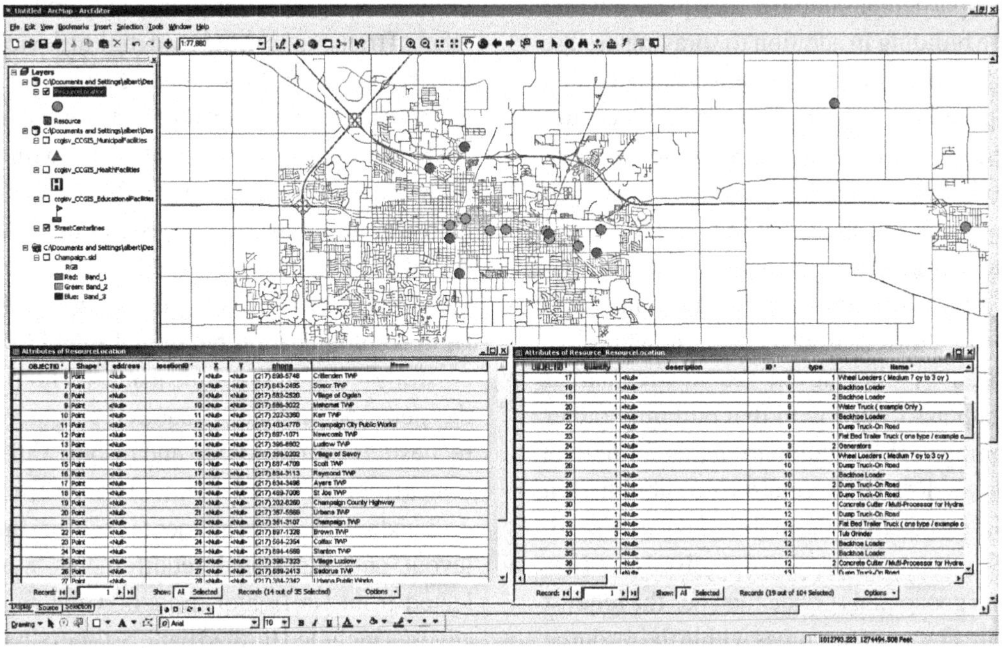

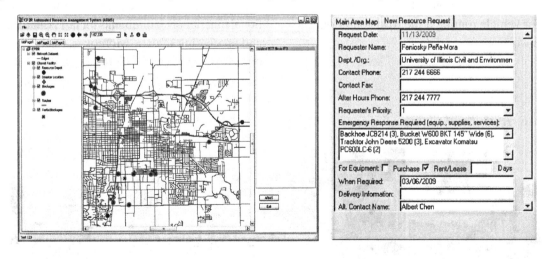

Figure 7. GIS visualization with fastest route analysis for resource deployment

Scenario

During disaster response, first responders, such as the US&R structural specialists or firefighters, are equipped with RFID tags and 802.11 networking enabled mobile devices, such as smart phones, PDAs or tablet PCs. Within the disaster zone, first responders set up a wireless ad hoc network with these mobile devices. When first responders arrive at a particular building, they place a RFID tag on the building in addition to spraying orange paint. The RFID tag stores assessment results for that particular building. The same procedure is carried out for each building within the area. Color-coded square markers show up on the map representing the assessed buildings (see Figure 4). If a responder nearby needs assessment information for a particular building, the responder can use the mobile device to retrieve the complete information by clicking on the square marker on the map associated with that building, rather than running to the building to search for orange spray marks. Resource requests are submitted in electronic format from the mobile devices as well.

When resource requests are received (see Figure 7), the geospatial service matches the request with the inventory of resources within the resource database. Once the resource is found, the service determines the most efficient routes from depots to the demand location. This potentially saves time for vehicle routing if the road conditions on the network are up-to-date.

TESTING

Field trials were carried out at the Illinois Fire Service Institute (IFSI) for system refinements. Field trials provide the opportunity to discover challenges not seen in laboratory settings. The system was tested at the IFSI training ground (see Figure 8) surrounded with simulated fire and smoke. Tests such as operation of BAS with the tester wearing full firefighting gear and measurement of transmission range of digital devices were carried out (Peña-Mora et al., 2009). The ARMS user interface was set up on a laptop computer and BAS was loaded on two tablet PCs with ad hoc network enabled. Three partially and totally collapsed buildings were assessed through Tablet PC A and RFID tags were placed onto the buildings for storage of assessment information. Tablet PC B, in communication range of Tablet PC A, successfully received assessment information through the network protocol. Required resources were requested through ARMS and resource allocation routes from resource depots to IFSI were produced.

Feedback on refinement of user interface layout and functional requirements were collected from firefighter trainers. Challenges discovered included limited battery time of handheld devices, limited range of signal for RFID tags and readers, screen visibility of mobile devices under sunshine, user interface improvement, and additional functional requirements for fast visualization and processing of data. Moreover, establishment of an ad hoc network requires a nontrivial proce-

Figure 8. Testing at the Illinois fire service institute

dure; it should not be assumed that users know how to carry out such a procedure. Therefore, a wraparound command script should be implemented for the network setup. In addition, there is a reasonable amount of uncertainty involved in using ad hoc networks for broadcast of critical information. Concrete slab or steel plates between mobile devices can interrupt the communication link. There is also potential risk of network and information overload when too many devices are on the network. The current data format for transmission and storage of the assessment information on RFID tags is in plain text. Although authentication is required to log into the system, packets sent on the network can still be intercepted. In addition, malicious writes to the RFID tags cannot be prevented in the current implementation. Further hardware and software improvements must be considered.

DISCUSSION

Because response actions are a sequence of interdependent operations, a more efficient resource supply for the initial US&R could greatly mitigate the damage caused by the disaster. Auf der Heide (1989) created the following example to show the interdependency of operations. Necessary medical care cannot be provided to victims unless the victim could be reached by emergency medical services (EMS) alive. When in serious condition, victims need to be sent to a hospital for surgery, which depends on the skills applied by private ambulance paramedics in the field. However, the EMS cannot gain access to the victim unless the victim is located in the rubble of collapsed or partially collapsed buildings, and in some conditions lifted under care out of the rubble to a safe and stable area. This requires the effort of US&R teams. A crane to remove the entrapping rubble or lifting of the victim is then required to carry out the rescue operations. This highlights

the importance of an efficient damage assessment and resource request and supply to save lives.

In the previous sections, information regarding damage assessment and resource distribution is streamlined in the digital format. A process for better information gathering, sharing, and compatibility is proposed for resource response to disasters. Electronic formats of assessment information and resource requests provide a better opportunity for a more efficient decision making. BAS could provide rescuers a better situational awareness of the working environment. Responders on the disaster site would have a better view of the conditions of surrounding buildings and locations of potential victims. The risk that responders are exposed to could be reduced and the productivity of search and rescue operations could be improved.

Because the initial assessment information of disasters is not always accurate (Auf der Heide, 1989), digital assessment that reduces the effort for information updates is especially useful. As discussed in previous sections, updating information, physically marked on structural components of buildings with the international orange spray paint, is not trivial. In addition, information dissemination in the paper format is difficult. BAS enables an easy update for the BMS, instantly viewable to responders on the network. This makes periodic updates of building assessment less costly in terms of time and effort. As a result, resource response to the disaster could be adjusted based on the updated needs.

ARMS provides decision makers better situational awareness of available resources. Understanding the resource needs and available resources could provide better information for decision making, which would make the on disaster site US&R operations more efficient. The standard NIMS format that ARMS has adapted makes it possible to match on disaster site resource requests with available resources in the digital inventory. Decisions such as vehicle/truck routing for loading and unloading of critical resources could potentially be scheduled to save time and effort to facilitate the logistical response to disasters.

However, there are still great challenges in the disaster response process including, but not limited to, protection of IT systems during and after disasters, resource inventory keeping, information fusion, multi organizational communication, security, and resource convergence (Holguín-Veras et al., 2007; Kendra & Wachtendorf, 2003; Auf der Heide, 1989). One of the major challenges is the process of logistics supply in the centralized command and control system. As disaster response is a highly complex and dynamic process, having too many layers of command makes the resource request process inefficient. Not only is receiving the request and locating the resource challenging, but the tracking of the request and the delay caused by the lengthy approval process also makes the logistical response difficult. In addition, management of volunteers and donation of resources challenge the centralized resource management. Management of high priority resources is sometimes hindered by the vast amount of resources in, or coming into, the command and control system. Mitigation of all of these difficulties in disaster response is critical.

CONCLUSION

In this paper an IT system with two system components, BAS and ARMS, is presented for on disaster site information gathering and resource allocation decision making to support US&R. BAS embeds structural triage and BMS used for post-disasters evaluation on a disaster site and resource request for US&R. Information critical to decision making is collected by first responders through BAS and is disseminated and retrieved electronically by decision makers. The resource management component, ARMS, processes and visualizes the collected information for automated decision making. With these data collection and data processing components, the system is expected to expedite decision making for a more effective

and efficient disaster response. Future research directions include incorporation of sophisticated logistics decision models and further system testing in on-field disaster response scenarios in collaboration with different levels of public emergency management agencies.

ACKNOWLEDGMENT

This paper is an extension of the conference paper included in the proceedings of the 7th International Conference on Information Systems for Crisis Response and Management (Chen et al., 2010). The authors would like to thank the National Science Foundation for their support of the grant of award number 0427089, Bill Keller (Champaign County EMA), Mark Toalson (Champaign County GIS Consortium) for his kind feedbacks, and Richard Jaehne (Director of IFSI) and Gavin Horn (Research Program Director of IFSI) for their help and guidance in the exercise at the IFSI. The authors also thank the Champaign County GIS Consortium for the geospatial data they have provided and the anonymous reviewers of ISCRAM and IJISCRAM for their invaluable comments.

REFERENCES

Applied Technology Council (ATC). (2001). ATC-20 Procedures used to evaluate Damaged Buildings near World Trade Center. *ATC News Bulletin*. Retrieved from http://www.atcouncil.org/pdfs/1101news.pdf

Auf der Heide, E. (1989). *Disaster Response: Principles of Preparation and Coordination*. St. Louis, MO: CV Mosby.

Aziz, Z., Peña-Mora, F., Chen, A. Y., & Lantz, T. (2009). Supporting Urban Emergency Response and Recovery Using RFID-Based Building Assessment. *Disaster Prevention and Management: An International Journal*, *18*(1), 35–48. doi:10.1108/09653560910938538

Bissell, A. B., Pinet, L., Nelson, M., & Levy, M. (2004). Evidence of the Effectiveness of Health Sector Preparedness in Disaster Response. *Family & Community Health, 27*(3), 193–203.

Centre for Research on the Epidemiology of Disasters (CRED). (2009). *2008 Disasters in Numbers.* Retrieved from http://www.reliefweb.int/rw/rwb.nsf/db900SID/LSGZ-7NJKJV?OpenDocument

Chen, A. Y., Peña-Mora, F., Mehta, S. J., Plans, A. P., Brauer, B. R., Foltz, S., & Nacheman, S. (2010, May 2-5). A GIS Approach to Equipment Allocation for Structural Stabilization and Civilian Rescue. In *Proceedings of the 7th International Conference on Information Systems for Crisis Response and Management (ISCRAM),* Seattle, WA.

Chen, A. Y., Peña-Mora, F., & Ouyang, Y. (in press). A Collaborative GIS Framework to Support Equipment Distribution for Civil Engineering Disaster Response Operations. *Automation in Construction.*

Dawes, S., Cresswell, A., & Cahan, B. (2004). Learning from Crisis: Lessons in human and Information Infrastructure from the World Trade Center Response. *Social Science Computer Review,* 52–66. doi:10.1177/0894439303259887

ELANSO. (2009). *Earthquake rescue – learning from disaster.* Retrieved from http://www.elanso.com/ArticleModule/HlVwUKSETDLcPzIsUf-VcPAIi.html

ESRI. (2009). *ArcGIS.* Retrieved from http://www.esri.com/products/indexb.html

Federal Emergency Management Agency (FEMA). (2003). *National Urban Search and Rescue Response System: Field Operations Guide.* Retrieved from http://www.fema.gov/usr/index.shtm

Gasiorek-Nelson, S. (2002). *DAU hosts 9/11 first responder: challenges and logistics of responding to Pentagon terrorist attack - Logistics Preparedness.* Retrieved from http://findarticles.com/p/articles/mi_m0KAA/is_5_31/ai_94771290/pg_3?tag=content;col1

Gentes, S. (2006). Rescue Operations and Demolition Works: Automating the Pneumatic Removal of Small Pieces of Rubble and Combination of Suction Plants with Demolition Machines. *Bulletin of Earthquake Engineering, 4,* 193–205. doi:10.1007/s10518-006-9006-1

Gill, G. (2007). *Will a Twenty-first Century Logistics Management System Improve Federal Emergency Management Agency's Capability to Deliver Supplies to Critical Areas, During Future Catastrophic Disaster Relief Operations?* Unpublished master's thesis, University of Phoenix, AZ.

Holguín-Veras, J., Perez, N., Ukkusuri, S., Wachtendorf, T., & Brown, B. (2007). Emergency Logistics Issues Affecting the Response to Katrina: A Synthesis and Preliminary Suggestions for Improvement. *Transportation Research Record: Journal of the Transportation Research Board,* 76-82.

International Search and Rescue Response (INSAR). (2005). *International Search and Response: Guidelines.* Retrieved from http://www.reliefweb.int/undac/documents/insarag/guidelines/Id-mark.html

Kendra, J. M., & Wachtendorf, T. (2003). Elements of Resilience after the World Trade Center Disaster: Reconstituting New York City's Emergency Operations Centre. *Disasters, 27*(1), 37–53. doi:10.1111/1467-7717.00218

Kevany, M. J. (2005). Geo-Information for Disaster Management: Lessons from 9/11. In van Oosterom, P., Zlatanova, S., & Fendel, E. (Eds.), *Geo-Information for Disaster Management* (pp. 443–464). Berlin, Germany: Springer. doi:10.1007/3-540-27468-5_32

McGuigan, D. M. (2002). *Urban Search and Rescue and the Role of the Engineer*. Unpublished master's thesis, University of Canterbury, New Zealand.

Mizuno, Y. (2001). *Collaborative Environments for Disaster Relief*. Unpublished master's thesis, Department of Civil & Environmental Engineering, Massachusetts Institute of Technology, Cambridge, MA.

National Commission on Terrorist Attacks Upon the United States. (2004). *9/11 Commission Report*. Retrieved from http://govinfo.library.unt.edu/911/report/911Report.pdf

Peña-Mora, F., Chen, A. Y., Aziz, Z., Soibelman, L., Liu, L. Y., & El-Rayes, K. (2010). A Mobile Ad hoc Network Enabled Collaborative Framework Supporting Civil Engineering Emergency Response Operations. *Journal of Computing in Civil Engineering*, *24*(3), 302–312. doi:10.1061/(ASCE)CP.1943-5487.0000033

Peña-Mora, P., Aziz, Z., Chen, A. Y., Plans, A., & Foltz, S. (2008). Building Assessment during Disaster Response and Recovery. *Urban Design and Planning*, *161*(4), 183–195. doi:10.1680/udap.2008.161.4.183

Sullum, J., Bailey, R., Taylor, J., Walker, J., Howley, K., & Kopel, D. B. (2005). *After the Storm Hurricane Katrina and the failure of public policy*. Retrieved from http://www.reason.com/news/show/36334.html

US Army Corps of Engineers. (2008). *Urban Search & Rescue Structures Specialist Field Operations Guide* (4th ed.). Washington, DC: Author.

This work was previously published in the International Journal of Information Systems for Crisis Response and Management, Volume 3, Issue 1, edited by Murray E. Jennex and Bartel A. Van de Walle, pp. 19-30, copyright 2011 by IGI Publishing (an imprint of IGI Global).

Chapter 3
WORKPAD:
Process Management and Geo–Collaboration Help Disaster Response

Tiziana Catarci
Sapienza Università di Roma, Italy

Massimo Mecella
Sapienza Università di Roma, Italy

Massimiliano de Leoni
Technical University of Eindhoven, The Netherlands

Alessandro Russo
Sapienza Università di Roma, Italy

Andrea Marrella
Sapienza Università di Roma, Italy

Renate Steinmann
Salzburg Research Forschungsgesellschaft mbH, Austria

Manfred Bortenschlager
Samsung Electronics Research Institute, UK

ABSTRACT

In complex emergency/disaster scenarios, persons from teams from various emergency-response organizations collaborate to achieve a common goal. In these scenarios, the use of smart mobile devices and applications can improve the collaboration dynamically. The lack of basic interaction principles can be dangerous, as it could increase the level of disaster or can make the efforts ineffective. This paper examines the main results of the project WORKPAD finished in December 2009. WORKPAD worked on a two-level architecture to support rescue operators during emergency management. The use of a user-centered design methodology during the entire development cycle has guaranteed that the architecture and resulting system meet end-user requirements. The feasibility of its use in real emergencies is also proven by a demonstration showcased with real operators. The paper includes qualitative and quantitative results and presents guidelines that can be useful in developing emergency-management systems.

DOI: 10.4018/978-1-4666-2788-8.ch003

INTRODUCTION

Due to the recent increase of safety threats like environmental disasters or terrorist attacks, Crisis Response has become a relevant application field for the development of new information technologies. In this context, team members need to collaborate in order to reach a common goal. The use of mobile devices and applications is valuable for the improvement of collaboration, coordination and communication amongst members of team(s) to achieve the desired goals. But there are also risks in the usage of mobile applications, e.g., decrease of performance. In emergency/disaster scenarios most of the tasks are highly critical and time demanding; for instance, in such scenarios the saving of minutes can result in saving people's lives. Therefore it is unacceptable to use systems that lack proper interaction principles.

The European project WORKPAD (www. workpad-project.eu) achieved to provide an architecture that intends to improve the collaboration in emergency management. According to the initial user requirements collection (de Leoni et al., 2007), and by the analysis of how Emergency Management is faced in the different European countries, the consortium learned that the most suitable architecture is two-level: a first level is deployed on the spot and a second level involves the servers of the different rescue organizations. There are several front-end teams on the field, each composed of several rescue operators. Rescue operators are equipped with PDAs and their work is orchestrated by a Process Management System (PMS) which is located on the team leader's PDA. In fact, based on several studies of emergency plans and end-user interviews, emergency plans can be seen as special cases of business processes (de Leoni et al., 2010). The Process Management System manages the execution of emergency-management processes by orchestrating the human operators with their software applications and some automatic services to access the external data sources

and sensors. The use of a PMS aims at improving the efficiency and effectiveness in dealing with the emergency's aftermath, thus reducing the event's consequence. At the back-end side data sources from several servers are integrated and the result is a single virtual data source that front-end devices can query, thus obtaining information aggregated from several sources.

The development of WORKPAD followed a methodology focused on user-centered design principles (Dix et al., 2003). This methodology relies on continuous involvement of users during the whole development cycle which guarantees that the final system meets user expectations. The section "Overview of the usability evaluation methodology" will provide detailed information about the different types of user tests performed within the scope of the project.

The feasibility of the WORKPAD system is demonstrated by a drill that the project consortium showcased on 18th June 2009 in the village of Pentidattilo (Calabria, Italy). In particular we simulated the occurrence of an earthquake and asked real users from different rescue organizations to deal with the situation by using WORKPAD. A video that illustrates the successful showcase is available at http://www.youtube.com/watch?v=48Hs5Qwg0ho. The section "Workpad Showcase" gives more details on a showcase storyboard and illustrates the interaction among the WORKPAD components. The section "Showcase Results" summarizes the showcase results and provides information on the user evaluation, whereas the section "Lessons Learned" mentions some guidelines which were established based on the evaluation results.

The existence of two levels in the architecture and the strong focus on user evaluation is a novelty compared with other relevant research projects in the area of emergency management, such as SHARE (http://www.share-project.org), FORMIDABLE, EGERIS (http://www.egeris.org) and ORCHESTRA (http://www.eu-orchestra.org).

BACKGROUND AND RELATED WORKS

System Requirements for Emergency Management

Jul (2007) reflects the American disaster management practices, investigating how the emergency size influences the response type and how collaboration should occur on the spot to deal with the aftermath of an emergency. Calamities are classified into three groups: (i) *local emergencies* that are short-lived event whose effects are localized in a single community; (ii) *disasters*, i.e., long-lived events affecting many communities, but community and response infrastructures are affected by few damages; and (iii) *catastrophes*, long-lived events affecting hundreds of communities, destroying almost every infrastructure and damaging the response systems. For example, the three categories comprise respectively a house explosion, the 9/11 terroristic attack, and Hurricane Katrina in 2005. From this moment on, we are not going to consider any longer local emergencies, since they do not require an extensive PMS support: the necessity of some collaboration is quite limited, due to the local nature of the happening. It is important to note that the American disaster management requirements reported in Jul (2007) are highly confirmed by the experience of the WORKPAD project.

As far as the Response and Communication Infrastructure, this can be characterized by local damages (medium-size disasters) or extensive destruction (large catastrophes). Even if it is not disrupted, past experiences suggest that the existing infrastructure should be used as less as possible. For instance, the Katrina catastrophe has shown that if all civil protection units use the existing infrastructure, it is destined to collapse or to experience a too low performance level, due to the overload. Indeed, it was not designed to support so many users at the same time. Therefore, it is advisable to opt for Mobile Networks (Requirement 1), which, generally speaking, are wireless networks in which hosts can act both as end points sending/receiving packets and as relays forwarding packets along the correct nodes' paths towards the intended recipients (Bertelli et al, 2008; Manoj & Baker, 2007).

The support for context awareness is crucial too: this is confirmed by (Jul, 2007): "context can be characterized by their similarity to environment known to the user, with individual contexts being very familiar, somewhat familiar or unfamiliar". Moreover "a given user, particularly in larger events, is likely to work in a variety of contexts, either because of physical relocation or because of changes in the context itself". Therefore, users cannot be assumed to have local knowledge of the geography and resources of the area. Consequently, PMSs should be integrated with Geographic Information Systems (GISs), which allow users to gain a deep knowledge of the area (Requirement 2).

As from Jul (2007), "context may be more or less austere" and "operations may be established in novel locations", as well as "response activities may be relocated". Hence, "uncertainty and ambiguity are inherent to disaster", from which it follows that response technology must allow for flexibility and deviation in their application, while imposing standard structures and procedures. So, PMSs should allow for large process specifications that are specialized time by time according to the specific happenings (Requirement 3), as well as they need to foresee techniques to adapt the process execution to possibly changing circumstances and contingencies (Requirement 4). Finally, PMS client tools must be extremely usable and intuitive. In fact, "the response typically involve semi-trained or untrained responders . . . and the proportion of semi-trained and untrained responders increases with the scale of the event, and they assume greater responsibility for response activities". As Emergency Management Systems are not used on daily basis but in exceptional cases, even experts could be not very trained: training sessions could be helpful, but a real emergency is totally different. From this, follows that Emergency Management Systems should be so intuitive that they can be easily mastered after few interaction sessions (Requirement 5).

Current-Day PMS Approaches for Emergency Management

Information systems are increasingly used for supporting emergency management activities, as they can significantly improve the effectiveness of the procedures and measures adopted for dealing with an emergency. Consider, for example, the use of Emergency Management Information Systems (EMISs) for supporting preparedness and mitigation activities (Turoff, 2002), Geographic Information Systems (GISs) for hazard and vulnerability mapping and analysis (Gunes & Kovel, 2000), Wireless Sensor Networks (WSNs) for monitoring risky areas (Lorincz et al., 2004), etc. However, when it comes to provide support for coordinating emergency operators during the response and short-term recovery activities, the availability of process-based information systems is still scarce.

Current approaches based on the adoption of PMSs in the emergency management domain mainly aim at providing support for the preparedness, response and recovery activities. In order to support emergency operators in quickly and efficiently defining process models, La Rosa and Mendling (2009) propose a domain-driven process adaptation approach based on configurable process models. Configurable process models capture and combine common practices and process variants related to specific emergency domains, which can be configured in a specific setting leading to individualized process models. In a configurable model, different process variants are integrated and represented through variation points, enabling process configuration. Variation points allow removing part of a large process specification that are irrelevant for the

current enactment, thus meeting Requirement 3 defined in the previous section. During process configuration, the requirements stemming from a specific emergency scenario reduce the configuration space, but due to the number of variation points and constraints, the model is in general too difficult to be manually configured. Process configuration is thus performed using interactive questionnaires, which allow process experts to decide each variation point and produce a configured model by answering specific questions.

A different approach based on design-time synthesis from scenarios is proposed in Fahland and Woith (2009). Under this approach, small processes, named *scenarios* and modeled as Petri Nets (van der Aalst, 1998), are dynamically merged upon request. Thus, a large emergency management process is synthesized by composing several fragments. This is a different approach for Requirement 3, but it is also useful for meeting Requirement 4: if needed, new scenarios can even be appended at run-time as a mechanism of adaptation of system and process behavior.

The support of process adaptation needs for emergency management is crucial (as from Requirement 4); a similar requirement already exists for classical business process management. Therefore, a large body of work has been devoted to this topic and different approaches have been proposed. When running processes need to be adapted to cope with changes in the execution environment, the adaptation approach can be either *manual*, envisioning an expert who is charge of modifying the instances to handle contextual changes and events, or *automatic*, where the schema of the running instances is automatically adapted to enable termination in the new execution environment. The only existing approaches that are applicable for emergency management are the *automatic* approaches, since *manual* ones would delay the execution in a way that can lead to serious consequences (e.g., death of people, collapsing of buildings, etc.).

Inside the category of the automatic approaches, the *pre-planned* strategies foresee that, when designing the process schema, the designers describe the policies for the management of all possible discrepancies that may occur. As a consequence, the number of possible discrepancies needs to be known a priori, as well as the way to deal with their occurrence. Therefore, *pre-planned* adaptation is feasible and valuable in static contexts, where exceptions occur rarely, but it is not applicable in dynamic scenarios where policies for too many discrepancies should be designed. On the other side, *unplanned* automatic adaptation approaches try to devise a general recovery method that should be able to handle any kind of event, including those unexpected.

Nowadays, commercial and open source PMSs use a *manual* adaptation approach, for example, ADEPT2 (Goser et al., 2007), ADOME (Chiu & Karlapalem, 2000), AgentWork (Muller et al., 2004), a *pre-planned* approach, for example, YAWL (van der Aalst & ter Hofstede, 2005), or both, and there exists no PMS using an *unplanned* adaptation approach (Mecella, 2008). An interesting previous case study for using PMSs for emergency management has been carried on using the AristaFlow BPM Suite (Lanz et al., 2010). AristaFlow, the commercial version of the ADEPT2 framework, allows for verification of the process structure and it features an intuitive approach to adapt process instances at run-time to deal with contingencies. This enables non-computer experts to apply changes and adapted processes are checked for soundness. But, unfortunately, the approach is still manual, even though interesting work has been conducted to simplify the procedures for adapting process instances. AristaFlow aims also at meeting Requirement 2: relevant information linked to tasks is visualized on geographic maps of the area where the emergency has broken out. In addition, AristaFlow provides a mobile version (Pryss et al., 2010) that can be installed on smartphones running Windows Mobile. The idea is that rescue operators connect to the server to retrieve the tasks they are assigned to and, later, they execute such tasks while disconnected from the PMS server. Finally or at any point in time, end users can synchronize their work with the server. The system supports the enactment of process fragments on mobile devices and semi-automatically handles errors and deviations that occur during executions. The system suggests the insertion and deletion of single activities (limited to human tasks) and the user manually and locally adapts process fragments running on the mobile device. The problem of this approach is that tasks are executed off-line and, hence, previous tasks can modify at any time the list of tasks that need to be executed afterwards. Consequently, users can be assigned to and carrying on tasks that, when synchronizing after finishing executing them, are learnt to be no more required. The most appropriate solution is that the server itself is constantly available on the spot running on mobile devices, so that off-line solutions are prevented.

An automatic process adaptation approach based on execution monitoring has been proposed in de Leoni, Mecella, and De Giacomo (2007) and de Leoni et al. (2009). According to the proposed approach, the PMS assigns tasks to resources considering execution context and resources' capabilities. For each execution step, an execution monitor aligns the internal virtual reality built by the system with the physical reality and data retrieved from the external world by sensors (intended as any software and/or hardware component able to get contextual information), possibly adapting the process to unforeseen exogenous events and producing an adapted process to be executed.

THE WORKPAD ARCHITECTURE

Two classes of users were identified: *Back-end* and *Front-end* users. The identification is based on the consortium's understanding of how Civil Protection works in Italy and other countries and on the collected user requirements (de Leoni et al., 2007). From an organizational perspective, front-end includes several teams of rescuers that are sent to area in order to manage an emergency, whereas back-end includes the control rooms/headquarters of the diverse organizations that have rescuers involved at front-end. These control rooms provide instructions and information to front-end teams to support their work. Typically, control rooms are provided with servers whose data need to be integrated in order to provide a unified view over the available information. At front end, every team is headed by a "leader operator", who coordinates the intervention of the other team members.

Figure 1 shows the overall WORKPAD architecture. The figure refers to one single (front-end) team with different operators who are coordinated in an emergency. The operators collaborate with the support of PDAs. Collaboration strictly depends on the possibility that operators and their devices can communicate with each other. Communication is executed on top of mobile networks (Requirement 1). Such mobile networks provide gateways to connect to Back-end Servers which are located at headquarters, where the operators store data and also receive further information for their work. Front-end teams are composed of several workers who are equipped with low-profile devices, i.e., PDAs. This fact poses several constraints in the development and deployment of the components, as memory is limited in size and CPUs are not very powerful (Satyanarayanan, 1996). Reduced screen size raises also new Human-Computer Interaction (HCI) issues, as the amount of information that can be visualized at the same time is not as much as on a laptop.

Figure 1. The overall WORKPAD architecture

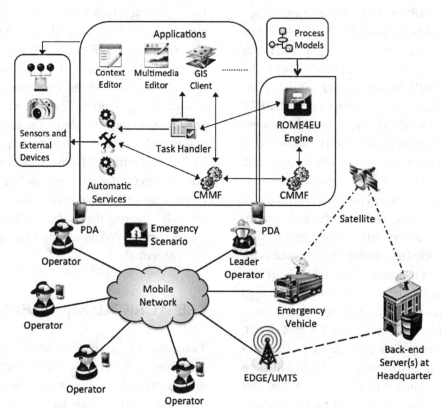

A Process Management System named RO-ME4EU is at the heart of the system. It manages the execution of emergency management processes, created from scratch or by customizing a previously defined process template (Requirement 3). The core component is the engine which assigns tasks to qualifying members. One of the workers in the team is the team leader. In addition to perform tasks, the leader supervises the work of the others. The engine is installed on the most powerful device which is typically the team leader's device. The engine performs task assignments on the basis of some preconditions over the process status. Preconditions can range from the completion of tasks to variables which have a value in a certain interval and to the availability of certain members skilled with specific capabilities (e.g., equipped with cameras or specific external sensors).

When a task has been assigned to a certain operator, the engine connects to the client running on the device named Task Handler. The Task Handler is an interactive GUI-based application that manages the interaction between rescuers and the engine, by providing intuitive features for a simplified management of the emergency (Requirement 5). The Task Handler is informed about each assignment made to the respective operator. The communication between the engine and the Task Handler relies on a Web service middleware. Each message is exchanged by a one-way invocation of a Web service end point. Once the Task Handler receives notification of a certain task assignment to respective users, it displays the name of the task together with relevant information. At any time users can decide to start a task by accepting the offer. In fact, task handlers do not execute process tasks: tasks are executed with the support of external applications. The Task Handler only takes care of mediating the interaction between users and the ROME4EU engine and starts the applications that support users in the execution of tasks. For instance, the

task "Build a medical tent" can be supported by the GIS-based application which shows the area, the terrain conditions and differences in altitude, as well as buildings and other objects of an area (Requirement 2). In this, the best location is identified where to build a tent. The fact that the ROME4EU's Task Handlers and the engine are fully operational on PDAs is an important novelty with respect to the current state of the art of PMSs and, more in general, Computer Supported Cooperative Work (CSCW). As a matter of fact, we motivate in (Battista et al., 2009) that almost all of the other Process Management Systems require the server engine to be running on a laptop/desktop. Furthermore, the ROME4EU engine follows an adaptation approach (pre-planned) based on process execution monitoring and driven by context awareness. Whenever discrepancies are detected thus leading to no successful process termination, the engine modifies the failing executing process into a new one (Requirement 4).

Some tasks may be automatic, i.e., no human intervention is required to carry them out. This kind of tasks is executed automatically by some special services running on a certain device. For instance, there exist some automatic services that retrieve environmental data from sensors and store them in the so-called Context Monitoring and Management Framework (CMMF). There are different kinds of sensors, such as the ones retrieving environmental data (e.g., humidity, temperature, precipitation) or others obtaining the state of devices (e.g., the battery level, the GPS position). CMMF is a P2P System for sharing information among PDAs (Juszczyk et al., 2009). The PDA on which the ROME4EU engine is installed is able to retrieve such environmental data from sensors installed on board of every team PDA. Almost all modern PDAs are equipped with GPS hardware and, hence, it is feasible to assume that every PDA is able to retrieve its own position. The information harvested by sensors is useful to monitor possible changes in the environment.

In the WORKPAD project the following applications have been developed to support the execution of emergency management tasks:

- The *Context Editor* component lets users enter additional contextual information that the sensors could not capture automatically. Context Editor stores all inserted data in the CMMF and retrieves them for the same component.
- The *Multimedia Editor* allows users to take and modify pictures of an area.
- The *GIS Client* (Bortenschlager et al., 2008) allows users to make an overview of the area and retrieve relevant information of the objects and buildings present in the area. The position of every team member is visualized to get a quick insight into the area where members are operating. Information of the relevant objects and buildings can also be updated. All the information is stored in a Back-End GIS Server and cached locally to the front-end team for quicker retrieval.
- The *Lightweight Sharing System* enables to share pictures, questionnaires and other files among all operators. In this way, every operator has the same knowledge of the situation s/he is facing.

The WORKPAD Front-end architecture is completely independent of the specific mobile-network technology e.g., it can be deployed on MANETs, or Mesh Networks but it works also with High Speed Downlink Packet Access (HSDPA) or Universal Mobile Telecommunications System (UMTS), even though this technology would delay the situation management due to the low bandwidth. For the showcase with the WORKPAD system in June 2009 we have deployed a Wireless Mesh Network (WMN) (Wang et al., 2008). A WMN is characterized by a backbone composed of several *Mesh Routers* that are connected with each other by multi-hop router paths. Each device connects to one of the mesh routers and can communicate with any other device that is connected to any of the routers. It is unfeasible to suppose mesh routers to be already deployed. Therefore mesh routers should be taken to the emergency area by the operators and power supply should be provided, as well.

The realistic solution is to equip civil protection jeeps with routers in order to have the necessary power supply. During the demo, rescue operators reached the area by jeeps equipped with such mesh routers, and they parked them in a way that the entire area is covered. Alternatively some of the rescue workers can be equipped with special bags that include a Mesh Router and a suitable battery.

WORKPAD front-end networks are connected to specific back-end systems (Vetere et al., 2009), which include a Web services platform to allow data exchange and integration. This platform is designed as a P2P network, in which each system (peer) can act as data provider, consumer and integrator. By plugging into WORKPAD's back-end network, a back-office system works as a WORKPAD back-end peer. This peer exports its ontology (i.e., a schema reflecting its conceptual model) which is mapped on a sort of global ontology. In this way peer data sets are integrated and exposed a single (virtual) data source that can be queried. During the showcase we have simulated several peers that accomplish the query "How many people were roughly located in building X?" We integrated data from the mobile-phone companies together with registry offices.

OVERVIEW OF THE USABILITY EVALUATION METHODOLOGY

The whole WORKPAD system results from applying a methodology that specializes the User-Centered software development process described in ISO Standard 13407. According to the ISO, users have been actively involved throughout the whole software development process, thus ensuring a user-driven development of the system. The results of the user tests were evaluated using qualitative evaluation methods. The evaluation results include improvement recommendations which the developers took into account. Tests allowed learning the correct requirements, thus obtaining a system that, from the one side, improves the task performance (efficiency), and, from the other side, ensures a higher level of usability. Figure 2 gives an overview of the usability evaluation methodology used in WORKPAD. Each evaluation step enabled an improvement of the prototype. We started with paper prototypes and performed the final evaluation of the WORKPAD system in the Pentidattilo drill in June 2009. In the following we list and explain the testing steps accomplished:

- **Online Pre-tests:** The first tests were conducted with online mock-ups. Users were asked to analyze animated images of how the system was envisioned. Then they had to answer a questionnaire on user satisfaction of the different components (i.e., task management, map overview, connection establishment, multimedia and context editor, file sharing). 13 users (8 male and 6 female) from Calabria region, 3 of age 46-60 and 10 of age 31-45, with different experience with PDAs participated in the test.

- **Controlled experiments:** After the development of a first software prototype, we conducted experiments on the different WORKPAD components in a laboratory under controlled conditions. These tests

Figure 2. Usability evaluation methodology in WORKPAD

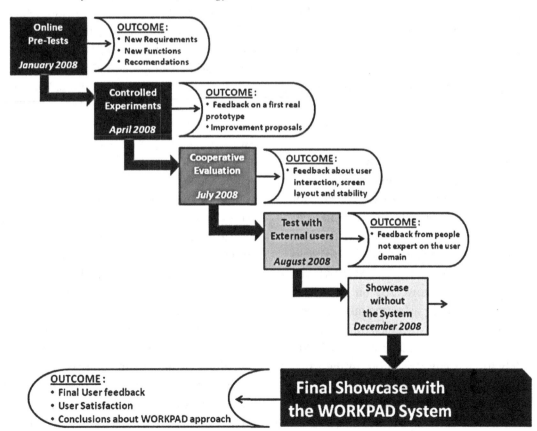

intended to observe users when using the system and wanted to discover open issues and areas of improvement. Special focus was given to the communication and the integration of the different components: users should feel the impression to work with a single system rather than with different components.

- **Cooperative evaluation:** This method is rooted in the notion that users typically prefer to get to know a system by using it rather than for example study a manual. Initially consortium members explained to the users the purpose of the WORKPAD system and the evaluation. Then users were asked to interact with the system in order to complete a specific task. Evaluators guided the users through the test and continuously interacted with them in order to gather information on user satisfaction. These tests were recorded by video cameras in order for us to analyze the level of the usability of the system off-line and look for recurrent usage patterns that possibly could be sped up.

- **Test with external users:** After the performance of the user test with expert users, the Consortium accomplished another usability test with people unfamiliar with the emergency topic in order to gain a different perspective on the usability issues.

- **Showcases:** Term "showcase" stands for the set-up of a concrete and realistic scenario deploying the system with the purpose of testing the feasibility, effectiveness and efficiency of a system. In WORKPAD, the system was showcased on 18th June 2009 by simulating an earthquake in a certain site with the involvement of real emergency operators. In December 2008 big celebrations were carried out in the same zone of our showcase in order to commemorate a real earthquake that devastated the cities of Messina (Sicily) and Reggio Calabria (Calabria) in Italy in 1908. It caused the death of more than 100.000 people. During celebrations simulations of rescue operations were executed in order to show how real emergencies are nowadays executed (without the WORKPAD's system). In particular we focused our attention on two of the showcases. The first one concerned "building some medicals tents" to provide health support to injured people. The second one dealt with "saving people entrapped in buildings" such as their own houses or offices. According to the experiences earned during their execution in December, we defined two accordant storyboards for the WORKPAD showcase in June.

THE WORKPAD SHOWCASE

The WORKPAD system was completely deployed in a realistic setting in accordance with the architecture previously described. And it was tested during a simulation of an earthquake that was supposed to occur in the small village of Pentidattilo (Reggio Calabria, Italy) on 18th June 2009. As previously said, two storyboards were executed, but the lack of space does not allow us to give details on both. We focus on the storyboard "Build medical tents".

Figure 3 depicts the structure of the process for this intent. In the first task emergency teams have to become aware of the area where tents need to be built. "Query about how many medical tents can be built" is an automatic task that is executed by an automatic service that connects to the Back-End and queries for obtaining the number of medical tents required. This query is evaluated by integrating data about the number of injured people and data about the maximum number of tents that are available in certain storage areas. At this stage the process splits in many branches that have the same structure. Each branch concerns one of tents

Figure 3. The process describing the storyboard: "establishing a medical point"

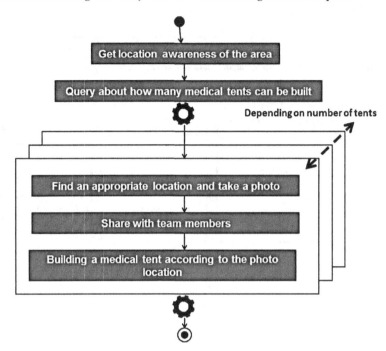

that need to be built up as from the query result. Once initiated these branches are independent of each other and work on local variables, which are synchronized when the branches join again. Every branch comprises the tasks (i) to find an appropriate location for a tent and take a photo of it; (ii) to share pictures with the other members; and (iii) to build the medical tents in the locations shown in the pictures shared at point.

Task (i) is carried out with the support of the Multimedia Editor, whilst task (ii) is supported by the Lightweight Sharing System. Task (iii) is slightly different, as most of the work required is not executed by the WORKPAD system: the rescuer uses the Multimedia Editor to watch the photo of the location of interest and then moves there where she builds manually the tent.

During the showcase, we aimed to gain feedback from people with diverse cultural background. In this way, we could improve the effectiveness and efficiency by leveraging on comments from a wider range of users, thus obtaining a final system that mediate different

needs. Therefore, the showcase involved rescuers from several organizations that are typically involved in emergency management. In particular, the following organizations were involved: Fire Brigades (in Italian: *Corpo Nazionale dei Vigili del Fuoco – VVFF*), the National Body for Alpine and Speleological Aid (*Corpo Nazionale Soccorso Alpino e Speleologico – CNSAS*), Service of Urgency and Medical Emergency (*Servizio di Urgenza ed Emergenza Medica - SUEM*), Italian Red Cross *(Croce Rossa Italiana – CRI)* as well as two voluntary organizations, i.e., *Europa Unita* and *Confraternita Misericordia*.

The next section provides a summary of the showcase's results and the user eventuation. This section describes the messages that are exchanged between the WORKPAD's component to execute certain tasks (such as method invocations, files sent/received, etc.). Figures 4 and 5 show two sequence diagrams: arrows are annotated with the size of messages in order to get an insight into the amount of data to be exchanged. The purpose is to demonstrate that we successfully achieved the

Figure 4. The interaction diagram of the WORKPAD components to enable the execution of task "Get Location awareness of the area"

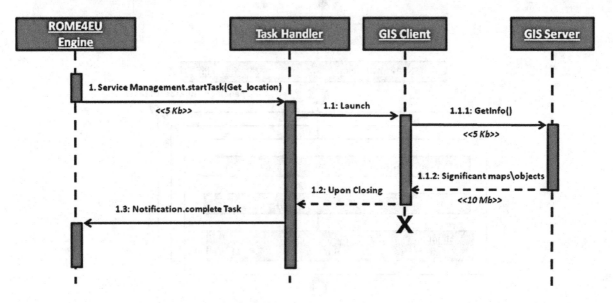

Figure 5. The interaction diagram of the WORKPAD components to enable the execution of task "Query about how many medical tents can be built"

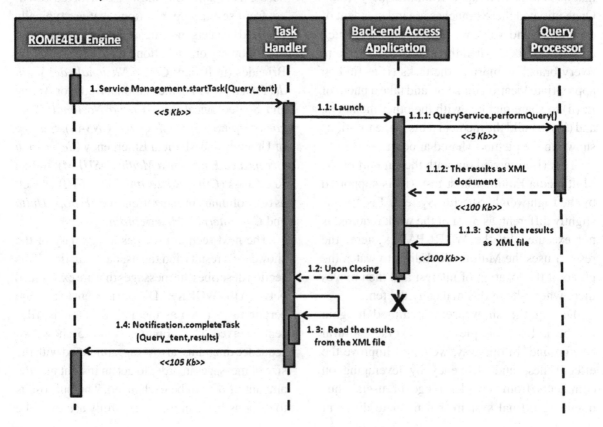

goal of keeping the exchanged data size relatively low and take the slowness of the available communication links into account.

As figures show, the interaction always starts from the ROME4EU engine that assigns tasks to appropriate workers/services. The worker is informed of the task s/he is assigned to through the Task Handler (TH) installed on his/her device, which displays the task name and other information relevant for its execution (Figure 6b shows the way these data are displayed on the GUI of Task Handler). When workers/services complete the execution of a certain task, Task Handler informs the engine about the event. The basic size of such messages is quite small, around 5 Kilobytes. Of course the message size may grow depending on the size of the input/output data enclosed in the messages. It is also worthy highlighting that at any time the team leader can overview the status of the other team members in terms of tasks assigned, running, completed. Figure 4 shows the interaction diagram to carry out task "Get Location awareness of the area". This task involves four components: the RO-

ME4EU engine, TH, GIS Client and GIS Server. The interaction diagram is depicted in Figure 4. The GIS Client is the support application for the task execution. The GIS Server is used by GIS Client to retrieve the map of the area. When this task is ready for assignment, ROME4EU chooses the member that guarantees the best performance. Afterwards, his/her Task Handler displays the information of the assignment so as to inform it. When the worker is ready, she/he clicks on the *Start* button. This task is associated to the GIS Client and, hence, after clicking on the button, such an application is launched. GIS Client needs to show some area maps, which are stored in some GIS Server at back-end. Therefore, GIS Client connects to that server, thus retrieving the orthophotos requested. Figure 6c shows how GIS Client looks like, once the map is obtained. When the GIS Client application is closed, the task is considered as completed and, hence, the engine gets a notification about the completion, thus enabling other tasks to be assigned.

Figure 5 depicts the interaction diagram for task "Query about how many medical tents can

Figure 6. Screenshots of some WORKPAD tools

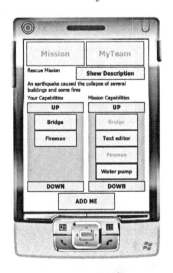

a) The outlook of the team members and their state.

b) The Task Handler after the assignment of a task to the user

c) The Client of Geographic Information System with some annotation

be built". This task involves four components: the ROME4EU engine, TH, Query processor and an ad-hoc service that queries the back-end for the right number of tents that need to be built up. This information involves an integration of data from different sources, such as to know the number of tents available in a certain storage area or the number of people available expert for building up the tents, the number of power generators available, etc. The description of the remaining interaction follows the same pattern as task "Get Location awareness of the area".

SHOWCASE RESULTS

This section describes the results of the showcase. During the showcase we considered how frequently users required assistance from consortium members during the execution of showcase tasks with the WORKPAD system. Further on we focused on the number of tasks which were correctly or wrongly carried out and measured the system robustness. Conclusions were drawn on possible problems that users encountered during system usage and on efficiency of such a kind of system for emergency management.

Each of the two storyboards was executed twice. In the first "trial" execution people from the

WORKPAD consortium were side by side with the rescue operators and explained them how to use the system to execute the tasks of each storyboard. Afterwards users moved to Pentidattilo which was simulated to be affected by an earthquake and took part in a second "real" showcase. Consortium members accompanied the emergency operators and filled in the task execution forms from which the indicators were derived. Figure 7 shows the results measured for such indicators. The first conclusion that can be drawn is that the execution of storyboard tasks was longer in the "trial" run than in the real showcase. From this we conclude that the users learned the WORKPAD system very quickly so as to be able to gain benefits just after one run. This high level of usability is resulted from applying a user-centered design approach throughout the whole system design and engineering process. The ease of use of the WORKPAD system is also confirmed by the other indicators. The "Number of assistance operations required per tasks" dropped from 2.0 in the trial to 0.7 in the showcase, and "Mean duration of task executions", which decreased on average by 0.6 after only one execution.

After each storyboard execution 12 users were interviewed with the goal to get information on user satisfaction and also to collect proposals for further improvement of WORKPAD. Firstly we

Figure 7. Summary of task execution form results

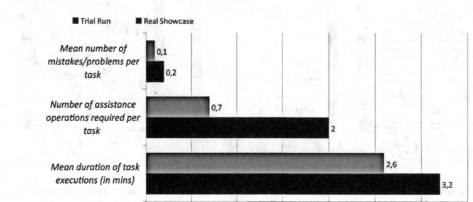

made clear to the users that WORKPAD is still a research prototype which can cause some problems in the usage of the system. For half of the test users it was the first time that they used a PDA. This aspect is particularly significant with regard to the good performance results measured during the drill. User feedback is summarized in Figure 8. The first question was whether they regard the system as intuitive and easy to use: four users fully and eight partially agreed (see Figure 8a). Some users added that they had problems with the visibility on the screen in the blazing sun. The problem of the sunbeams on the PDA's screen seems to be a critical issue; we already discovered the problem during some previous usability tests

and tried to improve the screen visibility in these conditions. But it seems that PDA screens are built using technologies that naturally do not address well the direct sunbeams on it. Other mentioned issues concerned the slowness of the system, but we are sure that further improvements of the WORKPAD system would solve the problem through new technological solutions. As depicted in Figure 8c, two third of the users repute WORKPAD can really improve the rescue operations. In particular, Picture Taking and Sharing, Geo-referenced Information and Process Management were the topics that mostly impressed the users (see Figure 8d).

Figure 8. Pie charts of some results of the user evaluation of the WORKPAD system

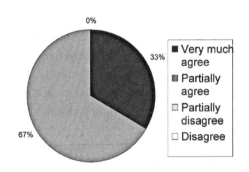

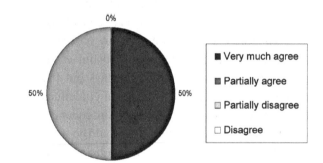

a) Is the WORKPAD system intuitive and easy to use?

b) The system efficiently supports users during their execution of emergency-management activities.

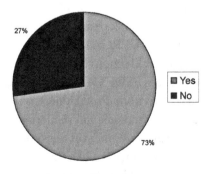

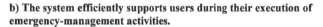
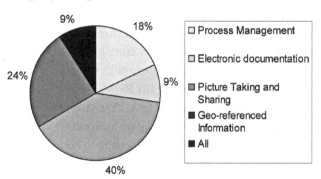

c) Does the WORKPAD system improve emergency management?

d) Which are the aspects that are more useful?

LESSONS LEARNED AND CONCLUSION

In the light of results obtained during the showcase in term of users' satisfaction and of improvement of the efficiency and effectiveness in emergency management, we are able to mention a few guidelines. These should be taken into account when an emergency management system is being developed:

1. Process Management Systems are useful during emergency management to improve the effectiveness (Hagebölling et al., 2009) and the process is a metaphor that is well understood by end users.

2. Task Handler and other client applications should be developed for the usage in extreme conditions and under sunbeams. The use of high-contrasting color should be considered (such as black on white, yellow on blue). This was also claimed in Agostini et al. (1996) for CRT monitor displays but the direct applicability of the same guidelines to PDAs does not follow as a consequence. In addition, the GUI widgets should be sized in the way that stylus pens can be avoided (e.g., buttons should be rather big).

3. As also claimed in Kellerer et al. (2005), Geo-referenced Information Systems are necessary to allow users to gain insight on the current situation. Indeed, they can be combined with ROME4EU in order to enable and improve the collaboration among team members. For instance, geographic information can be used to support ROME4EU to decide whom a certain task should be assigned to (e.g., by leveraging on locations where users are and where tasks need to be executed).

4. Data exchange and integration of various back-end systems located at different headquarters cannot rely on a hierarchical infrastructure. Different systems cannot rely on a common data schema ("global view") or a repository shared among all systems of all organizations; organizations would not trust such external entities. As in WORKPAD, different organizations need to keep using their (legacy) systems with private schemas, while, at same time, contributing to the global knowledge in a flexible and dynamic way. Moreover, due to rigid privacy policies, organizations need to control what parts of data they want to disclose, map and integrate in the global knowledge, thus making them available for the rescue operations.

In conclusion, the WORKPAD project has released a novel system, whose novelties lay in the interplay of process management and geo-awareness as basic tools for improving collaboration. This, to the best of authors' knowledge, is one of the few research projects effectively releasing an industrial mock-up, i.e., a running system which with a few efforts can be effectively marketed and made available to emergency operators. The authors should be very soon involved in an effort aiming at deploying the system in Calabria.

ACKNOWLEDGMENT

This work has been partially supported by the European Commission through the FP6-2005-IST-5-034749 project WORKPAD. The authors would like to thank all project partners, namely Università di Roma TOR VERGATA, IBM Italia - CAS, Salzburg Research, Technische Universitaet Wien, APIF Moviquity, SOFTWARE 602 and Regione Calabria – Settore Protezione Civile, for useful discussions and insights undertaken during the design and development of the system. A preliminary version of this work has been published as: Tiziana Catarci, Massimiliano de Leoni, Andrea Marrella, Massimo Mecella, Manfred Bortenschlager. The WORKPAD Project Experience: Improving the Disaster Response through

Process Management and Geo Collaboration. In Proceedings of the 7th International Conference on Information Systems for Crisis Response and Management (ISCRAM 2010), Seattle, USA, May 2010

REFERENCES

Agostini, T., & Bruno, N. (1996). Lightness contrast in CRT and paper-and-illuminant displays. *Perception & Psychophysics*, *58*(2), 250–258. doi:10.3758/BF03211878

Battista, D., Graziano, D., Franchi, V., Russo, A., de Leoni, M., & Mecella, M. (2009). *A Web Service-based Process-aware Information System for Smart Devices*. Retrieved from http://padis2. uniroma1.it:81/ojs/index.php/DIS_TechnicalReports/issue/view/198

Bertelli, G., de Leoni, M., Mecella, M., & Dean, J. (2008). Mobile Ad hoc Networks for Collaborative and Mission-Critical Mobile Scenarios: a Practical Study. In *Proceedings of the 17th IEEE International Workshops on Enabling Technologies: Infrastructure for Collaboration Enterprises* (pp. 147-152).

Bortenschlager, M., Goell, N., Haid, E., Rieser, H., & Steinmann, R. (2008). GeoCollaboration-Location-based Collaboration in Emergency Scenarios. In *Proceedings of the 17th IEEE International Workshops on Enabling Technologies: Infrastructure for Collaborative Enterprises*.

Chiu, D., Li, Q., & Karlapalem, K. (2000). A logical framework for exception handling in ADOME workflow management system. In *Proceedings of 12th International Conference on Advanced Information Systems Engineering* (pp. 110-125).

de Leoni, M., De Giacomo, G., Lespèrance, Y., & Mecella, M. (2009). On-line adaptation of sequential mobile processes running concurrently. In *Proceedings of the 2009 ACM Symposium on Applied Computing* (pp. 1345-1352).

de Leoni, M., De Rosa, F., Marrella, A., Mecella, M., Poggi, A., Krek, A., & Manti, F. (2007). Emergency Management: from User Requirements to a Flexible P2P Architecture. In *Proceedings of the 4th International Conference on Information Systems for Crisis Response and Management* (pp. 271-279).

de Leoni, M., Marrella, A., & Russo, A. (2010). *Process-Aware Information Systems for Emergency Management*. Paper presented at the International Workshop on Emergency Management through Service Oriented Architectures.

de Leoni, M., Mecella, M., & De Giacomo, G. (2007). Highly Dynamic Adaptation in Process Management Systems Through Execution Monitoring. In *Proceedings of the 5th International Conference on Business Process Management* (pp. 182-197).

Dix, A., Finlay, J., Abowd, G., & Beale, R. (2003). *Human Computer Interaction* (3rd ed.). Upper Saddle River, NJ: Prentice Hall.

Fahland, D., & Woith, H. (2009). Towards process models for disaster response. In *Proceedings of the Business Process Management Workshops* (pp. 254-265).

Goser, K., Jurisch, M., Acker, H., Kreher, U., Lauer, M., Rinderle, S., et al. (2007). Next-generation Process Management with ADEPT2. In *Proceedings of the BPM Demonstration Program at the 5th International Conference on Business Process Management*.

Gunes, A. E., & Kovel, J. P. (2000). Using GIS in emergency management operations. *Journal of Urban Planning and Development*, *126*(3), 136–149. doi:10.1061/(ASCE)0733-9488(2000)126:3(136)

Hagebölling, D., & de Leoni, M. (2009). Supporting Emergency Management through Process-Aware Information Systems. In *Proceedings of the Business Process Management Workshops* (pp. 298-302).

Jul, S. (2007). Who's Really on First? A Domain-Level User and Context Analysis for Response Technology. In *Proceedings of the 4th International Conference on Information Systems for Crisis Response and Management* (pp. 139-148).

Juszczyk, L., Psaier, H., Manzoor, A., & Dustdar, S. (2009). Adaptive Query Routing on Distributed Context - The COSINE Framework. In *Proceedings of the International Workshop on the Role of Services, Ontologies, and Context in Mobile Environments* (pp. 588-593).

Kellerer, W., Schollmeier, R., & Wehrle, K. (2005). Peer-to-Peer in Mobile Environments. In Steinmetz, R., & Wehrle, K. (Eds.), *Peer-to-Peer Systems and Applications* (pp. 401–417). doi:10.1007/11530657_24

La Rosa, M., & Mendling, J. (2009). Domain-Driven Process Adaptation in Emergency Scenarios. In *Proceedings of the Business Process Management Workshops* (pp. 290-297).

Lanz, A., Kreher, U., Reichert, M., & Dadam, P. (2010). Enabling Process Support for Advanced Applications with the AristaFlow BPM Suite. In *Proceedings of the Business Process Management Conference Demonstration Track*.

Lorincz, K., Malan, D. J., Fulford-Jones, T. R., Nawoj, A., Clavel, A., & Shnayder, V. (2004). Sensor networks for emergency response: Challenges and opportunities. *IEEE Pervasive Computing / IEEE Computer Society [and] IEEE Communications Society*, *3*, 16–23. doi:10.1109/MPRV.2004.18

Manoj, B. S., & Baker, A. H. (2007). Communication challenges in emergency response. *Communications of the ACM*, *50*(3), 51–53. doi:10.1145/1226736.1226765

Mecella, M. (2008). Adaptive Process Management. Issues and (Some) Solutions. In *Proceedings of the 2008 IEEE 17th Workshop on Enabling Technologies: Infrastructure for Collaborative Enterprises* (pp. 227-228). Washington, DC: IEEE Computer Society.

Muller, R., Greiner, U., & Rahm, E. (2004). AGENTWORK: a workflow system supporting rule-based workflow adaptation. *Data & Knowledge Engineering*, *51*, 223–256. doi:10.1016/j.datak.2004.03.010

Pryss, R., Tiedeken, J., & Reichert, M. (2010). Managing Processes on Mobile Devices: The MARPLE Approach. In *Proceedings of the CAiSE'10 FORUM*.

Satyanarayanan, M. (1996). Fundamental challenges in mobile computing. In *Proceedings of the 15th Annual ACM Symposium on Principles of Distributed Computing* (pp. 1-7).

Turoff, M. (2002). Past and future emergency response information systems. *Communications of the ACM*, *45*(4), 29–32. doi:10.1145/505248.505265

van der Aalst, W. M. P. (1998). The Application of Petri Nets to Workflow Management. *Journal of Circuits. Systems and Computers*, *8*(1), 21–66.

van der Aalst, W. M. P., & ter Hofstede, A. H. M. (2005). YAWL: Yet Another Workflow Language. *Information Systems*, *30*(4), 245–275. doi:10.1016/j.is.2004.02.002

Vetere, G., Faraotti, A., Poggi, A., & Salvatore, B. (2009). Information Management for Crisis Response in WORKPAD. In *Proceedings of the 6th International Conference on Information Systems for Crisis Response and Management.*

Wang, X., & Azman, L. (2008). IEEE 802.11s Wireless Mesh Networks: Framework and challenges. *Ad Hoc Networks*, 6(6), 970–984. doi:10.1016/j.adhoc.2007.09.003

Chapter 4
Experience Report:
Using a Cloud Computing Environment During Haiti and Exercise24

Brianna Terese Hertzler
San Diego State University, USA

Eric Frost
San Diego State University, USA

George H. Bressler
San Diego State University, USA

Charles Goehring
San Diego State University, USA

ABSTRACT

The events of September 11, 2001, the Indian Ocean tsunami in 2004, and Hurricane Katrina in 2005 awakened American policymakers to the importance of the need for emergency management. This paper explains how a cloud computing environment can support social networks and logistical coordination on a global scale during crises. Basic cloud computing functionality is covered to show how social networks can connect seamlessly to work together with profound interoperability. Lastly, the benefits of a cloud computing solution is presented as the most cost-effective, efficient, and secure method of communication during a disaster response, with the unique capability of being able to support a global community through its massive scalability.

INTRODUCTION

The last decade has endured multiple man-made and natural disasters. They have been physically devastating, such as Hurricane Katrina, they have been traumatic, such as 9/11 in America, and some have been widespread, like the Indonesian/Malaysian/Thai tsunami. One common characteristic that they all have in common is that the world stayed connected during and after the event through new technological capabilities.

DOI: 10.4018/978-1-4666-2788-8.ch004

The global community continues to be increasingly connected as technology evolves. Just as global economies are affected, both positively and negatively, by trading goods with each together, so has the emergency response system for wide-scale disasters. Therefore, as societies become more efficiently connected it is imperative that we respond faster, with greater effect, in order to reverse the potential damage caused to our economies, infrastructure, and personal lives. Through these events, catastrophic insufficiencies in communication and logistical coordination have become publicly apparent. The severe overloading of technological services within moments of the incidents was unavoidable.

With technological advancements, emergency managers have been able to better respond to disasters with groundbreaking success. Technologies such as geographic information systems, with visual technologies, and analytical reasoning capabilities are helping address these problems. However, when these services are run though a cloud computing environment, additional availability, performance, and reliability, make these services dramatically better. Cloud computing is a new application deployment approach through which to address emergency management. While the concept is young, its effects and benefits are far-reaching.

Examples of the cloud computing benefits, in response to the above insufficiencies, have already started to prove their worth. For example, the use of a cloud computing environment transitioned from hypothetical theory to fact in the wake of the Haiti disaster. Geographic information systems and social networks were used to provide a platform for the entire world to communicate and coordinate efficiently, and effectively, to meet the Haitian needs. In another example, the uses of cloud technology were further solidified though Exercise24, a two-day-long international and multidisciplinary crisis simulation. In this exercise, technologies were implemented to connect civilian and military organizations, in humanitarian assistance and disaster relief efforts, by using cloud-based applications.

Cloud computing is creating an environment for emergency managers and decision-makers to work and respond seamlessly. The breakthroughs and enhancements made in the cloud computing world have started to allow emergency managers to respond in a quick, effective, and efficient manner, changing the field of emergency management in drastic and profound ways.

This paper begins by evaluating some of the major technical areas of weakness presented during disasters. Many problems can be identified that need to be addressed and solutions found. Once there is an understanding of what is needed, steps can then be taken to evaluate currently available systems. This is followed by suggested improvements that can be implemented to address the noted problems. Furthermore, all new system proposed are evaluated to ensure that they operate within security requirements. The balance of the paper reviews how some of the new systems are being implemented. This includes exercises, such as Exercise24, where both its success, as well as some remaining open issues, is reviewed.

What is Cloud Computing?

There is potential for a lot of confusion surrounding the definition of cloud computing. In its basic conceptual form, cloud computing involves five primary fundamentals: shared resources, on-demand, elasticity, networked access, and usage-based metering (Craggs, 2009). Shared resources are the shared pool of IT resources, such as applications, processors, storage and databases. On-Demand allows users to call up resource from the cloud and use them as needed. When the user is finished with the resources they release them in a self-service fashion (Craggs, 2009). Elasticity, or flexibility that includes scalability, allows the cloud to be dynamic to the user's demand, allowing the cloud to satisfy peak demands and then release resources when demand subsides (Craggs, 2009). Networked access allows the cloud to be accessible widely, primarily though the internet. Lastly, the usage-based metering allows users of the cloud to pay for the services when needed and used, and to release them when they are no longer need, resulting in many benefits including cost and storage efficiency (Craggs, 2009).

The foundation of cloud computing is virtualization (Golden, 2010). Virtualization is the consolidation of servers and environment management. Cloud computing implementation has an underlying virtualization layer (Craggs, 2009). However, cloud computing is much more than virtualization. While the virtualization concept allows a decent amount of scalability for a particular task, cloud virtualization can demand extreme scalability. For example, virtualization takes a particular workload and allocates particular resources, when it exceeds those resources it requires an IT specialist to evaluate and implement the changes needed. In a cloud environment, if more resources are needed than what is currently available, the cloud offers a self-service approach. The user can requests additional capabilities, applications, or resources, and the requests can be satisfied without delay of configuration changes; the latter which can take weeks in a traditional computing environment (Craggs, 2009).

One fundamental tenant of cloud is widely available access to applications. Clouds involve user-based metrics that allow the cloud provider to allocate individual resources usage to the appropriate user when they are using the application.

Another fundamental tenant of cloud is that the cloud does not specify what resources are provided in the cloud or where the cloud is located. In contrast, traditional computing has corporate boundaries on specifically designed premises. Cloud computing may run applications that vary in resource and location.

It is important to note that conflicting definitions about cloud computing currently exist. The varying definitions often stem from attempting to define the different types of clouds, such as private cloud as compared to public cloud, or the various use cases of cloud, such as the multiple array of applications from Tweeter at one extreme to an internal NASA application on the other; yet both can be deemed, *cloud computing*. This ultimately confuses the fundamentals that separate them. Therefore, the definitions above are designed to represent the fundamentals of cloud computing and not how cloud is being implemented; forming generally accepted way to talk about clouds (Reese, 2000).

BENEFITS OF THE CLOUD

Cloud computing offers an array of benefits that traditional computing systems cannot provide. Cloud has the ability to break dependency on hardware and allows the user to instantly access resources from wherever, and whenever, they are needed. The user no longer needs to hoard resources when they are not needed, and cloud allows access to more dynamic resources and applications as needed very quickly (Golden, 2010). Furthermore, cloud allows users to fluctuate optimally, or flex the amount of resources, as needs and demands increase and decrease.

Cloud computing turns the machine into a virtual software or application image, which resides on some physical server in the cloud's hosting environment (Golden, 2010). However, that virtual software image can be diverted between hardware resources, breaking the hardware dependency associated with external hosting. This is a primary benefit, and concern, within cloud computing. With hardware dependency no longer an issue, a user system is insulated from hardware breakdowns (Golden, 2010). If a hosting system falters, or becomes overloaded, the cloud provider would automatically move the user software image to another piece of hardware while it fixes the original hardware. Additionally, if an application starts to demand more resources then what are assigned to it, then more resources can easily be added to ensure the system does not suffer (Golden, 2010).

Cloud connects the hosts with the users though the web. This allows the user to be able to access the resources as long as they have access to the internet (Craggs, 2009). This exponentially increases the connectivity and accessibility from previous systems

Another benefit of cloud computing is that it addresses resource management in profoundly better ways. Previously, regardless if a user was currently using a particular resource or not, they would have to store their data on additional resources. However, in a cloud environment the user no longer needs to be responsible for identifying resources for storage. If a user wants to store more data they request it from the cloud provider and once they are finished they can either release the storage, by simply stopping the use of it, or move the data to a long-term lower-cost storage resource. This further allows the user to effectively use more dynamic resources because they no longer need to concern themselves with storage and cost that accompany new and old resources.

Lastly, the new resources requested by a user can be delivered much faster. New equipment and IT managers are no longer need for each and every new application allowing the user to use their new resources much quicker. This also allows for the applications and resources to be better tailored to the user's needs.

INFRASTRUCTURE MANAGEMENT BENEFITS

Cloud computing offers an array of infrastructure management benefits. Through a cloud computing environment users can save and eliminate cost of services, personnel, and IT infrastructure. Reducing cost is one of the major attractions to cloud computing. In traditional infrastructures, users must concern themselves with all of the hardware and software maintenance costs; that can comprise of up to 40% of IT personnel and asset resources.

Users in a cloud environment can eliminate costs though standardization and reuse of IT infrastructure. Due to systems being hosted in a cloud, problems that arise are handled, and dealt, with internally by the hosting server, usually with zero interruptions or knowledge to the user. Access to cheaper resources also saves money as cloud computing has inherently less risk of human error. For example, in capacity estimation a cloud provider can offer many times more computing resources than an individual IT organization, so that when demand goes up, resources are naturally available. Lastly, the cloud computing usage-based verses capital expenditure financial model, of both hardware and software, can provide substantial financial savings.

USAGE-BASED COST: WHAT DOES IT LOOK LIKE?

Usage-based cost works similar to electricity. For example, when you turn your lights on at your house you get billed for the amount that you use. Once you are done using your lights you can switch the light switch off and you no longer get billed for the electricity while you are no longer using your lights. Cloud computing costs continue past "on" and "off" switches. Take a taxi for example; as the taxi moves you get charged a flat rate and continuous fee. When the taxi stops at a light, the fee slows down because you are not using the car and the gas as much. Once you get out of the taxi you stop paying all together.

Before there was cloud computing capabilities, companies would continuously be billed whether they were using the systems or not. It would be like being continuously billed for electricity 24 hours a day 7 days a week. Or in the taxi example, it would be like buying the car even if it sat in the garage all year. The old system of servers is inefficient and costly. Of course, organizations do not want to pay, or submit for approval, for services that are not currently in place.

CLOUD FLAVORS

Cloud computing is not simply offloading IT workloads from traditional methods to cloud methods. Cloud computing can be more or less complex depending on key variables; thus clouds can be deployed and accessed in numerous ways. Deploying, or migrating, an application from traditional computing to cloud computing primarily depends on whether resources are on or off premise, and it depends whether access is restricted to one organization or if access is granted to multiple user types. The main trade-off between these differences is the cost reduction compared to the risks, including security risks.

Depending on the needs of a user, the cloud will fall in one of four categories, public, community, private, or a hybrid cloud. In a public cloud, IT resources are owned by a third party that is off the premises and who hosts the services needed. These clouds can be accessed by anyone on a metered based service. In a community cloud, IT resources and services are owned and maintained on behalf of a community of users. Arguably, a community cloud is also a public cloud, though it is treated as distinct due to special characteristics it encompasses. In a private cloud, IT resources and service are owned or leased by a single company for its own private use. Private clouds can also have internal aspects where all resources remain on-premise. Lastly, the hybrid cloud combines two or more cloud styles (Craggs, 2009). This latter type is not common.

In a public cloud the user does not need to concern itself with capacity management and investments associated with hardware and IT costs, the hosting server takes responsibility for these issues (Craggs, 2009). These issues and responsibilities could also involve software licenses, another

cost benefit of the cloud. It is important to note that a public cloud, that presents the option of a private cloud, is not the same as a private cloud. The private option, within the public cloud, is referring to security software within the public cloud that can provide extremely high security measures; however it still remains separate from a true private cloud because it is operated with in a public cloud (Craggs, 2009). These private clouds, running on public clouds, are often referred to as a virtual private network (VPN).

The community cloud offers large incentives for companies or groups that wish to share IT cost, communicate, and share resources more seamlessly. The cloud location could be external, or internal, within the company or grouped users. It is characteristically different than a public cloud because access to the resources is limited to the community members of the cloud.

Private clouds are very attractive to users that are concerned about information security and data transmission security. The private cloud reduces cost savings, but increases security concerns. Although scalability is still available, the speed at which it can occur depends greatly on its own IT infrastructure, instead of being able to leverage large resources and IT pools (Craggs, 2009). Not all traditional applications will run efficiently in a private cloud, and so applications often have to be re-written and designed for a private cloud in order to be migrated from traditional computing to cloud computing.

Lastly, the hybrid cloud is best described from the end user perspective, because from this perspective the user perceives the cloud as a single cloud, rather than a hybrid. Under a hybrid cloud a user can fuse two or more cloud models to the particular functions needed. For example, a company could use the private cloud model for sensitive information and data, and use the public cloud for low risk data and communication. This helps reduce cost where the risk is low and still protect concerning data (Craggs, 2009).

CLOUD SECURITY

Recently, *InformationWeek* surveyed 152 federal IT professionals who are using or assessing cloud services. 57% indicated it was unlikely that their agencies would be tapping into commercial cloud services. This is attributed to that fact that 93% percent cited security concerns as a primary reason for resisting the implementation of Clouds (Biddick, 2010).

Security is the aspect of cloud computing that enthusiasts often want to avoid. Such negligence is often due to complexity, effort, and expense to ensure that access to and protection for data for the cloud application is secure. Security is an important aspect of cloud computing that poses numerous hurdles to be addressed. Security refers to a broad set of policies, technologies, and controls deployed to protect data, applications, and the associated infrastructure of the cloud computing environment (Biddick, 2010). With greater benefits of application flexibility, accessibility, and adaptability come greater security risks using a cloud-based infrastructure.

In regards to government use of clouds, security concerns fall into one of three main categories of vulnerabilities, secure data transmission, insider tampering, and data loss or leakage.

First, at the forefront of the conversation surrounding cloud computing security is securely transmitting data. If companies and government agencies cannot depend on the cloud environment for secure transmission, it will limit the use of many applications on the cloud environment, regardless of costs that can be saved. Usually transmission encryption is utilized for communication transmission between the user and the application running in the cloud. There are many types of encryption methods, but this topic is beyond the scope of this paper.

Second, insider security concerns remain at the forefront of issues surrounding the cloud environment. This concern pertains to the hardware and software. Clouds need to provide users with a secure physical environment where their data and resources are being hosted. This includes the physical location and the IT personnel with access to the systems. Although the government requires cloud hosting contractors to meet government security standards, many are only at the minimum security levels.

Furthermore, the cloud software must be secure from those that want to hack the systems or intercept data information, which involve firewalls and other access protection systems, including dual authentication.

Finally, clouds must provide security of the data itself from data leakage. Particularly pertaining to the government and disaster management, the data must not only be readily accessible, it must remain secure even when it is not used.

GOVERNMENT AND CLOUD COMPUTING

The buzz around cloud computing does not pertain simply to the private sector. In recent years the government has been aggressively perusing ways to incorporate cloud computing. In a turbulent economy, with the necessity for efficiency in the ever increasing fast paced technological world, cloud has the potential to meet government's needs in groundbreaking ways.

The Obama Administration has advocated and committed to the use and expansion of cloud computing. President Obama nominated Vivek Kundra to the position of Federal CIO and it was interpreted as a sign that the U.S. government was beginning to incorporate cloud computing through its IT policies and strategies (Microsoft, 2010). Kundra has built a career on a reputation of driving unprecedented change in federal IT. Concurrently, government IT stakeholders have been preparing the foundation for a broader adoptions of cloud services within government agencies. The National Institute of Standards and Technology's computer science division produced a draft definition of cloud computing including five key characteristics, on-demand self-service; ubiquitous network access; location-independent resource pooling; rapid elasticity; and pay-per-use (Microsoft, 2010). This also included ideas for government clouds, which was seen as innovated and ground breaking

Today, the government now has a Federal Cloud CTO, Patrick Stingley who has taken on the role within the General Services Administration (Microsoft, 2010). Together with Kundra, the government decided to move forward with cloud by transferring government websites to cloud systems.

On February 23, 2009 the General Services Administration (GSA) hired Terremark World-Wide to provide hosting, storages, and disaster recovery services for the Webportal USA.gov. GSA now is able to access services for applications and data via a cloud though the internet (Beizer, 2009). One of the driving factors to transition to cloud computing for the GSA was cost. GSA predicted it could cut its web management cost by 50% by using cloud technology. USA.gov, and its Spanish-language companion site, receive over 140 million visits per year, according to GSA (Beizer, 2009). They also hope that it would be the foundation for further online services to save costs. Tom Freebairn, the director of USA.gov Technologies remarked on the new advances that since this is still relatively new territory for the federal government, there is the expectation of sharing our experiences (Beizer, 2009).

The government spends approximately $76 billion annually on IT, $19 billion toward infrastructure maintenance (Claburn, 2009). In a speech at the NASA Ames Research Center in California in 2009, Federal CIO Vivek Kundra announced that the government would be committed to cloud computing to cut cost and to further efficiency, continue being innovative, and to be environmentally responsible. "Why should the government pay for and build infrastructure that is available for free?" he said. "In these tough economic times, the federal government must buy smarter" (Claburn, 2009).

In 2009 the GSA operated Web site launched Apps.gov for government agencies to buy and deploy cloud computing applications (Foley, 2009). Both Google and Microsoft have launched government clouds that are in the final stages of approve (Hoover, 2010). Google has already been approved as a FedRAMP launch vendor, and has submitted its final materials in hopes to certify the Google government-specific cloud. Similarly, on February 24, 2010, Microsoft took cloud security and privacy for government agencies by unveiling numerous new enhancements and certifications for the Microsoft Business Productivity Online Suite (BPOS) (Microsoft, 2010). Microsoft Online launched a new dedicated government cloud based on the BPOS. Furthermore, Microsoft extended identity federation services to Microsoft Live@edu, which supports interoperability while providing reliable security within a cloud setting (Microsoft, 2010). With the new Business Productivity Online Suite, Federal claims to meet the security, privacy, and compliance needs of U.S. federal government agencies. In addition, future capabilities and certifications that will continue to raise the bar in cloud security and privacy, including two-factor authentication, also known as dual authentication, enhanced encryption, and the expected attainment of Federal Information Security Management Act (FISMA) certification, a critical requirement for government information technology, all of which are moving government cloud capabilities forward (Microsoft, 2010).

Most recently, members of Congress and other federal officials met with chief technology officers from 13 leading software and computer hardware companies to discuss the integration of cloud computing for the federal government (Duvall, 2010). The primary benefits being cost cutting and improved performance.

The government is enticed by cloud technology for the same reasons as the private sector; that is, cost. The pay-as-you-use model is exciting for any company or government operating on a reduced budget. General IT cost can be dramatically reduced depending on usage. By reducing cost through the elimination of infrastructure management it allows the public sector to focus on delivering valuable programs to the public (Claburn, 2009).

WHERE IT ALL IS LEADING

Government data specialists project that federal, state, and local government agencies will spend more than $1.4 billion on cloud services by 2013 (Foley, 2009). Government agencies have "begun the march" to cloud computing in 2010. Although concerns continue to circulate around using cloud computing in the government, there is still a "can-do" mind set among government technology mangers, portraying a "when," not "if" attitude. The desire to implement cloud services is strong among the government, including areas that most would never have expected. Regardless of the real security concerns, government decision makers agree that the benefits outweigh the risks, as long as the risk can be mitigated though careful planning (Foley, 2009). This is verified in the consistent and relentless push from the government to overcome the hurdles preventing integration of cloud computing.

MEETING SECURITY REQUIREMENTS IN CLOUD COMPUTING

As mentioned, security is a primary concern regarding applications running on a cloud computing infrastructure. There are solutions possible today to secure a cloud environment. In order to do so effective encryption, multiple points of authenticity and standardization need to be implemented.

The government and private companies have the ability to implement encryption for all network traffic between users and the applications running in the cloud services. This produces a more secure cloud environment.

Another important aspect of cloud computing security is to require all users to use multiple points of authenticity. Though many systems are better secured from hackers, there is nothing to prevent a hacker from "getting in" if they have the user's login credentials, except if two or more authentications are required. This is one of the most important security measures that the cloud environment could implement. By requiring a user name/password, and an additional form of authentication (e.g., finger print, retina scan, or a RSA token) no hacker could access the application, and it would be a more secure cloud computing environment. The cost and the time to implement these changes can be minimal particularly compared to the benefit.

CLOUD COMPUTING BENEFITS FOR CATASTROPHIC DISASTERS

Cloud computing is the best solution to the needs and desires of the government, organizations, and individuals responding to catastrophic disasters. The availability, scalability, cost, speed of communication, and potential security offer many solutions to current dilemmas within the emergency response and relief work community.

Cloud computing services are more readily available for a response to a catastrophic event. Since the applications are hosted at geographically dispersed locations the application is not at risk of going down if one of the facilities fails.

Cloud computing is effective because they can scale when user load dramatically increases. Although we can never fully predict the full magnitude of communication requirements during a disaster, clouds allow us to be flexible because they can expand quickly as the application demands increase.

Cloud computing is diverse by connecting users over the internet. This provides the ability for users to communicate coordination efforts between those in the field and those outside the field. With cloud computing if one has access to the internet, whether though cell phone or a computer, they can connect with the cloud.

Lastly, clouds are secure, and though adaptation of dual authenticity, encryption, and meeting security software regulation, large concern about secure can be put aside. Furthermore, the cloud is not in one place, meaning the risk of systems failures substantially decreases.

Using Cloud Computing: Haiti

The lessons learned in previous disasters such as 9/11, the Indonesia Tsunami, and Katrina, demanded numerous areas of improvement for emergency managers and relief efforts. The advancement in technology, including cloud computing, provided tools for emergency managers to use in response efforts that had previously failed or proven inadequate to handle catastrophic levels. These advancements were applied and utilized in humanitarian assistance and disaster relief efforts during the Haiti Earthquake response.

Using InRelief as a case study we can better understand how a Cloud environment works in catastrophic disasters. Inrelief.org was created as part of the solution for seamless unclassified communication that the government could use to interact with other groups and individuals responding to disasters (Giasson, 2010). InRelief was able to provide layered interactive maps, a virtual warehouse, and verified information and communication though a cloud environment operating out of the InRelief.org website.

Layered interactive maps provided valuable up-to-date imaging for responders in Haiti (Disastersrus, 2006). These maps were able to utilize various groups' information and make comprehensive images. This included roads, building, refugee camps, food distribution, runways and critical infrastructure. With the maps having the ability to be layered you can look at one map while applying additional information at the same time. For example, you can look at roads while adding the ability to see where the refugee camps or food distribution locations are at the same time. Furthermore, responders could simultaneously zoom in on particular sites.

The Cloud provides the ability to utilize technology to respond in profoundly effective ways. The interactive maps mean that responders no longer need to fly "blind" as they did in the Indonesia Tsunami. The flexibility, scalability, and adaptability of the Cloud allow maps such as the one above to be created quickly, allowing available resources to be utilized immediately resulting in saved lives.

The InRelief website also provided a platform for a virtual warehousing during the Haiti response. This allowed the resources available to be placed into a mass spread sheet and for maps to be created of operational warehouse assets (Giasson, 2010). Groups, such as the Red Cross, can benefit immensely with virtual warehouses in the Cloud environment because as long as one as access to the internet they can access the resources available, even from a cell phone in the field. This means that responders can be more efficient in using supplies and making decisions because they have more real time information available to them.

Lastly, InRelief.org provided accurate verified information during Haiti. This addresses a large problem recognized in previous disasters such as Katrina. Information simply posted to the web may or may not be accurate. However, in Haiti, InRelief demonstrated the power of attaching a trusted organization to the information. Responders could look to InRelief for trusted information and to verify or dispel rumors. This is a great solution to the mass amount of information that floods the internet during a crisis. The cloud environment allows organization like InRelief to monitor information circulating the internet and address, as a reliable source, concerning rumors and to disperse important information such as evacuation orders. This solution addresses major issues we faced during 9/11 and Hurricane Katrina where information could not be communicated, or too much information flooded the internet with little accountability or verification.

Cloud computing is still in its infancy and has not had many catastrophic events to prove its full abilities. Haiti however, showed the Cloud is not a theory, and further, that the theory works.

EXERCISE24 (X24)

The government, particularly the Department of Defense and the Navy, are interested in finding better and more effective ways to communicate and disseminate unclassified information during disasters. Although the Navy, and even less so, the DOD are rarely called upon to assist in disasters, when they are needed they need to have systems in place that allow them to communicate as seamlessly as possible in the unclassified domain.

Exercise24 (X24) was created to begin the process of finding, and perfecting, a solution for the government's unclassified communication need. Through exercises like X24, Inrelief.org and other stakeholders around the world can test the ideas and theories behind a cloud computing platform to obtain unclassified communication on a mass level. Doing exercises, such as X24, allows all the players to understand what works and to find new solutions for what does not work before a catastrophic event happens. This way those involved will understand, be familiar with, and know what works in gathering and dispersing unclassified information.

The exercise was hosted and managed by the Immersive Visualization Center (VizCenter) at San Diego State University and was funded by the US Navy which included the collaboration of a cloud computing platform "InRelief.org". The exercise produced a virtual humanitarian assistance and disaster relief simulation on September 24-25, 2010. To date, X24 was the largest online integration of real life events into a simulation for the collaborated efforts of humanitarian assistance and disaster relief.

Exercise24 was a two-day international multidisciplinary simulation of the capabilities that support communication, logistics coordination, and response to catastrophic disasters and crisis that created an off shore oil spill, tsunami, displaced communities requiring immediate mass sheltering, damage to critical infrastructure inland, and numerous environmental impacts.

Through using social networking tools such as Google, Twitter, Facebook, U-Stream, and the InRelief website the X24 staff created a real time scenario:

The scenario initiates with an earthquake off the coast of Huntington Beach, California, USA generating a catastrophic subsurface and surface oil spill. A large inland earthquake occurs within the first hour resulting in reports of deaths and injuries, damage to roadways, power lines, and other key resources and critical infrastructures in Southern California, USA and Northern Baja California, Mexico. A series of aftershocks, fires, loss of power, disease concerns, and other challenges continue throughout the exercise to facilitate participant objectives (Giasson, 2010)

HOW IT WORKED

The scenario was initiated though injects on the InRelief website and by linking technology such as smart phones, back-end cloud computing and visualization, to help nurture the feeling of a real life event and time line. Through the information provided in the injects, participants could simulate how they would respond, resources they would need, resources they could provide, and problem solve to create solutions for complex emergencies in difficult settings.

The function of a "real-play" format, where the intensity of emergencies and controlled chaos takes place, was crucial to foster conversations of global players to analyze and better prepare for the next catastrophic disaster (Bressler, 2010).

Exercise24 had three main objectives in creating the scenario and executing the simulation. First, to use a cloud computing environment to connected and integrate geographically dispersed experts from the onset of the simulated international crisis (Bressler, 2010). Through smart technologies and social networks, community leaders of the affected areas could seamlessly communicate, coordinate, and be supported on topics such as mitigation, response, recovery and resumptions of societal normalities (Bressler, 2010). The second objective was to utilize smart phones, ultra-lights (in the United States), and air systems in Mexico to create further threat and damage assessment of the simulation that would provide information of damage infrastructure, which would lead to decision making information such as sheltering locations or evacuation routes. The last objective of X24 was to accelerate the coordination of powers of NGO's, faith based groups, involved governments, international groups, corporate groups, and social networks to identify and efficiently send aid to South California and Baja California; the affected areas in the simulation.

Each of the objectives focused on areas that previously failed or proved too weak to handle massive disasters of the past, such as 9/11 or Katrina. The objectives specifically focused on planning and testing new ways of addressing issues common to every global disaster.

All injects that were placed into the scenario were specifically requested by participants and driven from experience, concern, and areas known to be problematic that individuals sought to work though. Once the injects were fused into a coherent, scientifically realistic, and globally impacting event they were mapped and posted. Images were also added to the injects to enhance the real life real time atmosphere. Although time compression was needed to meet all objectives the scenario, the exercise provided a real life catastrophic event that allowed the global community to effectively interact and prepare for the next large-scale disaster.

PARTICIPANTS AND PARTICIPATION INFORMATION

Over 150 people physically attended the VizCenter at San Diego State University, which allowed them to observe and connect with thousands of people around the world using the Internet and social-networking tools via Google, Twitter, Facebook, and U-Stream. Furthermore, 79 different countries, speaking 33 different languages, located on every continents of the planet, except Antarctica, participated virtually in the exercise (Giasson, 2010). These visitors reached the Exercise24 websites using mobile devices as well as personal computers, which further simulated a real life event and communication information flow (Giasson, 2010). Statistics were able to be collected using BuzManager. They concluded that 84% of the visitors to the Exercise24 websites were first time visitors, not return or long-time users. The overall percentage of single-page visits, or visits in which the person visited only the home page, was very small, 1.27% (Giasson, 2010). Referrals from other sites, excluding Inrelief.org related sites, accounted for 74% of the visitors to the Exercise24 websites. Significantly, referrals from CNN's website accounted for 50.5% of the non-inrelief.org-related, suggesting that traditional media still has a role in Humanitarian/Disaster response, if only to spark interest for more detailed information. The BuzzManager snapshot report indicates that visitors to the websites stayed long enough to develop opinions regarding the conduct of the exercise (Giasson, 2010).

X24 TEAM

The Exercise24 team was comprised of the InRelief.org staff who are funded by the United States Navy, the VizCenter staff, and undergraduate and graduate students from San Diego State University with varying majors.

X24 COST

X24 did not have a budget outside of existing salaries and systems. Essentially, X24 was functionally a zero cost simulation, which was the intended goal of the X24 team. By operating on a low, or zero cost, budget it simulated how it could be done in real life and function from the onset of an incident without needing to wait approval of funds from the private or public sectors.

WHAT X24 ACCOMPLISHED

Exercise24 was the start of a longer journey to find effective, efficient, and secure platforms of communication. It is important to note that X24 demonstrated its primary objective that using a cloud platform, such as inrelief.org, works for mass communication and coordination. Further, that it can be done securely. With thousands of people participating and 79 countries, it proved that a cloud platform does allow global communication seamlessly while being cost effective, and further, this is where emergency management is moving.

One of the advantages X24 demonstrated was the flexibility of its account systems. Previous platforms of communication had a more simplistic user approach. Account holders would communicate on mass levels, but tailoring to the needs, such as single user to another single user, or one mediator sending a message to a particular group, or smaller groups to groups, was not possible within the systems formation. Inrelief.org provides the ability to communicate to anybody on any scale needed by the user. A large advantage because the systems allows a user or moderator to use the same platform to communicate mass information on a public or individual scale.

X24 also demonstrated the ability to communicate on a cloud platform and its effectiveness for activities such as virtual warehouses. A cloud computing platform, utilized in a disaster, also groups to share resources that ultimately create a virtual warehouse. Groups and individuals can log into a cloud environment and enter in what resources are available. Through these mapping capabilities leaders can have a real time spread sheet of all the available resources, even on a global scale, in one location or spread sheet. This allows leaders to respond with resources effectively on monumental ground breaking levels.

Perhaps the most important piece X24 demonstrated was that the government and emergency management is transitioning to the cloud. There is no longer and argument for why it is needed, but rather how to most effectively use it. By incorporating government leaders and emergency managers with other leaders and emergency managers globally, the message was clear, those that make the decisions among all governments want to find the best way to utilize the cloud environment; because it works.

CONCLUSION

This paper gave a background to some of the previous communication and system failures experienced during the last decade's disasters and presented the needs, and desires, that systems must obtain in order to be effective and efficient. Through the clarification and understanding of current systems there is an understanding that the capacities in regards to adaptability, scalability, and reliability are proven insufficient for effective emergency response. A clear understanding of cloud computing, while adequately addressing security concerns, provides a cost effective solution to the needs and desires of leaders and responders to disaster coordination and communication relief efforts.

For further research we recommend security, authenticity, priority, roles and emergency operation center complications to be evaluated. Although this paper addressed security concerns, more research could be conducted to evaluate security software advances within cloud computing.

The need for authenticity for security purposes is vague within the disaster management arena, and further how it could be implemented. Although the majority of information communicated in disasters is unclassified, secure systems are still needed for classified information; authenticity of the user would be a great way to address security concerns.

The past decade has shifted the priority of security in profound ways. However, there is still resistance in preventive measures that often require large funds, regardless of the fact that preventive measures often save more money than they cost. More research could be done on the priority of preventive measures and how better to communicate the need.

Finally, roles of leaders and emergency operation centers during national and global disasters are unclear. For example, in a disaster who is responsible for setting up the cloud environment? Inrelief is an example of one Cloud that works, however, what if there is ten InRelief type website operating at the same time? Should this responsibility and leadership fall on government organizations such as FEMA, or should it be left to education systems like the VizCenter at San Diego State? These questions require more research and understand to provide the best solutions.

In summary, using a cloud computing environment is the most effective system because of its ability to adapt in dynamic ways to any disasters regardless of scope and scale. Though using encryption and dual authenticity and continuing to focus on the software for security rather than the hardware, clouds are made to be as secure as the user desires, including meeting government standards of security.

Clouds provide improved physical protection, higher availability, easier access, and faster performance and are cost effective. With the government endorsing and working rapidly towards and secure cloud environment for agencies and emergencies management, questions should now revolve around how to do it better, rather than questioning the use of a cloud environment.

The effective and efficient use of a cloud environment for disaster management is no longer a theory. Though examples such as Haiti and Exercise24 cloud computing has proven it is no longer a theory of an effective tool, it is the best solution to disaster response, that is cost effective, adequately meets security concerns, and provides a system that provides immediate response that save infrastructure, economies, and lives.

REFERENCES

ActionAid. (2006). *Tsunami Response: Human Rights Assessment*. Retrieved September 27, 2010, from http://www.actionaid.org/docs/tsunami_human_rights.pdf

Beizer, D. (2009, February 23). USA.gov Will Move to Cloud Computing. *Federal Computer Week*. Retrieved September 11, 2010, from http://fcw.com/articles/2009/02/23/usagov-moves-to-the-cloud.aspx

Biddick, M. (2010, August 6). In Government, Private Clouds May Trump Google Apps. *Information Week*. Retrieved October 5, 2010, from http://www.privatecloud.com/2010/09/17/in-government-private-clouds-may-trump-google-apps-2

Bressler, G. (2010). *Exercise24* [Lecture notes]. San Diego, CA: San Diego State University.

Claburn, T. (2009, September 15). Government Embraces Cloud Computing, Launches App Store. *Information Week*. Retrieved October 1, 2010, from http://www.informationweek. com/news/government/cloud-saas/showArticle. jhtml?articleID=220000493

Comer, D. E. (2006). *Internetworking with TCP/ IP - Principles, Protocols and Architecture*. Upper Saddle River, NJ: Pearson Prentice Hall.

Craggs, S. (2009). *Cloud Computing Without the Hype; an Executive Guide*. Lustratus Research Limited.

Disastersrus. (2006). *Hurricane Katrina Reports*. Retrieved October 3, 2010, from http://www. disastersrus.org/katrina/

Duvall, M. (2010). *CTOs Meet with Fed Officials to Discuss Cloud*. Retrieved October 3, 2010, from http://www.information-management. com/news/government-cloud-computing-push-BSA-10018810-1.html?msite=cloudcomputing

Energy Ideas. (n.d.). *Centralized versus Distributed - Computer Analogy*. Retrieved October 1, 2010, from http://energyideas.ca/Central_vs_Distributed_Computers.php

Foley, J. (2009, July). How Government's Driving Cloud Computing Ahead. *Information Week*. Retrieved September 5, 2010, from http:// www.informationweek.com/story/showArticle. jhtml?articleID=218400025

Giasson, J. (2010). *About the Exercise*. Retrieved September 19, 2010, from https://sites.google. com/a/inrelief.org/24/

Giasson, J. (2010). *Exercise24* [Lecture notes]. San Diego, CA: San Diego State University.

Giasson, J. (2010). *Injects and Updates*. Retrieved October 1, 2010, from https://sites.google.com/a/ inrelief.org/24/injects

GoGrid. (2010). *The Cloud Computing Pyramid*. Retrieved September 1, 2010, from http://pyramid. gogrid.com/?ref=servepath

Golden, B. (2008). *Cloud Virtualization*. Retrieved October 6, 2010, from http://advice.cio.com/ bernard_golden/cloud_virtualization

Hoover, J. N. (2010, May 26). Gov 2.0: Google Readies Government Cloud. *Information Week*. Retrieved October 1, 2010, from http://www. informationweek.com/news/government/cloud-saas/showArticle.jhtml?articleID=225200270

IBM. (n.d.). *Get the Facts on IBM vs the Competition - The Facts about IBM System Z 'Mainframe'*. Retrieved October 3, 2010, from http://www-03. ibm.com/systems/migratetoibm/getthefacts/ mainframe.html

Jevans, D. (2008). Cloud Security: The Need for Two-Factor Authentication in Cloud Computing. *Cloud Computing Journal*. Retrieved October 8, 2010, from http://cloudcomputing.sys-con.com/ node/644838

Microsoft. (2010). *Microsoft Unveils New Government Cloud Offerings at Eighth Annual Public Sector CIO Summit*. Retrieved September 24, 2010, from http://www.microsoft.com/presspass/ press/2010/feb10/02-24ciosummitpr.mspx

Murphy, V. (2005, October 4). Fixing New Orleans' Thin Grey Line. *BBC News*. Retrieved October 3, 2010, from http://news.bbc.co.uk/2/ hi/4307972.stm

National Commission on Terrorist Attacks Upon the United States. (2004). *9/11 Commission Report*. Retrieved September 22, 2010, from http:// govinfo.library.unt.edu/911/report/911Report_ Ch9.htm

New York Magazine. (2002, September 11). *9/11 by the Numbers*. Retrieved September 20, 2010, from http://nymag.com/news/articles/wtc/1year/ numbers.htm

News, B. B. C. (2004, December 31). *UN Urges 'Special' Wave Response*. Retrieved September 28, 2010, from http://news.bbc.co.uk/2/hi/asia-pacific/4136153.stm

News, B. B. C. (2005, January 27). *Tsunami Aid: Who's Giving What*. Retrieved September 27, 2010, from http://news.bbc.co.uk/2/hi/asia-pacific/4145259.stm

Office of the Director of National Intelligence. (2008). *Information Sharing Strategy*. Retrieved September 10, 2010, from http://www.dni.gov/reports/IC_Information_Sharing_Strategy.pdf

Pierre, R. E., & Gerhart, A. (2005, October 5). News of Pandemonium May Have Slowed Aid. *Washington Post*. Retrieved October 1, 2010, from http://www.washingtonpost.com/wp-dyn/content/article/2005/10/04/AR2005100401525.html

Rappaport, E. (1993). *Preliminary Report for Hurricane Andrew*. Miami, FL: National Hurricane Center. Retrieved October 1, 2010, from http://www.nhc.noaa.gov/1992andrew.html

Reese, G. (2000). Distributed Application Architecture. In *Database Programming with JDBC and Java* (2nd ed.). Sebastopol, CA: O'Reilly Media. Retrieved October 1, 2010, from http://java.sun.com/developer/Books/jdbc/ch07.pdf

Rhome, R., Brown, J., & Brown, D. P. (2005). *Tropical Cyclone Report: Hurricane Katrina: 23–30 August 2005*. Miami, FL: National Hurricane Center. Retrieved September 20, 2010, from http://w5jgv.com/downloads/Katrina/TCR-AL122005_Katrina.pdf

Shah, A. (2005). *Asian Earthquake and Tsunami Disaster*. Retrieved September 27, 2010, from http://www.globalissues.org/article/523/asian-earthquake-and-tsunami-disaster#Greatesteverpeacetimereliefoperationunderway

TelaScience. (n.d.). *Haiti Earthquake*. Retrieved October 3, 2010, from http://hyperquad.telascience.org/haiti3/?zoom=13&lat=18.57027&lon=-72.32214&layers=BT

Wellnomics. (n.d.). *The Wellnomics® System Architecture*. Retrieved September 24, 2010, from http://wellanomics.com/pages/it_architecture.aspx

WindowSecurity. (2002). *Security: Secure Internet Data Transmission*. Retrieved October 1, 2010, from http://www.windowsecurity.com/whitepapers/security_secure_internet_data_transmission.html#WhatIsTransmissionSecurity

Word, I. Q. (2010). *Mainframe Computer*. Retrieved October 2, 2010, from http://www.wordiq.com/definition/Mainframe_computer

This work was previously published in the International Journal of Information Systems for Crisis Response and Management, Volume 3, Issue 1, edited by Murray E. Jennex and Bartel A. Van de Walle, pp. 50-64, copyright 2011 by IGI Publishing (an imprint of IGI Global).

Chapter 5
Multi–Layers of Information Security in Emergency Response

Dan Harnesk
Luleå University of Technology, Sweden

Heidi Hartikainen
Luleå University of Technology, Sweden

ABSTRACT

This paper draws on the socio-technical research tradition in information systems to re-conceptualize the information security in emergency response. A conceptual basis encompassing the three layers—technical, cognitive, and organizational—is developed by synthesizing Actor Network Theory and Theory of Organizational Routines. This paper makes the assumption that the emergency response context is built on the relationship between association and connectivity, which continuously shapes the emergency action network and its routines. Empirically, the analysis is based on a single case study conducted across three emergency departments. The data thus collected on information security, emergency department routines, and emergency actions is used to theorize specifically on the association/connectivity relationship. The resultant findings point to the fact that information security layers have a meaning in emergency response that is different from mainstream definitions of information security.

INTRODUCTION

One dimension of information security that is increasingly enjoying attention in emergency organizations is that of emergency responders maintaining information security while operating various types of technical devices and complying with organizational security frameworks, such as

DOI: 10.4018/978-1-4666-2788-8.ch005

the information security guidelines. The context of emergency response (ER) is a domain which involves rapid and intricate cross-boundary socio-technical communications to manage undefined situations (Turoff, 2002), often under pressure while engaged in the act of rescuing people and saving lives (Harrald, 2006). In contrast to mainstream information security (Dhillon & Backhouse, 2001; Siponen & Oinas-Kukkonen, 2007) emergency response challenges responders to exploit their cognitive ability so as to avoid disclosure of sensitive information to unauthorized individuals. In the emergency response literature, cognition is seen as: "*..execution of a set of behaviors that an individual is expected to be able to perform*" (Mendonça et al., 2007, p. 47), and deemed as a critical construct for the capacity to manage the nonlinear structure of emergency response (Comfort, 2007).

As enacting information security increasingly manifests itself as a fundamental part of communication patterns in emergency response, it is appropriate to ask how information is shared and acted upon, rather than how it is protected. To this end, we define information security in the context of emergency response as a process of three interoperable layers; technical, cognitive, and organizational to render a secure emergency response environment. This definition is consistent with previous research on secure communications that incorporates technical, conceptual, and organizational views on access to IT, secure communication, security management, and security design (Siponen & Oinas-Kukkonen, 2007).

To develop the argument for multi-layers of information security, we draw on Actor Network Theory (Bardini, 1997; Latour, 2005; Walsham, 1997) to conceptualize how actors associate with one another and other on-site emergency actors. That is, how they utilize cognitive capability to maintain secure information sharing channels during emergency response operations. Second, we use theory of organizational routines (Pentland & Feldman, 2007) to characterize the organization of emergency response routines, which inform the entire community of emergency responders about rules and obligations in emergency response. These two theoretical bodies were used as an initial guide to design a single case study with data collection undertaken from the three emergency response departments; Fire Dept, Police Dept, and Para Dept. We closely observed how information security was appreciated, considered, and managed during emergency response.

The paper proceeds as follows. First, we review selected research contributions in the field of emergency response to characterize its context. In this review we discuss how emergency actor networks shape response capability to enact the associations defined by constitutive actors. We end the review with diagnosing the formal organization and its generally accepted premises for emergency response. Third, information security is introduced and its relevance in the context of emergencies is discussed. Fourth, we present the result of the analysis and finally we offer an alternative understanding of information security that is meaningful in the emergency response context.

THE DUAL CONTEXT OF EMERGENCY RESPONSE

The context of emergency response has been characterized as a collaborative adhocracy consisting of inter-related individuals, groups, and organizations who have different backgrounds and levels of expertise (Mendonça et al., 2007). Responding effectively and rapidly to emergency events is likely to require complex decision making about issues that are grounded in different language, cognition, and routines. This results in the re-production of multiple logics of response. Given the adhocracy view, improvisation, adaptability, and creativity as vital means for effective problem solving during emergencies emerge as critical non-technical factors that leave imprints on the relative success of response operations (Comfort, 1999; Harrald, 2006). To ensure success, emergency departments typically engage in defining coordinated actions and communication patterns that are sought to be maintained during emergency events. In practise, this is particularly difficult to achieve as emergency responders operate in two different types of situations: Firstly- as networks in which they are responding to an emergency, and must communicate quickly, reliably, and accurately with other responders (Mayer-Schoenberger, 2002), often under pressure (Crichton et al., 2000), Secondly- in preparation periods where planning activities for the next event are critical for future response success (Manoj & Hubenko Baker, 2007).

Such contextual aspects have been identified in the literature (Turoff et al., 2004; Waugh & Streib, 2006) and include assumptions of dealing with capability issues such as compliance with formal procedures while also reacting to unpredictable emergency events, and simultaneously building up reliable and trusted relations with specialists from other organizations. However, this is not easily achieved since emergency actors acting during response events have limited resources to build concerted knowledge because each emergency is different and poses different problems. For example, the staff involved in any situation might shift from emergency to emergency as well as during emergencies. As noted by Turoff et al. (2004), role transferability is a key problem in emergency response as it is virtually impossible to pre-define what particular roles actors will undertake during an emergency. Thus, one problematic task in defining and setting up emergency action networks revolves around drawing a common cognitive map of network constituents to exploit the full potential of existing knowledge (Carlsson, 2001).

Emergency responses, as collaborative adhocracy, have been ascribed numerous features that challenge and endanger the information security. As emergency actors during the events need to access information regarding scale, location, damage, and available human and physical resources from multiple organizations, this makes the response arena highly interactive (Dantas & Seville, 2006; Horan & Schooley, 2007). The process of prioritizing information at different nodes within the emergency actor network aims to minimize the risk of responders acting on incomplete and erroneous information. While information prioritizing is critical for decision making and shaping of representations in a common operating picture (Carver & Turoff, 2007)

the assumption that it could sustain organizational boundary, as each node in the network acts on the premise of its own focus and responsibility, has been critiqued. For instance, Comfort (2005) claims the main problem in maintaining security in complex environments such as emergencies and disasters, is the understanding of uncertainty during response. However, as Comfort notes, the real problem, is that emergencies as contrasted to organizational activities, not only concern the emergency organization, but also other public, private, and non-profitable organizations. Thus, the scope of information flows related to an actual incident or emergency contain both objectively bounded interpretations as well as subjective non-operational opinions of the situation. As a result, emergency response may produce large amounts of information, a fact which may be counterproductive when used in response actions due to information overload in certain nodes of the network. As Turoff et al. (2004) note, "..people responding to an emergency are working 18-24 hours a day and have no tolerance or time for things unrelated to dealing with the emergency" (p. 7). The concept of interoperability has been used to explain the effects of uncertainty on emergency response. In contrast to ordinary organizational interoperability issues where incompatibility between technical systems consumes most resources, the emergency response arena has to be in operation even when the technical systems fail. In the context of emergency response, it is information that is the primary resource to adaptive performance. If systems fail, communication procedures cannot assure adequate information security levels (Comfort, 2005) because information may be changed or distributed unintentionally to responders or even persons outside the system.

Comfort and Haase (2006) demonstrate that fragility of communications infrastructure determines the level of inter-organizational performance in actual disaster operations by referring to the hesitant intergovernmental response to the Hurricane Katrina. Coordination of actions has been the outstanding recipe for success over the years for inter-organizational emergency communications and performance. The emphasis has been to explain how the enactment of pre-defined action plans serves as a means to achieve controlled and concrete actions during response. In contrast to traditional command and control models (Dynes, 1994), Mendonça and Wallace (2004) propagates that response organizations must possess agility and discipline to respond to extreme events. In particular, the many ways to share information available through modern technology, within dedicated response teams and public agencies chiefly challenge the coordination of response actions. Harrald (2006) observes that "..coordinated outcome can be achieved only if the evolving emergency management structure is an open organization, aware of and adjusting to the rapidly changing external environment showing the importance of improvisation, adaptability, and creativity to the management of this transition from chaos to stability" (p. 261).

Other researchers have noted that coordination is only successful if emergency responders and managers are able to lessen the gap between planning and practise. Comfort (2007) states that "cognition is essential to activating the response process..//...//..this means achieving a sufficient level of shared information among the different organizations and jurisdictions participating in disaster operations at different locations, so that all actors readily understand the constraints on each and the possible combinations of collaboration and support among them under a given set of conditions" (p. 191).

Chen et al. (2008) developed a framework to analyze trends occurring in the emergency response life cycle that explain the coordination patterns: On-site response coordination and Remote response coordination (by commanding officers). Without developing an understanding of the relationship between on site coordination and remote coordination, there will be always remain

a lack of clarity of the rapid coordination decisions that responders must make when exploring the solution domain.

Effective emergency response routines constitute the heart of any emergency department. Written response plans based on planning processes contain specific objectives aimed at achieving emergency preparedness (Perry & Lindell, 2003). The typical aim with plans and guidelines is engineered to initiate responders into action, but in real situations different contingencies often force responders to follow or deviate from these guidelines. This happens because the cognitive roadmap may change over time during informed actions (Waugh & Streib, 2006). Extreme events such as the Hurricane Katrina have led many researchers to study how planned emergency routines are enacted during actual response. For example, Comfort and Haase (2006) concluded on basis of results from a social network analysis of the hurricane Katrina that the human cognitive capacity and the technical and organizational means displayed obvious discrepancies between formal plans and actual practise. One line of criticism which postulates against too much reliance on formal plans is the inherent difficulty for emergency responders to make sense of the plans. Murhen et al. (2008) refer to this as - discontinuity is the rule in emergencies, and continuity is an exception. As further noted by (Murhen et al., 2008), it is very difficult to define the beginning and the end of an emergency, and therefore it is difficult to consciously enact emergency routines.

Institutionalization of routines is in many ways a function of selecting between different available methods to establish well defined procedures. To this end Comfort (2005) suggests that emergency departments should conduct long-term policy analysis and also generate alternative scenarios alongside as a method to internalize target goals. This method, Comfort (2005) argues, warrants careful consideration, as it is grounded in systematic, rigorous, and detailed analysis of the characteristics and interactions of the actors/ agents involved in the selected system. It differs from traditional methods of forecasting in that it considers large ensembles of scenarios; seeks robust, rather than optimal strategies; employs adaptive strategies; and designs analysis for interactive exploration among relevant stakeholders. Its weakness Comfort (2005) notes, is that it cannot predict which of the potential scenarios is most likely to actually happen.

Given the characterization of emergency response as a collaborative adhocracy, emergency actors may experience being associated to a certain emergency response progress along an array of trans-situational actions, rather than reacting mindfully to institutionalized knowledge in their respective organizations. For example, viewing exchange of information as critical for allowing continuous process of updating the changing status of an emergency community under stress, Comfort and Haase (2006) describe emergency management as a complex socio-technical communication process aimed at performing informed collective actions. As Mendonça et al. (2007) note, "..when attempting to consolidate information to obtain a shared situational awareness, there is a very real possibility that information that is relevant to one or more parties could be inadvertently left out" (p. 46).

Situational awareness of the events during response is the result of having access to specific information to prioritize certain measures before other measures. The picture can then be shared with other responders so that coordinated actions can be performed with a desirable outcome. However, the basis of this premise is that responders are connected to their backbone, i.e., their own organizationally defined emergency routine. This involves translating the rules for response into an operative state where responders understand the effects of action regardless of whether they applied controlled actions or creative actions during emergency response operations (Mendonça et al., 2007).

Enacting Emergency Response

Enactment of emergency response is explained in the literature as the assurance that inscribed action procedures are interpreted and followed by responders through the means of response plans and information technology (Murhen et al., 2008). This is described in ANT as a course of co-evolvement to achieve actability (Latour, 1987). Indeed, actability is essential in emergency response and when responders arrive at the emergency scene, they commence with the critical assignment to form working alliances (Mendonça et al., 2007). The reinforcement of actability to operate collectively originates from the various actors' ability to recognize patterns from prior responses and superimpose those onto ongoing emergencies (Comfort, 2007). However, the complexity of dependencies between actors in emergency response entail a clear hazard of hiding organizational routines from view, and thereby concealing the way that actability is represented (Holmström & Robey, 2005). As noted by Comfort (2007) "..the challenge is to build the capacity for cognition at multiple levels of organization and initiate coordinated action in the assessment of risk to vulnerable communities" (p. 189).

In order to build such capacity, Actor Network Theory proposes that stable identities should be developed through non-negotiable elements that translate individual perceptions into collective intentions, thus setting the network into operation (Callon, 1991). In emergency response, this happens through the shift from a residing state into operative state whereby actors assume well defined roles and commence response activities. In ANT terms, this proceeds along a translation process where emergency responders first enter the obligatory passage point of constructing common definitions and meanings, e.g., rules for response (Turoff et al., 2004), second, they define representatives, for e.g., defining certain roles with assigned task responsibility (Harrald, 2006), third, they co-opt each other in the pursuit of individual and collective objectives, e.g., re-produce cognitive abilities (Comfort & Haase, 2006).

The assembling of parts to form such a unity comes about through a series of events, which, in the ANT perspective, are not about the separation between the social and technological, but as an assemblage of negotiated and reinforced ties that do not look like regular social ties (Latour, 2005). While Latour (2005) discusses association as a central tenet in ANT to describe how actors connect in a network, Bardini (1997) clearly differentiates between association and connection. In Bardini's view, humans use association capability to instantly grasp items that enforce change in a given trail of actions. Connection, on the other hand, is concerned with intelligibility of action content as communicated between actors. Thus, in an emergency response environment, we can think of the actor using his ability to associate and individually making decisions while at the same time following emergency response routines.

Given the evolving nature of emergencies and the necessity of responders to shift between formal routines and creative actions, the situation calls for an understanding of the commonalities that the routines prescribe and the communicative content that is shared by responders. In respect, a synthesis between actor network theory and theory of organizational routines is appropriate because of the common performative tenet both theories contain. In fact, (Feldman & Pentland, 2005) draw their advance of the theory of organizational routines on work on power of association (Latour, 1986), arguing that performances are situated actions carried out by specific actors at specific times and places. More importantly, actions are not just situated, they can also be characterized as purposive action and mediated actions, as Goldkuhl (2005) argues, "..to act is to do several things at the same time" (p. 159).

Organizing Enactment of Emergency Response

Connectivity is a concept used to explain how actors interact in networks. It indicates that every actor (human or technical) can be reached (Feldman & Rafaeli, 2002). Similarly, emergency departments define how response actors are expected to interact on the premise of defined standard operation procedures, and how they collectively enact those procedures during a variety of response situations (Perry & Lindell, 2003). In order to meet response goals, this becomes an issue of organizing for performativity and ensuring that features of the emergency hierarchy unmistakably inform actors of the enactment of emergency response routines. Orlikowski (2007) demonstrates that performativity is constituted by available technology and that performance results in the assemblage of deliveries that temporarily bind together a heterogeneous assembly of distributed agencies, which for the duration of the particular activity are provisionally stabilized. Consider the temporal dimension in emergency response and the critical variable which time plays during response. This is a fundamental premise that explains why emergency departments are typically organized in a hierarchy with formal structures and formally assigned responsibility for actions (Perry & Lindell, 2003). Mendonça et al. (2007) note that when emergency calls activate responders, the response context is highly heterogeneous and it must be recognized that not all actors involved in the response to an extreme event will require access to all relevant information. To this end, Jul (2007) point out that technological solutions that assemble information offered by users, tasks and context of emergency response is one way to reproduce the understanding of performativity. However, a controversy in emergency response is the tendency responders have of falling back on formal plans and rules when stressed with reduced control over a given emergency (Turoff et al., 2004). This means there is a gap between what is formally constituted through policies and what is enacted during emergencies – a discrepancy between 'routines as-expressions' and 'rules-to-be-executed'(D'addeiro, 2008). As Pentland and Feldman (2007) observe 'when people in organizations use tools to do tasks, they most often do so as part of an organizational routine' (p. 786). Considering the two pathways of enacting emergency response; preparing for response and activating the response, the cognitive ability is critical to avoid having both the components of emergency response coordination and communication operating in a static and disconnected mode (Comfort, 2007).

THE INFORMATION SECURITY CHALLENGE IN ER

Information security research have drawn from several reference disciplines, such as organization theory, strategy, psychology, design, mathematics, communication to define and explain different aspects of information security (Dhillon & Backhouse, 2001; Siponen & Oinas-Kukkonen, 2007). Theoretical developments in those fields have been adopted into information systems security approaches to understand and articulate concepts of information security. For example, Siponen (2005) addressed socio-technical contingencies and analyzed the organizational roles attributed to information systems security and noted that orientations like technical, socio-technical, or social are pertinent to security design.

With the increased knowledge about information security in general, and the need to secure information sharing across team borders and organizational boundaries, technical information security solutions are a must for the effective flow of information between systems (Bishop, 2003). Mainstream technical information security views authentication as a central principle for access to IS in that it ensures the subject to be the one the subject claims to be and thereby the prevention of

unwanted actions is assured (Kong et al., 2001). However, in the context of emergency response, information security is achieved on technology level, rather than on software control levels. This incline a number of security challenges. For example, dedicated networks are used as temporary communication channels while acting within emergency response. The practical application of such technologies has tended to be organization specific. For example, a group of policemen can talk among themselves, but not with fire fighters or paramedics (Kristensen et al., 2006). This makes data gathering, analysis, and decision-making in an emergency situation extremely complex as conditions will differ from the last response (Jennex, 2007). However, the fundamental difference between normal organizational information management and emergency management context is that the responders are those who share information, and simultaneously make rapid decisions while time critical actions are performed. In such an intricate situation, where people involved have to compile heterogeneous information sources to act effectively, there is always the risk of exposing confidential data. For example, communication over the GSM net transmits encrypted messages between the mobile station and the base transceiver station but in the fixed network, these are not protected the message is transmitted in plain text most of the times (Milanovic et al., 2004).

Emergency response is highly governed by action response plans as means for translating organizational objectives into collective intentions, and thus setting the emergency network into operation. The actions carried out are not only means for relief operations during emergency response, but also help to get feedback for improving future response actions (Nathan, 2004). Although there might be differences in specific details of the action concept, Goldkuhl (2005) showed that actors perform actions based on situational grounds where they assess specific situational conditions, and reflect on the conditions using knowledge in order to produce a communicative outcome. This is particularly relevant for this research as emergency response contains many, if not all, variants of actions due to its extremely dynamic nature. For example, institutionalized action, controlled action, trans-situational action, interactive action, purposeful action, and creative action (Goldkuhl, 2005) are particularly important instances of action in the typical emergency response situation. These action types reflect the levels of formality in defined response routines, and also highlight the need to delegate action responsibility between on-site emergency actor networks to facilitate role transferability (Turoff et al., 2004). For example, institutionalized action and controlled action are interrelated as per the rules of regulation as stated by national emergency authorities. Institutional actions are actions that have become a part of custom and collective habit (Goldkuhl, 2005). A typical instrument available to emergency responders in order to establish practise is the decision process as it is followed on any emergency call. Such process include rules of response with pre defined tasks, and serves as a means to perform actions for which the emergency responders have responsibility (Comfort et al., 2001).

Furthermore, the application of information security as an organizational routine is concerned with aligning formal approaches of information security to information technology infrastructures. Therefore, the underlying premise of information security in the organizational domain is; to ascertain the means by which information can be protected against hackers and other types of *mala fide* behaviours (Baskerville, 1991; Dhillon & Backhouse, 2001; Siponen & Oinas-Kukkonen, 2007). In Siponen and Oinas-Kukkonen (2007) view, organizational security policies are the most common organizational-level solutions to ensure security throughout organizational processes. However, Choobineh et al. (2007) found that *"Once a system has been implemented, it was a norm to follow a checklist to address whether any of the security 'holes' remained unplugged"* (p. 961). Frequently adopted approaches to ensure

information security on the organizational level are risk management and standardization (Dhillon & Backhouse, 2001; Doherty & Fulford, 2006). These approaches are described in policy documents as tools for information security managers. One major issue with such procedures is that security preferences are not well translated into the observed behaviour of users' (Stahl et al., 2008). Information security considerations can therefore conflict in terms of underlying requirements, values and objectives to the extent that their integration with organizational processes can also fail (Dhillon & Torkzadeh, 2006). The problem in practising information security management in the context of emergency response is dependent upon the management of the life-cycle of emergency. Policy development is of the essence, but time lapse between planning, training and actual emergency response limits the translation of policy preferences into actual security behaviour (Kotulic & Clark, 2004).

In Table 1 we have summarized the premises of information security in terms of means and objectives as discussed in this section to illustrate the difference to the traditional organizational context.

RESEARCH METHOD

A qualitative case study (Yin, 1994) with interpretative purpose which aimed at gaining an in-depth understanding of the impacts of information security on cross-boundary communication was carried out in three emergency organizations. Subsequently, we selected the emergency response context as a study object for two reasons. First-Information security is not well researched in that context. Second- The emergency response arena represents an extreme position of actions and interpretations of events that substantially differs from normal business oriented processes. Our proposition was that the emergency context with rapid flow of information, accuracy in emergency response, and the constant re-production of action

Table 1. Summary of security means/objectives

Domain	Means	Examples	Objectives
Technology	Business context: • Software authentication Emergency context: • Technology authentication	• User identifications • Virtual private networks	• Secure information systems • Secure communications
User/Actions	Business context: • Maximize security awareness Emergency context: • Decision process	• Training programmes • Rules for response	• Increased security knowledge • Secure actions
Organizational	Business context: • Security policy planning Emergency context: • Security policy planning	• Security standards and checklists. • Emergency scenarios	• Security compliance • Robust strategies

logic would support our layered approach to information security more reliably than a traditional business context with isolated security artefacts under the control of different users. Ontologically, we have infused our research with the philosophical assumptions of ANT to avoid the use of ANT as merely a method for data collection (Cordella & Shaikh, 2003), which has been recognized as problem in information systems research. ANT allows us to interpret the emergency response context as an intermingling arena of collective/individual actions and technology in use. In particular, ANT is apt for the identification of constitutive actors and the interplay that emerges, develops, and diminishes over time (Callon, 1986). The course of events in emergency response clearly display such process while at the same time indicating that emergency departments have difficulties identifying the interface between on-going response and accomplished response (Murhen et al., 2008).

Data Collection and Analysis

Miles and Huberman (1994) explain that qualitative researchers usually work with small samples of people, and that the samples are purposive, rather than random. Altogether 8 interviews were carried out with representatives at the different emergency departments. The representatives were chosen due to their familiarity of the subject matter, i.e., planning and performing emergency response. The interviews were carried out during a period of 1.5 months in 2009-2010. All interviews were recorded and transcribed to allow for content analysis. To ensure authenticity of the research, the transcripts and preliminary version of the results were sent to interviewees for credibility check. The emergency actors at the three different departments were aware of information security in general. However, specific attributes thereof were not particularly recalled by the actors who had only received basic education, as some cases had taken place 10-20 years ago. The focus of the interviews was to understand how the emergency actors perceived information security as a socio-technical facet. Based on the available literature, we wished to expand enactment of information security to include a cognitive dimension, in addition to techno-organizational aspects.

The first step in our analysis was to organize the data according to the main categories: technology in use, action relations, and organizational issues that were the guiding themes during interviews. We also wanted to illuminate the duality of emergency response as being both action oriented and routine oriented (see Table 2).

Table 2. Examples of data coded in the analysis

Layer → Domain ↓	Technology in use	Action relations	Organizational issues
Emergency Action	"We need to have the loud speakers of ComApp on, Sometimes when we are in a public place we forget that- meaning that outsiders can hear something that they are not supposed to hear."	"We develop different kinds of collaboration patterns with other rescue teams"	"We are careful not to make any personal data public, or even exchange that information over our own radio network."
Emergency Routine	"In the ComApp network, the traffic is so intense that people have trouble getting their messages through.	"We are bound by the confidentiality rules concerning our patients, and watchful sharing across emergency actors is important"	"There are some guidelines for radio traffic. But in our work, since our tasks involve patients, we must ensure data confidentiality"

RESULTS

Access to information after the alarm call is enabled is via the ComApp radio technology, which is used for communications between in house responders as well with other rescue departments. As only a limited number of ComApp devices are available, these are dedicated for the use of response officers and team leaders:

There is one ComApp device for every employee who is on duty. Then every vehicle has their own ComApp radio and there are a few extra. So those that are called in for medium or large situations can bring those even if it is not their shift. But when you go in for a really big situation, it might be that there are not enough ComApp devices for everyone. Then the group leaders or working pairs share one device. It depends on the situation (Fire Chief).

Given the type of technology used by rescue departments, secure communications are critical and certain measures have been taken to ensure security:

In practice we use many channels or talking groups, which can be separated into different folders. For example, we cannot listen in on the

every- day communications of the ambulance or the police. But they do have common talking groups with different organizations that enable collaboration between authorities. There are also dedicated talking groups for people in lead positions (Fire Officer).

With the use of ComApp there is possibility to share information between actuators along the pre-defined use practise. During fire and rescue operations, actuators could have access to injured people's personal information, and leakage among unauthorized actuators regarding peoples' privacy is clearly a risk:

I would say that if used right – and if people know what to do, ComApp is a very good tool and it enables very good and efficient leadership – in principle no matter the scale of the emergency. The problem usually is the lack of skills. People do not know how to use it. They do not know how to change folders, how to set home groups. On the other hand, even if they knew how to do that, our user rights are sometimes very limited. For example in the fire fighter level, they cannot set the home group from another folder. This can be a potentially very significant constraint in resource consuming collaborative assignments (Fire Officer).

The process by which the rules for response are decided upon is governed by situational conditions in any emergency. In an incident like this, all of the actions are planned ahead as far as possible; the command leader will be giving "pre-orders" that emotionally prepare the crew for a difficult rescue mission, so that everyone has a common idea of where they are about to go and what they have been preparing for. The three main response bodies operate individually, but mutually align with situational conditions to assure successful rescue. First responders have to cope with complex information sharing situations. Responders' main objective is ensuring a successful rescue operation, which is dependent upon controlled actions that lead to desirable outcome. Controlled actions are a direct result of institutionalization of action patterns formalized in rules for response:

We have a chain of command when we have a larger incident but – there is this L4, who is the general leader in medicinal rescue work. Then we have this triage leader who assesses the condition [health] of the patient and their need for care. Then we have the care leader who coordinates the care taking activities of patients and lastly the transport leader who coordinates the transportation of the patients (ParaMed).

However, situational conditions require rapid adaptation of actions to new and unforeseen conditions at the scene of the emergency. An arena of trans-situational dependencies emerges that challenge the chain of secure actions as foreseen in rules for response:

for example when using different tools, do people know who they can share information with and what kind of data they can share, so that they do not let anything slip by mistake? (ParaMed).

Interactive actions and trans-situational actions are common to all three rescue departments. For example, the fire department communicates with the ambulance almost daily in situations where there has been a car accident. The unit supervisor, who is the on duty officer at the department, contacts the ambulance units. In some situations, the leader of the rescue work, the so called P3, is the commanding officer in contact with the ambulance and also the police. However, the commonality in collaboration is sometimes a problem:

...even during events when I'm not assigned as command leader, I might still get direct calls, just because people have gotten used to calling the manager in city B for information. So in a way there is no common communications guideline that can be applied to every situation. And in a way that would not make any sense, because the police do not necessarily know what resources the other rescue department has sent in, and sometimes rescue departments don't reveal whether data is passed on to them or not (FireOfficer).

There are confidential matters concerning administration during response. Some things concern the site of the incident, some relate to assurance of specific rescue department security and any other involved third party:

...there are confidential things concerning each municipality's preparedness plans that must not be leaked. We obtain a lot of confidential data during preparation of operations (Fire Chief).

The paramedics felt a need that common guidelines concerning information security should be gone through more often. The superior usually scrutinizes these when they work out contracts and sign the confidentiality agreements. They also talk about these kinds of things between the group members but repeated and regular training would be appreciated. The need for continuous training programs is requested also by the fire fighters and the police, especially since technology can be misused in critical situations:

Nowadays people have cell phones with a camera and the tabloids pay for spectacular pictures, which is a risky situation for both our staff as well as people we try rescue in different emergency operations (Police officer).

Robust strategies are normally developed through exercises that help to gain knowledge from real situations. In that, rescue departments work out consistent actions:

Normally when we have exercises, we have time to prepare for them. We might know what we have to do beforehand or at least we know that we will be put into some sort of situation (Police Officer).

The main premise of emergency response is that actions taken during any response should be coordinated. Given the scenario described, when the first team goes in they take a look at the situation and give a general picture to other teams arriving, and it is likely that a command centre would be established. It is important that the police, fire department and paramedics have full situational awareness. Depending on the situation as many people as are deemed necessary are called out to assist in the emergency:

...the organizations that would be involved in the response would be the rescue authorities, the fire departments, the volunteer fire departments, ambulances, possibly some other groups from the health authorities, the police, border guard and rescue helicopters services (Police Officer).

One type of intriguing situations where responders have to respond instantly is when situational awareness is fragmented, and the available communication means fail to support responders:

So in a way there is no common communications guideline that can be applied to every situation. And in a way that would be kind of impossible, because the police do not necessarily know what resources the fire department has sent in. Or the ambulance might not know as all of our assignment data is not passed on to them (Police officer).

...ComApp still has some dead zones in the periphery of our region. In that case we use GSM technology. However, sometimes we have to operate without any kind of available communication infrastructure (Fire chief).

DISCUSSION

This paper has focused on information security and how it plays a role in the actual enactment of emergency response. To this end, we learned that effective emergency response does not only depend upon the knowledge and application of technical expertise and emergency operating procedures, but also on the non-technical skills of the actors involved. Thus, the social systems in emergency response are formed by patterns of the enacted conduct i.e., the repeated forms of social action and interaction by emergency responders. The technical and social system functions according to their respective internal structure and this in turn advices the operating actors of their duties, and the accepted conduct while performing those duties.

Our study shows that while one group may ensure confidentiality through the use of technology, other groups share information face to face during emergency actions, which creates ambiguity regarding security during on-going communications. This sets information security in emergency response in a deeply intertwined position between technical and social activities (*viz.* Latour, 2005) as the actions performed are formed along an array of pre-defined action patterns. As noted by Turoff (2002) and Comfort and Haase (2006), emergency responders have to perform accountable actions while staying prepared for the unforeseeable. To this end, our study shows that responders act securely even though the technology they use does not involve any authentication procedures, which make responders cautious in sharing information outside commonly agreed channels. Indeed, information security is deeply embedded in the actions performed by responders while they are responding to emergencies. That is to say, emergency responders use their cognitive capacity to assess and act on situated information (Comfort, 2007). The literature demonstrates that the concept of action is fundamental to cognition and is of utmost importance in the context of emergency response since the ability to meet response goals largely depend on behavioural activity that is creative but purposeful (Comfort, 2007; Mendonça & Wallace, 2007). Indeed, the complexity of actions among emergency responders and how their intentionality controls action leaves an imprint on information security alignment between organizational routines and emergency action networks. Our perspective on information security differs from those that presents information security as an application of risk management and uncertainty in the avoidance of risks (Baskerville, 1991; Comfort, 2005) by articulating that the cognitive dimension of information security, i.e., responders' capacity to maintain information sharing channels is fostered by characteristics of technology in use and preferences of security policies. This is carved out in the entanglement of rapid association and formal aspects to a decision situation, which was explicitly identified in our case where responders operate on fragmented situational awareness, and the available communication means fail to support response actions. Theoretically, this establishes a link between actor network theory and organizational theory by framing association and connectivity within the emergency response instance of what ANT defines as a translation process. Towards this end, information security may be thought of as an associative (Bardini, 1997) and connective (Feldman & Rafaeli, 2002)

relationship between actors. These two constructs convey how humans perceive content of actions. Association operates through the grasp of items, instantly implying what is likely to happen next or is what is suggested, by the association of thoughts. Connectivity means that a system has rules and formal structures formed by decision makers and designers which actors must comply with.

Information security can be maintained during emergency response using technology for secure communications together with robust security strategies. However, it is displayed in our case that such view has limitations. Emergency response rely heavily on trans-situational actions, which stress responders' capacity to maintain information sharing channels without revealing any confidential data, which might risk its integrity or imperil privacy. Without the cognitive layer, the technical and organizational aspects of information security remain static or disconnected to the actions performed during emergency response. This perspective differs from that of the existing information security research that considers information security as the premier safeguard to informational assets (e.g., Siponen & Oinas-Kukkonen, 2007). The cognitive layer makes technological and organizational security considerations entangled, in the sense that information security is not just a socio-technical construct; rather it is socio-technical relationships that shape emergency response security.

Our case illustrates the linkage between technical, cognitive, and organizational security layers as a relationship between association and connectivity, to use Latour's (2005) and Bardini's (1997) terms. The information security layers rest on the socio-technical foundation of emergency responders being able to deliberately associate the emerging needs during response, and at the same time being able to connect those associations with what has been formally decided upon, internally as well as with concerned rescue organizations.

CONCLUSION

We have suggested a layered structure of information security elements with the purpose of developing better understanding of information security in a context wherein information security has not been well researched. Our contribution is twofold. First we show that information security is not only a technical but an organizational concern. Second, we present advice on a theoretical synthesis that frames the socio-technical reality of information security by relating Actor Network Theory and Organizational Routines showing the response duality of emergencies. The approach of combining two theoretical accounts in this paper details the enactment of information security in emergency response so as to understand how cognition ties technical security features with organizational security issues. Theoretically our approach contributes constructively to describe an alternative approach to information security research to address the gap between formal and informal criteria of information security (Dhillon, 2007). In particular, we describe a way to support development of robust and heterogeneous collection (Latour, 2005; Feldman & Rafaeli, 2002) of security agendas in a context that largely rest on non-technical factors and the reaction to unpredictable emergency events (Comfort, 2007; Harrald, 2006; Turoff et al., 2004; Waugh & Streib, 2006).

As with all research there are limitations due to researchers' assumptions of the context. In positivist research there might be issues of self-bias influencing data analysis. In qualitative and interpretive research the number of cases is often subject to critique as generalization of findings is impossible if few cases are selected. To this end, our study is limited in that it presents a single case study across three emergency organizations only. It is limited to the extent that, the result represents a partial view on information security, and clearly points to the fact that further studies are needed. However, there are two methodological aspects that are important to mention in regards to limiting

factors. First, the choice of case site was based on Miles and Huberman's (1994) suggestion to develop familiarity with the phenomenon and the setting under study before embarking on inquiry to increase credibility in qualitative research. We have developed familiarity with the contexts of emergency response and information security in previous research projects. Second, qualitative research approaches are suitable in contexts where emerging aspects of a phenomenon are likely to influence interpretation due to *in situ* conditions. Although emergency response largely draw on prescribed action patterns response always has to adapt to emerging conditions during events.

We contend in this concluding discussion that our theoretical framework together with a qualitative method has much to offer research on socio-technical components of information security.

Given the result of our study we are inclined to challenge the current view of emergency response as a collaborative adhocracy as we found emergency response rather to be a collaborative bureaucracy; an assemblage of technocratic performances with semi-effective outcomes. This means that emergency organizations have clear descriptions and understanding of what technical information security means and also what the organizational security aspects are. However, in the relationship between association and connectivity where action determines outcome information security becomes much more cognitive than perceived in mainstream accounts of information security. As such, we consider this positioning a worthwhile subject for further studies related to information security conceptualizations in the context of emergency response.

REFERENCES

Bardini, T. (1997). Bridging the gulfs: From hypertext to cyberspace. *Journal of Computer-Mediated Communication, 3*(2).

Baskerville, R. (1991). Risk analysis: An interpretive feasibility tool in justifying information systems security. *European Journal of Information Systems, 1*, 121–130. doi:10.1057/ejis.1991.20

Bishop, M. (2003). *Computer security: Art and science*. Reading, MA: Addison-Wesley.

Callon, M. (1986). Some elements of a sociology of translation: Domestication of the scallops and the fishermen in St. Brieuc's bay. In Law, J. (Ed.), *Power, action and belief: A new sociology of knowledge?* (pp. 196–219). London, UK: Routledge & Kegan Paul.

Callon, M. (1991). Techno-economic networks and irreversibility. In Law, J. (Ed.), *A sociology of monsters: Essays on power, technology and domination* (pp. 132–161). London, UK: Routledge.

Carlsson, S. A. (2001). *Knowledge management in network contexts*. Paper presented at the the 9th European Conference on Information Systems.

Carver, L., & Turoff, M. (2007). Human-computer interaction: The human and computer as a team in emergency management information systems. *Communications of the ACM, 50*, 33–38. doi:10.1145/1226736.1226761

Chen, R., Sharman, R., Rao, H. R., & Upadhyaya, S. J. (2008). Coordination in emergency response management. *Communications of the ACM, 51*(5), 66–73. doi:10.1145/1342327.1342340

Choobineh, J., Dhillon, G., Grimalla, M., & Rees, J. (2007). Management of information security: Challenges and research directions. *Communications of the AIS, 20*, 958–971.

Comfort, L. K. (1999). *Shared risk: Complex seismic response*. New York, NY: Pergamon.

Comfort, L. K. (2005). Risk, security, and disaster management. *Annual Review of Political Science, 8*, 335–356. doi:10.1146/annurev.polisci.8.081404.075608

Comfort, L. K. (2007). Crisis management in hindsight: Cognition, communication, coordination, and control. *Public Administration Review, 67*, 189–197. doi:10.1111/j.1540-6210.2007.00827.x

Comfort, L. K., & Haase, T. W. (2006). Communication, coherence, and collective action: The impact of hurricane Katrina on communications infrastructure. *Public Works Management Policy, 10*(4), 328–343. doi:10.1177/1087724X06289052

Comfort, L. K., Sungu, Y., Johnson, D., & Dunn, M. (2001). Complex systems in crisis: Anticipation and resilience in dynamic environments. *Journal of Contingencies and Crisis Management, 9*(3), 144–158. doi:10.1111/1468-5973.00164

Cordella, A., & Shaikh, M. (2003). Actor network theory and after: What's new for IS research? In *Proceedings of the 11th European Conference on Information Systems*.

Crichton, M. T., Flin, R., & Rattray, W. A. R. (2000). Training decision makers - Tactical decision games. *Journal of Contingencies and Crisis Management, 8*(4). doi:10.1111/1468-5973.00141

D'addeiro, L. (2008). The performativity of routines: Theorising the influense of artefacts and distrubuted agencies on routine dynamics. *Research Policy, 37*, 769–789. doi:10.1016/j.respol.2007.12.012

Dantas, A., & Seville, E. (2006). Organisational issues in implementing an information sharing framework: Lessons from the Matata flooding events in New Zealand. *Journal of Contingencies and Crisis Management, 14*(1). doi:10.1111/j.1468-5973.2006.00479.x

Dhillon, G. (2007). *Principles of information security: Text and cases*. Hoboken, NJ: John Wiley & Sons.

Dhillon, G., & Backhouse, J. (2001). Current directions in IS security research: Towards socio-organizational perspectives. *Information Systems Journal, 11*(2), 127–153. doi:10.1046/j.1365-2575.2001.00099.x

Dhillon, G., & Torkzadeh, G. (2006). Value-focused assessment of information system security in organizations. *Information Systems Journal, 16*(3), 293–314. doi:10.1111/j.1365-2575.2006.00219.x

Doherty, N. F., & Fulford, H. (2006). Aligning the information security policy with the strategic information systems plan. *Computers & Security, 23*(1), 55–63. doi:10.1016/j.cose.2005.09.009

Dynes, R. R. (1994). Community emergency planning: False assumptions and inappropriate analogies. *International Journal of Mass Emergencies and Disasters, 12*, 141–158.

Feldman, M. S., & Pentland, B. (2005). Organizational routines and the macro-actor. In B. A. H. Czarniawska, T (Ed.), *Actor-network theory and organizing* (pp. 91-111). Copenhagen, Denmark: Copenhagen Business School Press.

Feldman, M. S., & Rafaeli, A. (2002). Organizational routines as sources of connections and understandings. *Journal of Management Studies, 39*(3), 309–331. doi:10.1111/1467-6486.00294

Goldkuhl, G. (2005). Socio-instrumental pragmatism: A theoretical synthesis for pragmatic conceptualisation in information systems. In *Proceedings of the Third International Conference on Action in Language, Organisations and Information Systems*, Limerick, Ireland.

Harrald, J. R. (2006). Agility and discipline: Critical success factors for disaster response. *The Annals of the American Academy of Political and Social Science, 604*(1), 256–272. doi:10.1177/0002716205285404

Holmström, J., & Robey, D. (2005). Inscribing organizational change with information technology. In Czarniawska, B., & Hernes, T. (Eds.), *Actor-network theory and organising*. Copenhagen, Denmark: Copenhagen Business School Press.

Horan, T. A., & Schooley, B. (2007). Time-critical information services. *Communications of the ACM, 50*(3). doi:10.1145/1226736.1226738

Jennex, M. E. (2007). Modeling emergency response systems. In *Proceedings of the 40th Hawaii International Conference on System Sciences*.

Jul, S. (2007). Who's really on first? A domain-level user, task and context analysis for response technology. In *Proceedings of the ISCRAM Intelligent Human Computer Systems for Crisis Response and Management Conference*.

Kotulic, A. G., & Clark, J. G. (2004). Why there aren't more information security research studies. *Information & Management, 41*(5), 597–607. doi:10.1016/j.im.2003.08.001

Kristensen, M., Kyng, M., & Palen, L. (2006). Participatory design in emergency medical service: Designing for future practice. In *Proceedings of the CHI Conference on Participatory Design*.

Latour, B. (1986). The power of association. In Law, J. (Ed.), *Power, action and belief: A new sociology of knowledge?* (pp. 264–280). London, UK: Routledge & Kegan Paul.

Latour, B. (1987). *Science in action*. Milton Keynes, UK: Open University Press.

Latour, B. (2005). *Reassembling the social: An introduction to actor-network theory*. Oxford, UK: Oxford University Press.

Manoj, B. S., & Hubenko Baker, A. (2007). Communication challenges in emergency response. *Communications of the ACM, 50*(3). doi:10.1145/1226736.1226765

Mayer-Schoenberger, V. (2002). *Emergency communications: The quest for interoperability in the United States and Europe*. Boston, MA: John F. Kennedy School of Government Harvard University.

Mendonça, D., Jefferson, T., & Harrald, J. R. (2007). Collaborative adhocracies and mix-and-match technologies in emergency management. *Communications of the ACM, 50*(3), 44–49. doi:10.1145/1226736.1226764

Mendonça, D., & Wallace, W. A. (2004). Studying organizationally-situated improvisation in response to extreme events. *International Journal of Mass Emergencies and Disasters, 22*(2), 5–29.

Mendonça, D., & Wallace, W. A. (2007). A cognitive model of improvisation in emergency management. *IEEE Transactions on Systems, Man, and Cybernetics: Part A.*

Milanovic, N., Malek, M., Davidson, A., & Milutinovic, V. (2004). Routing and security in mobile ad hoc networks. *Computer, 37*(2), 61–65. doi:10.1109/MC.2004.1266297

Miles, M. B., & Huberman, M. A. (1994). *Qualitative data analysis*. Thousand Oaks, CA: Sage.

Murhen, W., Van Den Eede, G., & Van de Walle, B. (2008). Sensemaking as a methodology for ISCRAM research: Information processing in an ongoing crisis. In *Proceedings of the 5th International ISCRAM Conference*, Washington, DC.

Nathan, M. L. (2004). How past becomes prologue: A sensemaking interpretation of the hindsight-foresight relationship given the circumstances of crisis. *Futures, 36*, 181–199. doi:10.1016/S0016-3287(03)00149-6

Orlikowski, W. J. (2007). Sociomaterial practices: Exploring technology at work. *Organization Studies, 28*(9), 1435–1448. doi:10.1177/0170840607081138

Pentland, B., & Feldman, M. (2007). Narrative networks: Patterns of technology and organization. *Organization Science, 18*(5), 781–795. doi:10.1287/orsc.1070.0283

Perry, R. W., & Lindell, M. K. (2003). Preparedness for emergency response: Guidelines for the emergency planning process. *Disasters, 27*(4), 336–350. doi:10.1111/j.0361-3666.2003.00237.x

Siponen, M. T. (2005). Analysis of modern IS security development approaches: Towards the next generation of social and adaptable ISS methods. *Information and Organization, 15*(4), 339–375. doi:10.1016/j.infoandorg.2004.11.001

Siponen, M. T., & Oinas-Kukkonen, H. (2007). A review of information security issues and respective research contributions. *The Data Base for Advances in Information Systems, 38*(1), 60–80.

Stahl, B. C., Shaw, M., & Doherty, N. F. (2008). *Information systems security management: A critical research agenda*. Paper presented at the Association of Information Systems SIGSEC Workshop on Information Security and Privacy.

Turoff, M. (2002). On site: Past and future emergency response information systems. *Communications of the ACM, 45*(4), 29–32. doi:10.1145/505248.505265

Turoff, M., Chumer, M., Van de Walle, B., & Yao, X. (2004). The design of a dynamic emergency response management information system (DERMIS). *Journal of Information Technology Theory and Application, 5*(4), 1–36.

Walsham, G. (1997). Actor-network theory and IS research: Current status and future prospects. In Lee, A., Liebenau, J., & DeGross, J. (Eds.), *Information systems and qualitative research* (pp. 466–480). London, UK: Chapman Hall.

Waugh, W. L., & Streib, G. (2006). Collaboration and leadership for effective emergency management. *Public Administration Review, 66*, 131–140. doi:10.1111/j.1540-6210.2006.00673.x

Yin, R. (1994). *Case study research*. Thousand Oaks, CA: Sage.

This work was previously published in the International Journal of Information Systems for Crisis Response and Management, Volume 3, Issue 2, edited by Murray E. Jennex and Bartel A. Van de Walle, pp. 1-17, copyright 2011 by IGI Publishing (an imprint of IGI Global).

Chapter 6
Cell Phone Use with Social Ties During Crises:
The Case of the Virginia Tech Tragedy

Andrea Kavanaugh
Virginia Tech, USA

Steven Sheetz
Virginia Tech, USA

Francis Quek
Virginia Tech, USA

B. Joon Kim
Indiana University-Purdue University, USA

ABSTRACT

Many proposed technological solutions to emergency response during disasters involve the use of cellular telephone technology. However, cell phone networks quickly become saturated during and/or immediately after a disaster and remain saturated for critical periods. This study investigated cell phone use by Virginia Tech students, faculty and staff during the shootings on April 16, 2007 to identify patterns of communication with social network ties. An online survey was administered to a random sample pool to capture communications behavior with social ties during the day of these tragic events. The results show that cell phones were the most heavily used communication technology by a majority of respondents (both voice and text messaging). While text messaging makes more efficient use of bandwidth than voice, most communication on 4/16 was with parents, since the majority of the sample is students, who are less likely to use text messaging. These findings should help in understanding how cell phone technologies may be utilized or modified for emergency situations in similar communities.

DOI: 10.4018/978-1-4666-2788-8.ch006

INTRODUCTION

April 16, 2007 was a Monday. The weekend before had been Admitted Students days on the VT campus. Seung-Hui Cho shot two students early Monday morning (about 7:15) in a dormitory, and the police had taken someone into custody within an hour. During this time, Cho had gone by the post office and dropped off a package to NBC news with a videotape of himself telling his story of the killings he was yet to carry out. About 9:30 he barricaded the doors of Norris Hall on campus, and went from classroom to classroom, shooting 29 students and faculty, before shooting himself. A total of 32 people, including Cho, died that morning.

Cell phone networks were overwhelmed during the tragic shootings at Virginia Tech on April 16, just as they were during Katrina and 9/11 (May, 2006; Weiser, 2006). Many notification forms during emergencies overburden not only landlines, but also wireless (or ubiquitous) communication among individuals. We know from various studies that the first responders in an emergency are often the immediate individuals on site who can provide information to authorities (May, 2006). Just as authorities do, citizens need to be able to communicate effectively and reliably under such circumstances. Therefore, it is important that we understand communication patterns and system usage in emergency situations.

Many studies of emergency communication usage have examined public sector communication systems, such as those used by police, fire, and rescue workers (Mehrotra, 2007; National Research Council, 2007; Schooley, Marich & Horan, 2007; Weiser, 2006). A few studies have focused more recently on a combination of public sector agencies and lay persons' use of communication networks, including citizen use of government websites, and fixed and mobile telephone services (Owen, 2005; Steinberg, 2006; Palen & Liu, 2007; Palen &

Vieweg, 2008). We build on studies that focus on ordinary citizens' use of communication networks. Specifically, we examined fixed and wireless voice communications, text messaging, electronic mail notifications, and web-based information sources for Virginia Tech affiliates (students, faculty and staff) during the tragic events of April 16th. We focused our study primarily on cell phone use.

Mobile communication (particularly, cell phone usage) by citizens is of special interest given that it is the communication channel that ordinary people are most likely to have close at hand in the moment of an emergency (Horrigan, 2007). Moreover, in future crises, the cell phone is likely to be used increasingly for high bandwidth services, as some people caught in an emergency will also attempt to receive information over their mobile devices from the Internet and other networked sources (e.g., to obtain web-based news over their cell phone).

For these interrelated reasons, we think that the use of text messaging is an important component as we plan ahead for crisis management. Text messaging could play a vital role in emergency response for several reasons: 1) it is very low bandwidth; 2) it is asynchronous; and 3) it is currently underutilized by subscribers in the United States, because providers typically charge extra for text messaging services. In situations where the citizenry forms a vital cadre of "first informers" (which is increasingly more common owing to the ubiquitous presence of mobile communication devices), text messaging can be the most efficient form of communication over a given network. Moreover, cell phone penetration has reached deep into the socio-economically disadvantaged segments of the US population (Horrigan, 2007). Hence, cell phones (along with the digital services that come with them) represent a compelling medium for broad access to digital information and communication.

PRIOR RESEARCH

Emergency (or crisis) communication management pertains to the use of communication channels and messages to coordinate activity and convey information among citizens, rescue workers, government agencies and others. Our study seeks to build on an area of work known as emergency information management and, more specifically, on a growing research area referred to as 'crisis informatics' -- a term coined by Hagar (2007) and extended by Hughes et al. (2008) and Palen et al. (2009), among others. Speaking on the role of information and communication technology (ICT), Palen et al. (2009) note that 'crisis informatics' includes "empirical study as well as socially and behaviorally conscious ICT development and deployment. Both research and development of ICT for crisis situations needs to work from a united perspective of the information, disaster, and technical sciences." It is the sharing of information in a disaster that is especially crucial, and different types of ICT facilitate sharing to a greater or lesser degree based on a variety of circumstances. The recent literature on crisis communication includes analyses of information seeking behavior following such disasters as September 11, 2001 (Schneider & Foot, 2004), Hurricane Katrina (James & Rashed, 2006), the southern California wildfires in 2007 (Sutton et al., 2008; Schlovski et al., 2008), and the Virginia Tech shooting tragedy April 16, 2007 (Palen et al., 2009; Vieweg et al., 2008; Sheetz et al., 2009).

The Hurricane Katrina disaster that began August 29, 2005 produced several lessons regarding communication and information exchange (May, 2006). Among these are: 1) the importance of citizens to provide information (and misinformation) to each other and to rescue workers and related authorities; 2) there cannot be over-reliance on a single communication infrastructure, such as wireless or terrestrial communications; and 3) individuals possess unprecedented capacity to access, share, create and apply information (May,

2006; Peskin & Nachison, 2005). During Katrina, the public sector communication used by police, fire and rescue workers, often failed due to incompatibility among systems and across geographical districts (Le et al., 2006). In response, researchers at Virginia Tech and their affiliated collaborators have been developing 'smart radio' communication systems that automatically determine the frequency and other necessary specifications in order to interconnect with diverse networks. "Smart radio" (cognitive, adaptive, and software defined radio) offers the promise of greatly alleviating spectrum scarcity and increasing the diversity of users and applications. Radios that can locate unused spectrum should significantly reduce congestion, and encourage competitive, commercial and community entry. Such technologies will have significant impacts on existing stakeholders, on market structures and on network performance when they become available.

In previous studies of the events of April 16 at VT, Palen et al. (2009) examined the use of the Internet, including social network sites, for communication during the crisis. For example, they studied the Facebook group named "I'm OK at VT" and found that students joined the group not only to make their own safety known, but also to assist in searches for other students (Vieweg et al., 2008). In their study of ordinary people coordinating disaster relief on the Internet, Torrey et al. (2007) found that people quickly organized themselves into small blog communities or ongoing discussion forums. They found that a large number of dispersed individuals were able to collaborate on time-critical tasks. White et al. (2009) also found effective use of online social networks for communication and information exchange in response to crisis.

These works have reported that much citizen enabled communications that take place during disaster take place online. Our work focuses on cell phone use specifically, and finds that most of the communications that took place on 4/16 by students was via mobile phones.

METHODOLOGY AND PLAN OF WORK

We used mixed methods to solicit and preserve data, including one-on-one interviews, online surveys, and emails through the university system. We designed and then administered an online survey from November 2007 through January 2008 requesting participation by emails through the university system. We conducted follow up interviews with a subset of survey respondents in February and March 2008.

We contracted with the VT Center for Survey Research to provide us with ID numbers blind coded to e-mail addresses (to preserve confidentiality as approved by the VT IRB) of a random sample of students, faculty and staff from the total VT population (about 36,000). We sought a sample that was proportional to the campus population; that is, in 2007, there were about 29,000 full-time students (undergraduate and graduate students) or about 80% of the total VT population; there were about 3200 full and part-time faculty (9%), and about 3900 full and part time staff who made up about 11% total of the total VT population.

We contacted 4000 students, 350 faculty and 350 staff by email and invited them to participate in our study of cell phone use on 4/16. We used the Dillman method (1978, 1999) in administering the online survey. Specifically, we sent an initial email to the randomly selected sample alerting them to the study, and letting them know that we would be sending a link in the next email to the online survey. A few days later, we sent another email to them with a link to the survey. We sent two follow up reminders about two weeks and four weeks later, with the link to the survey, but only to those individuals in the sample who had not already completed it. In the first email alerting them to the study, we asked potential respondents (faculty, staff and students) not to throw away their non-university phone bills for April – the purpose is to 'freeze' the information as soon as possible to facilitate collection. The sample did not include the entire freshman class since they would not have been on campus the prior Spring semester when the tragedy occurred; and Spring semester seniors, since they would have graduated and left campus to destinations unknown to us.

Random Sample Survey and Interview Questions

The online survey and follow up interviews asked about general communication on that day (e.g., land line phones, cell phones, email, and Internet). While we did not ask specifically about Twitter use, we had an open category for 'other' under forms of communication used that day. No one wrote in Twitter. We asked detailed questions about cell phone use, such as, how and when the respondent communicated, with whom (type of relationship, not specific individual by name), duration of the communication, whether the communication was incoming or outgoing, and the location of their contact (i.e., local or non-local). The categories we used for "type of relationship" were: parent, spouse or romantic partner, child, other family member, close friend, friend, acquaintance, and other. Thus, our data are about the 'kinds of relationship' with whom the subject communicated (and not the specific identity of the communicant).

We followed a standard social network analysis of individuals' relationships with others with whom they communicated on that day: specifically, strong and weak tie relationships. The strength of a tie is a combination of the amount of time, emotional intensity, intimacy (mutual confiding) and reciprocal services that characterize the tie (Berkowitz, 1982; Fischer et al. 1977; Granovetter, 1973; Marsden & Lin, 1982). Strong ties are characterized by (Wellman, 1992, pp. 211-212): 1) a sense of the relationship being intimate and special, with a voluntary investment in the tie and a desire for companionship with the tie partner; 2) an interest in being together as much as possible through frequent interactions in multiple social contexts over a long period (perhaps this definition excludes the teenager to parent relationship, as many students may not want this at this point in their lives – although generally child-parent is a strong tie); and 3) a sense of mutuality in the relationship, with the partner's needs known and supported. Conversely, weak ties are more instrumental than strong ties – providing informational resources rather than support and exchange of confidences. In an emergency situation, individuals are likely to contact their strong ties to assure them that they are okay or that they need help; they are likely to contact weak ties, such as, fellow members of groups with which they are affiliated after the emergency has subsided in order to bolster needed help, emotional support or companionship.

There were two main parts to the online survey. Part A consisted of yes/no, multiple choice and Likert scale types of questions about the respondent's communication behavior on April 16, 2007, including use of other communication media, such as television, desktop internet access, radio and face-to-face interactions. We also asked in Part A for standard demographic and VT status information (e.g., gender, age, and whether the respondent was a student, faculty or staff person). Part B consisted of questions presented in a table format about each specific cell phone communication/calls of the respondent during the day of April 16. The items collected for each communication included kind of relationship with the contact, whether the communication was voice, text or Internet access, and by time of day.

The Part B section of the survey also included questions for each communication about the duration of the communication and the location of the social contact (for example, within 100 miles, farther than 100 miles, as measures of local and non-local contacts). The respondents provided these data for each individual communication, starting with the first use (if any) of their cell phone that day. For example, the first use was typically a phone call to "spouse/significant other" or a parent. So, all the information about

that call was included in the table. Given that the shooting occurred in April 2007 and we conducted the survey between November 2007 and January 2008, we encouraged respondents to use their actual cell phone bill when completing Part B in order to recall more accurately their cell phone use that day. To facilitate this we included links to the billing websites of cell phone providers.

At the end of the online survey, we asked respondents whether they would be willing to be contacted for a follow-up interview. For the interviews, we created a stratified sample of 31 individuals from the pool of 470, including faculty, students and staff, again roughly proportional to the campus population. Specifically, we interviewed 21 students (undergraduate and graduate), 5 faculty and 5 staff. To elaborate on the surveys, in the interviews we solicited more of the rationale and other motivating factors about their cell phone use on April 16. In addition, the interview protocols allowed respondents to describe where they were at the time of the shootings and throughout the day. The main results of the interview portion of the study are reported in Sheetz et al. (2009).

RESULTS

We summarize the findings in several sections, including demographics, technology use, social networking factors and network performance problems. Many of the findings are consistent with other studies and are not surprising. However, our interest is in how these different modes of communication and types of relationships interact in an emergency situation. We expected and found that people communicated with their strongest social ties earliest, most often, and for longer duration than weaker ties. These are some of the defining characteristics of strong ties, i.e., someone with whom a person communicates more frequently. We were also interested in cell phone use for text messaging and the implications for its use in crisis situations.

Demographics

From our pool of all VT faculty, undergraduate and graduate students (except for freshmen and seniors who would not have been enrolled in 2008 when we conducted the surveys and interviews), the VT Center for Survey Research created a random sample of 4000 students, 350 faculty and 350 staff, as noted above. The number of respondents and response rates for each Part of the survey by different types of VT affiliates are shown in Table 1.

More people completed Part A only than completed both Parts A and B of the survey. (Respondents could not submit Part B if they had not already completed Part A.) The total number of respondents for Part A was 979 (that is, an overall response rate of 21%); the total number of respondents who completed both Parts A and B was 496 (an overall response rate of 9%). A higher response rate offers more reliable data analyses, although a 10% response rate for surveys is fairly typical. For Part A, the response rate ranged from 19% (students) to 31% (staff), with an overall response rate of 21%. For Part B, the response rates were lower, ranging from 8% (students) to 14% (staff), and an average of 9%. In Part A, as noted, respondents replied to questions in a multiple choice response format. In Part B, respondents provided detailed information on calls they made or received that day. The respondents were equally divided on gender, and, of course, the majority of respondents were college age students. Specifically, the bulk of students were undergraduates, with 369 students aged 19-20 and 206 aged 21-22.

Modes of Communication Used on 4/16

Regarding various modes of communication used on 4/16, over 70% of all groups (students, faculty and staff) used both cell phone for voice calls and

Table 1. Survey respondents and response rates

VT Status	Sample	Number of Respondents (N)		Response Rate	
		Part A	Part B	Part A	Part B
Student	4000	776	332	19%	8%
Faculty	350	95	41	27%	12%
Staff	350	108	48	31%	14%
Total	4700	979	421	21%	9%

email from personal computers (see Figure 1). Students were the greatest users of text messaging from cell phones and instant messaging (IM) from computers. Faculty and staff had greater access to land lines, since many of them were at offices. The vast majority (85%) of respondents indicated that having a cell phone on 4/16 was important; 64% reported it was extremely important.

Students also used Facebook for communication far more often than faculty or staff (Sheetz et al., 2009).

Cell Phone Use

Of all the different types of communication they used via cell phones (voice, text, internet, and email), the vast majority (82%) was voice; only 11% was text messaging (see Figure 2). A very small percentage of all cell phone communication by all respondents was either email (N=63 or 2% of all cell phone communication) or internet (N=58 or 1% of all cell phone communication).

Respondents listed a total of 4,481 cell phone communication entries (Part B of the survey) for that day. For each entry, they also listed what type of social tie the communication was with, such as, parent, sibling, close friend, acquaintance, etc., and other information noted earlier. Just over half of respondents contacted 10 or fewer people by cell phone. The average number of cell phone communications of all types by all respondents was only 6, although some respondents had over 100 cell phone communication entries.

Figure 1. Modes of communication by type of user

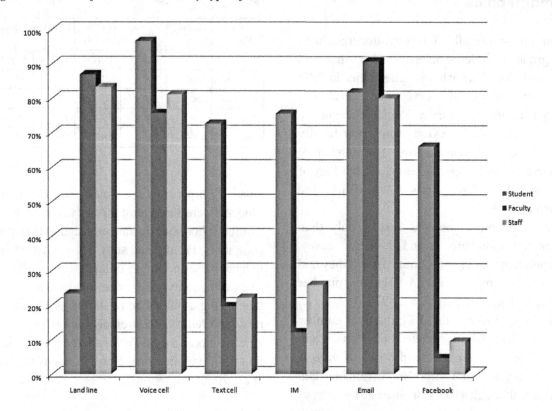

Social Network Ties and Cell Phone Use

We examined strong versus weak social ties by type of user and cell phone communication type (voice, text, email, internet). We coded strong social ties as those with family (i.e., parents, children, siblings, spouse or romantic partner, other family members) and close friends. Half of all people with whom respondents communicated by cell phone were family; another 25% were close friends, thereby together representing 75% of all social ties reached that day by cell phone (see Table 2).

Overall, respondents preferred to use cell phone voice communication with all relationship types except acquaintances (the weakest type of social network tie). Respondents' use of social networking sites was higher for acquaintances and friends than for stronger social relationships (see also Sheetz et al., 2009).

First Calls and Call Duration

As the majority of respondents were students, and just over half (51%) of first calls were to a parent(s); the next biggest percentage of first calls were to a spouse or romantic partner (15%). There was a significant negative correlation between the strength of the tie and time from the shootings around 10 am (that is, as time passed, respondents communicated with weaker ties).

The overall pattern in the sequence of communication by cell phone with respondents' social ties showed the following progression: first communication was with parents, then spouse or romantic partner, then close friends, then friends, and then siblings. Siblings may come this far down along the progression because parents contacted the siblings of the respondents. This was revealed in the interviews. Some interviewees reported extensive calling trees, such that, the person the interviewee contacted reached many other people

Figure 2. Type of cell phone communication

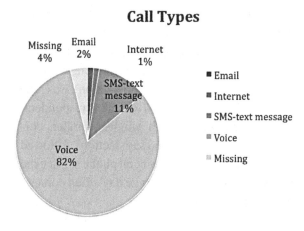

Call Types

Text Messaging

Of all the cell phone communications reported for 4/16 (4,481 total entries by all respondents), 11% was text messaging, as noted earlier. Students did most of the text messaging (see Table 3). Not surprisingly, age and text messaging were significantly correlated. The number of text messages respondents reported sending on an average day was also significantly associated with age (r=-.77, p<.00); the negative correlation indicates that older users sent fewer text messages on an average day.

On 4/16, undergraduate students used text messaging more than graduate students, faculty, or staff. Figure 3 shows the number of text messages sent from cell phones by different age groups. The youngest respondents are clearly the most prolific texters. We grouped ages as they relate to status as lower and upper level undergraduate (19-20 and 21-22), younger and older graduate student (23-26 and 27-33) and middle age faculty and staff (34-38) and over 38. The specific status is not as important as age in relation to experience with text messaging.

More females sent text messages on 4/16 than males. Texters reported they were more extroverted than non-texters and contacted almost 3 times as many people as non-texters. Students not only sent more text messages, they sent them primarily to other young people in their social networks (i.e., friends, close friends and siblings).

A cross-tabulation of the strength of social tie (e.g., parent, friend, acquaintance, etc.) by cell phone communication type (e.g., voice, text) showed significant differences. Students used text messaging primarily with strong ties within their age cohort (see Table 4), such as their romantic partner or spouse, close friends and friends, and siblings.

A graph of the number of text messages by time from the shootings mirrors that of voice, meaning that text suffered from network saturation also.

in their social network (i.e., siblings, other family members and friends).

The average duration of voice communication by cell phone was the same over time (about 3 and a half minutes). Nonetheless, a one-way ANOVA indicated significant differences in call duration across the level of social tie (F=29.162, *df*=8/4037, p<.001). People talked longer to a close family member, such as a parent (M=3.15, SD=1.29) than to close friends (M=2.63, SD=1.28) and talked with close friends longer than with friends (M=2.40, SD=1.15).

Table 2. Call frequency by strength of social tie

Strength of Tie		Frequency	Percent
Strongest ties	Parent	1044	23.3
	Child	68	1.5
	Sibling	357	8.0
	Spouse/Partner	410	9.1
	Other Family Member	373	8.3
	Close Friend	1101	24.6
	Friend	778	17.4
	Acquaintance	153	3.4
	Other	142	3.2
Missing		55	1.2
Total		4481	100%

Table 3. Who used text messaging on 4/16?

	Frequency (number of texts)	Percent
VT Students	469	92.5%
VT Faculty	11	2.2
VT Staff	11	2.2
Other	16	3.1
Total	507	100%

Satisfaction with Network Service

While 88% of respondents reported they were satisfied with their service on an average day, only 51% were satisfied on 4/16. Texters were more satisfied with their network service on an average day than non-texters, yet were less satisfied on 4/16

than non-texters. We conducted multiple regression tests to examine the primary determinants of satisfaction with cell phone performance on 4/16 (see Table 5).

The model explains 38%, p < .001, of the variance of satisfaction. The variable with the most influence (reducing satisfaction) was 'the number of people that tried to call the respondent, but could not get through' (higher numbers corresponded with decreased satisfaction), followed by 'how satisfied with their cell phone service the respondent was on an average day', then connection problems.

Almost three-fourths of respondents reported they experienced problems over an extended period with their cell phone network performance on 4/16. Not until 3 PM did less than a majority of respondents report problems as shown in Figure 4.

Figure 3. Text messaging by age

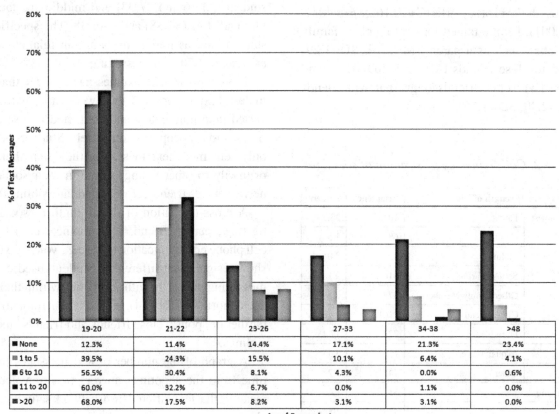

	19-20	21-22	23-26	27-33	34-38	>48
■ None	12.3%	11.4%	14.4%	17.1%	21.3%	23.4%
■ 1 to 5	39.5%	24.3%	15.5%	10.1%	6.4%	4.1%
■ 6 to 10	56.5%	30.4%	8.1%	4.3%	0.0%	0.6%
■ 11 to 20	60.0%	32.2%	6.7%	0.0%	1.1%	0.0%
■ >20	68.0%	17.5%	8.2%	3.1%	3.1%	0.0%

Age of Respondents

Table 4. Text messaging with social ties

With Whom	Text Messaging by:							Total	
	Students		Faculty		Staff		M		
Acquaintance	28	6%	0	0%	0	0%	2	30	6%
Child	0	0%	3	27%	0	0%	0	3	1%
Sibling	54	12%	0	0%	1	9%	1	56	11%
Close Friend	170	36%	0	0%	7	64%	6	183	36%
Friend	143	30%	4	36%	3	27%	6	156	31%
Other Family	23	5%	0	0%	0	0%	0	23	5%
Parent	13	3%	0	0%	0	0%	0	13	3%
Spouse/Partner	35	7%	4	36%	0	0%	1	40	8%
Other	3	1%	0	0%	0	0%	0	3	1%
Total	469	100%	11	100%	11	100%	16	507	100%

Table 5. Satisfaction with cell phone network service on 4/16

Cell Phone Satisfaction	Standardized coefficients (Beta)	Sig.
How many people were not able to get through	-0.416	0.000
How satisfied with cell phone service on average day	0.246	0.000
Connection problems	-0.171	0.000
How many calls were dropped	-0.142	0.000
How many different people contacted by cell phone	0.101	0.002

There is one main wireless cell phone service provider in the Blacksburg area, Verizon Wireless. Most residents subscribe to Verizon Wireless. Other providers have a presence, including T-Mobile, AT&T, Nextel and Cingular, but they have less coverage and reliability. Verizon Wireless increased its capacity throughout the late morning and early afternoon as its network became saturated, but it was not able to accommodate total demand during the peak period of the crisis between 10:30 am and 3 pm.

DISCUSSION AND IMPLICATIONS

The communication of VT students, faculty and staff showed heavy use of voice and text messaging from cell phones and email from computers, as part of a mix of communication and information media during the tragic day of 4/16. Most communication was with users' strongest social ties, such as family members and close friends. They communicated earlier and more frequently with stronger ties than with weaker ones. We expected to see these kinds of patterns. We were particularly interested in the use of text messaging from cell phones, as this is one of the most bandwidth efficient forms of communication. Efficient use of bandwidth is especially critical during emergencies, when networks become quickly saturated and many calls can fail to connect or get dropped shortly after connecting.

It is clear in this study that text messaging was used representing 11% of all contacts, however it was voice communication dominated the use of cell phones. Students, particularly undergraduates,

Figure 4. Connection problems over time

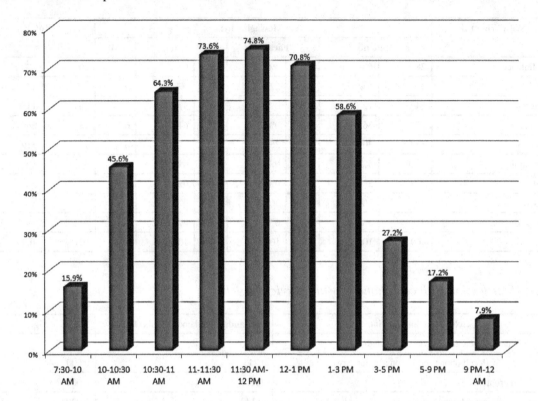

used text messaging the most. Moreover, they used it not with all their strongest ties but only with their strong ties that also generally use text messaging, that is, other young people (their siblings, close friends, friends and romantic partners or spouse).

The significance of this finding is that there seems to be a need to increase familiarity and acceptance of texting. In terms of bandwidth usage, we could improve the network performance by greater use of text messaging in emergencies than voice communication. However, since older adults are less familiar with text messaging, they are less likely to use it in general. While there is some tendency among older adults to use new technology less than younger adults, the use of text messaging in the United States seems to be further inhibited by the fees charged by cell phone service providers for individual text messages.

It is important to note that the current high charges for text messaging (e.g., 20 cents per mes-

sage whether sending or receiving with Verizon Wireless) are a disincentive to subscribers to learn this simple but potentially essential communication capability. Our thinking is that it is important that there be incentives for cell phone subscribers to learn how to use text messaging (such as, a certain number of free text messages per month with any subscription), so that in the event of an emergency, subscribers are not trying to use text messaging for the first time to notify kin or rescue authorities. Another possibility is advances in voice recognition, but this is not the focus of our research. Ultimately, we might recommend that cell networks be designed for emergency loads or preferential access to emergency responders; however, we do not want to foreclose the communication that is vital among ordinary citizens in situ during an emergency.

Texters in our study differed significantly from non-texters on many variables, not only the amount of texting on an average day but also the amount of dropped connections or other network problems on 4/16. These differences, however, are largely due to the predominance of students in the user population. Nonetheless, a similar university environment should expect to have the same population distribution and cell phone usage patterns, since students are the predominant user population on campuses.

The tragic events at Virginia Tech on April 16, 2007 can be considered a combination of crisis informatics and social convergence events. It is a crisis in that some people were under attack from a lone gunman; the subsequent attempts to obtain information and account for people's safety are a social convergence event, insofar as a large number of people are communicating with each other and with people who are in other locations (home, other parts of the city, other cities, and abroad) who have an interest in what the individual participant is seeing and doing. Given the finding of Hughes and Palen (2009) that people who adopted twitter during a crisis or social convergence event continued to use the service, and given the unfortunate recurrence of campus shootings, we can expect to see an increasing use of text messaging by older users (e.g., university and college faculty, staff and parents of students) for crisis informatics and social convergence in similar situations.

ACKNOWLEDGMENT

We are grateful to the National Science Foundation (IIS-0738390) for support for this research. We would also like to thank Virginia Tech students, faculty and staff who participated in the study. We appreciate very much the assistance provided by Jason Browning with the technical preparation of this paper.

REFERENCES

Berkowitz, S. D. (1982). *An introduction to structural analysis: The network approach to social research*. Toronto, ON, Canada: Butterworths.

Center for Technology in Government. (2004). *Learning from the crisis: Lessons from the World Trade Center response*. Retrieved from http://www.ctg.albany.edu/publications/reports/wtc_symposium

Dillman, D. A. (1978). *Mail and telephone surveys: The total design method*. New York, NY: Wiley-Interscience.

Dillman, D. A. (1999). *Mail and Internet surveys: The tailored design method* (2nd ed.). New York, NY: John Wiley & Sons.

Fischer, C., Jackson, R., Stueve, C., Gerson, K., McCallister, L., & Baldassare, M. (1977). *Networks and places: Social relations in the urban setting*. New York, NY: Free Press.

Granovetter, M. (1973). The strength of weak ties. *American Journal of Sociology*, *78*(6), 1360–1380. doi:10.1086/225469

Hagar, C. (2007). The information and social needs of farmers and use of ICT. In Nerlich, B., & Doring, M. (Eds.), *From mayhem to meaning: Assessing the social and cultural impact of the 2001 foot and mouth outbreak in the UK*. Manchester, UK: Manchester University Press.

Horrigan, J. (2007). *A typology of information and communication technology users*. Washington, DC: Pew Internet & American Life Project.

Hughes, A., & Palen, L. (2009, May). Twitter adoption and use in mass convergence and emergency events. In *Proceedings of the 6th International ISCRAM Conference*, Gothenburg, Sweden.

Hughes, A., Palen, L., Sutton, J., Liu, S., & Vieweg, S. (2008, May). "Site-seeing" in disaster: An examination of on-line social convergence. In *Proceedings of the 5ᵗʰ International Conference of the Information Systems for Crisis Response and Management*, Washington, DC.

James, A., & Rashed, T. (2006). In their own words: Utilizing weblogs in quick response research. In Leifeld, J. A. (Ed.), *Learning from catastrophe quick response research in the wake of Hurricane Katrina* (pp. 71–96). Boulder, CO: University of Colorado.

Le, B., Rondeau, R., Maldonado, D., Scaperoth, D., & Bostian, C. W. (2006, November 13-16). Signal recognition for cognitive radios. In *Proceedings of the Software Defined Radio Technical Conference*, Orlando, FL.

Marsden, P., & Lin, N. (Eds.). (1982). *Social structure and network analysis*. Thousand Oaks, CA: Sage.

May, A. (2006). *First informers in the disaster zone: The lessons of Katrina*. Washington, DC: The Aspen Institute.

Mehrotra, S. (2007, May 20-23). Information technologies for improved situational awareness. In *Proceedings of the Digital Government Research Conference*, Philadelphia, PA (p. 333).

National Research Council. (2007). *Improving disaster management: The role of IT in mitigation, preparedness, response and recovery*. Washington, DC: National Academies Press.

Owen, J. (2005, July 11). London bombing pictures mark new role for camera phones. *National Geographic News*.

Palen, L., & Vieweg, (2008). The emergence of online widescale interaction in unexpected events: Assistance, alliance & retreat. In *Proceedings of the ACM Conference on Computer Supported Cooperative Work* (pp. 117-126).

Palen, L., & Liu, S. (2007). Citizen communications in disaster: Anticipating a future of ICT-supported public participation. In *Proceedings of the SIGCHI Conference on Human Factors in Computing Systems* (pp. 727-736).

Palen, L., Vieweg, S., Liu, S., & Hughes, A. (2009). Crisis in a networked world: Features of computer-mediated communication in the April 16, 200 Virginia Tech event. *Social Science Computer Review*, 27(5), 1–14.

Peskin, D., & Nachison, A. (2005). *We Media 2.0: Landfall synapse*. Retrieved from http://www.mediacenter.org/synapse/wemedia20_synapse_screen.pdf

Schneider, S., & Foot, K. (2004). Crisis communication & new media: The Web after September 11. In Howard, P., & Jones, S. (Eds.), *Society online: The Internet in context* (pp. 137–154). Thousand Oaks, CA: Sage.

Schooley, B., Marich, M., & Horan, T. (2007, May 20-23). Devising an architecture for time-critical information services: Inter-organizational performance data components for emergency medical services. In *Proceedings of the Digital Government Conference*, Philadelphia, PA (pp. 164-172).

Sheetz, S., Kavanaugh, A., Quek, F., Kim, B. J., & Lu, S.-C. (2009, May). The expectation of connectedness and cell phone use in crises. In *Proceedings of the 6ᵗʰ International Conference on Information Systems for Crisis Response and Management*, Gothenburg, Sweden.

Shklovski, I., Palen, L., & Sutton, J. (2008). Finding community through information and communication technology in disaster events. In *Proceedings of the ACM Conference on Computer Supported Cooperative Work* (pp. 127-136).

Steinberg, L. (2006, May 19-22). E-Government and the preparation of citizens for natural disasters (National Science Foundation digital government grant # 429240). In *Proceedings of the Digital Government Conference*, San Diego, CA.

Sutton, J., Palen, L., & Shklovski, I. (2008, May). Backchannels on the front lines: Emergent uses of social media in the 2007 southern California wildfires. In *Proceedings of the 5th International Conference on Information Systems for Crisis Response and Management*, Washington, DC.

Torrey, C., Burke, M., Lee, M., Dey, A., Fussell, S., & Kiesler, S. (2007). Connected giving: Ordinary people coordinating disaster relief on the Internet. In *Proceedings of the 40th Annual Hawaii International Conference on System Sciences* (p. 179a).

Vieweg, S., Palen, L., Liu, S., Hughers, A., & Sutton, J. (2008, May). Collective intelligence in disaster: Examination of the phenomenon in the aftermath of the 2007 Virginia Tech shooting. In *Proceedings of the 5th International Conference on Information Systems for Crisis Response and Management*, Washington DC.

Weiser, P. (2006). *Clearing the air: Convergence and the safety enterprise*. Washington, DC: The Aspen Institute.

Wellman, B. (1992). Which ties provide what kinds of support? *Advances in Group Processes*, 9, 207–235.

White, C., Plotnick, L., Kushma, J., Hiltz, S. R., & Turoff, M. (2009, May). An online social network for emergency management. In *Proceedings of the 6th International Conference on Information Systems for Crisis Response and Management*, Gothenburg, Sweden.

This work was previously published in the International Journal of Information Systems for Crisis Response and Management, Volume 3, Issue 2, edited by Murray E. Jennex and Bartel A. Van de Walle, pp. 18-32, copyright 2011 by IGI Publishing (an imprint of IGI Global).

Chapter 7
Mining Geospatial Knowledge on the Social Web

Suradej Intagorn
USC Information Sciences Institute, USA

Kristina Lerman
USC Information Sciences Institute, USA

ABSTRACT

Up-to-date geospatial information can help crisis management community to coordinate its response. In addition to data that is created and curated by experts, there is an abundance of user-generated, user-curated data on Social Web sites such as Flickr, Twitter, and Google Earth. User-generated data and metadata can be used to harvest knowledge, including geospatial knowledge that will help solve real-world problems including information discovery, geospatial information integration and data management. This paper proposes a method for acquiring geospatial knowledge in the form of places and relations between them from the user-generated data and metadata on the Social Web. The key to acquiring geospatial knowledge from social metadata is the ability to accurately represent places. The authors describe a simple, efficient algorithm for finding a non-convex boundary of a region from a sample of points from that region. Used within a procedure that learns part-of relations between places from real-world data extracted from the social photo-sharing site Flickr, the proposed algorithm leads to more precise relations than the earlier method and helps uncover knowledge not contained in expert-curated geospatial knowledge bases.

INTRODUCTION

When disaster strikes, people are increasingly turning to social media sites to reach out to friends and family, post information, including images and videos, about current conditions, and receive updates about shelters and safe harbors. Humanitarian relief community could potentially integrate this data with geo-spatial information contained in maps, satellite imagery and news reports, to monitor the unfolding situation, assess damage and help coordinate relief efforts. As an example, images posted on the social photo-sharing site Flickr both before and after a tornado devastated a

DOI: 10.4018/978-1-4666-2788-8.ch007

town, could by combined with eyewitness accounts and missing persons reports on the microblogging service Twitter to create a detailed view of the affected area and its population. First-hand accounts on Twitter could then be used to monitor the availability of shelter and critical supplies, such as fresh water. The challenge, however, is to link places people talk about in their posts to the actual geo-spatial entities, since ordinary people are highly unlikely to use terms from a predefined geo-spatial vocabulary, or may refer to places that are not formally defined, such as, neighborhoods and landmarks.

We address this problem by automatically mining social media content to learn about places and relations between them. In addition to creating rich content in the form of text documents, images, and videos on the Social Web sites such as Flickr and YouTube, people often annotate content with keywords, called tags that they use to label and categorize content, as well as geographic coordinates, or geo-tags. Although social metadata lacks a controlled vocabulary and predefined structure, it reflects how a community organizes knowledge, including geospatial knowledge. A corpus of social metadata created by large numbers of people can be mined to reveal concepts (Plangprasopchok, 2004) including places and relations between them (Keating, 2005). Community-generated knowledge[1] that is automatically extracted from social metadata can complement expert-curated geospatial knowledge (Keating & Montoya, 2005; Kavouras et al., 2006), such as Geonames (http://geonames.org) or Yahoo! GeoPlanet (http://developer.yahoo.com/geo/geoplanet/). Community-generated knowledge is more likely to stay complete and current, since it is learned from metadata that is distributed and dynamic in nature (Golder & Huberman, 2006). It is also more likely to reflect colloquial *folk* knowledge that people use to talk about places.

Recently we proposed a method for aggregating geo-tagged data created by thousands of users of the social photo-sharing site Flickr to learn places and relations between them (Intagorn et al., 2010). The method represents a place by the coordinates of the geo-tagged images Flickr users labeled with the place name and uses geospatial subsumption to learn relations between places. Our key challenge is to efficiently and accurately represent places. In the original work, we used convex hulls to represent places, but found they did a poor job, since places were often concave. To address this problem, we present a simple, computationally efficient algorithm to find a possibly concave contour of a planar shape. Our method starts with a bounding box that subsumes all points and gradually erodes it until the boundary converges to a polygon that best represents that shape. We evaluate the method on data set consisting of US zipcodes. We then apply it to learn the boundaries of places extracted from social metadata. We show that the new method enables us to learn more precise relations between places using geospatial subsumption. Some of what we learn includes novel relations not found in the formal directories, for example, that Wild Animal Park is in San Diego. While not technically correct, such expressions of folk knowledge are still quite useful.

SOCIAL METADATA AND THE CHALLENGES OF LEARNING FROM IT

Tagging has become a popular method for annotating content on the Social Web. When a user tags an object, for example, an image on Flickr, she is free to select any keyword from an uncontrolled vocabulary to describe it. In addition to tags, some social Web sites, such as Delicious, and Flickr, provide the ability for users to organize content hierarchically. While the sites do not impose any constraints on the vocabulary or the semantics of the hierarchical relations, in practice users employ them to represent subclass relationships ('paris' is a kind of 'city') and part-of relationships ('yosemite' is a part of 'california') (Plangprasop-chok and Lerman, 2009). Social metadata offers rich evidence for learning how people organize knowledge. While illustrate with examples from Flickr, similar functionality is offered by other Social Web sites.

Tags: Figure 1(a) shows a Flickr image, along with metadata associated with it. Tags that describe this image include useful descriptors ("lighthouse" or "parola" in Tagalog), features ("archway", "jetty", "dock") and colors ("white", "red"), in addition to where the image was taken ("Lapu-Lapu", "Cebu", "Philippines").

Sets and collections: Flickr allows users to group photos in folder-like *sets*, and group sets in *collections*.[2] Both sets and collections are named by the owner of the image. While some users create multi-level hierarchies, the majority create shallow hierarchies consisting of collections and their constituent sets. The image in Figure 1(a) was grouped with other images taken around the Philippine province of Cebu in the eponymous set (see Figure 1(b)). This and sets describing other places around Philippines were grouped together in a collection "the Philippines" (see Figure 1(c)).

Geo-tags: In addition to keywords, users can attach geospatial metadata to photos in the form of geographic coordinates. This allows images to be displayed on a map. Figure 1(d) shows images (purple dots) in the "Cebu" set displayed on a map.

Figure 1. Examples of data and metadata created by a Flickr user, (a) Tags assigned to an individual image (geotags are not shown), (b) images in the set "Cebu", (c) sets in a collection called "the Philippines" created by the user, (d) geo-tagged images in the "Cebu" set displayed on the map

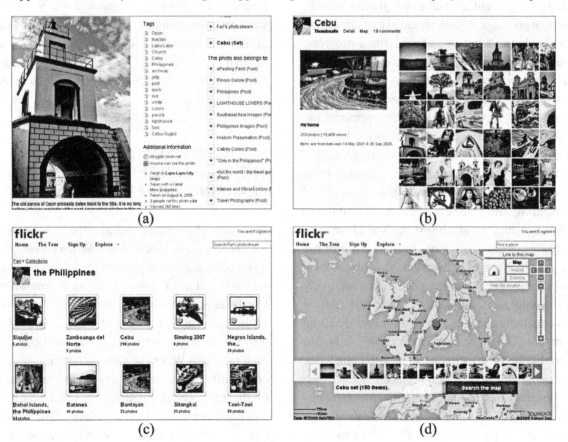

(a)

(b)

(c)

(d)

Extracting geospatial concepts from social metadata presents a number of challenges. Individuals vary in their level of education, expertise, experience and enthusiasm for creating and annotating content. As a result, user-generated data is sparse and uncertain. Annotation terms are often ambiguous (Mathes, 2004; Golder & Huberman, 2006): for example, "Victoria" could refer to a place in Canada or Australia. Data is also very noisy. In addition to misspellings and idiosyncratic naming conventions, there are geo-referencing errors and mistakes people make when tagging images. Geo-tagged data is also highly variable in quality. Expert users provide detailed annotations at different granularity levels (Kang & Lerman, 2010), while novice users specify generic terms (Golder & Huberman, 2006). For example, an expert user may tag an image with "golden gate bridge" (landmark), "San Francisco" (city), "California" (state), and "usa" (country), while a novice may tag a similar image with "San Francisco" only. Data is also non-uniform. Whereas a few tags will be used thousands of times, the majority will be used far less frequently. Similarly, while a few popular places will be geo-tagged thousands of times, others will have only a few data points associated with them.

Despite these challenges, geo-tagged social metadata provides a valuable source of evidence for learning how people conceptualize places and relations between them. In a recent work we described an *ad hoc* method that aggregates geo-tagged photos on Flickr to learn relations between places (Intagorn et al., 2010). Our method has two main steps: (1) representing places, e.g., 'cebu' and 'phillipines' and (2) using geospatial subsumption to learn part-of relations between them, e.g., 'cebu is part-of philippines'. One of the key challenges in this work is to accurately represent places using points, i.e., geo-tagged photos users assigned to that place. In the next section we describe an efficient algorithm that can learn arbitrary boundaries from a collection of points in the plane, and later, how this method is used to learn geospatial knowledge.

LEARNING PLACE BOUNDARY

Consider a set of points S sampled from a region shown in grey in Figure 2(a). The simplest boundary finding algorithm computes the bounding box of S, the so-called minimum bounding box (MBR), which is given by the minimum and maximum coordinates of the points. The bounding box of S is shown in blue in Figure 2(b). Another popular method computes the boundary of a convex region that contains all points in S. This is called the convex hull, and it is the shape that a rubber band would take when stretched over all points in S. Unfortunately, neither of these methods is well suited for representing boundaries of geospatial entities, such as zipcodes, states, or neighborhoods, which are often concave. For example, representing the boundary of the state of California by a convex hull will incorrectly include portions of the state of Nevada.

While several boundary finding algorithms exist for constructing polygons that are not necessarily convex, they are either computationally complex, require specialized libraries for computing them, or make unrealistic assumptions. The state-of-the-art alpha hull method (Edelsbrunner et al., 1994), for example, can find the shape of a set of points at a tunable level of detail, controlled by the parameter alpha. The optimal value of the alpha parameter has to be chosen experimentally depending on the density of points (Djurcilov, 1999). Since points in our data set are non-uniformly distributed, alpha hull is a poor choice for our purpose.

Erosion Algorithm

We propose a simple, efficient algorithm that starts with a bounding box and gradually erodes it until it finds a possibly concave polygon that best describes the boundary of S. The erosion algorithm starts with k equally spaced points placed along each segment of the bounding box of S, as shown in Figure 2(c) for $k = 2$. For each of the k cut points, we find a point $s \in S$ that is closest to

Figure 2. Boundary finding example. (a) Set of points (black) sampled from an unknown region (grey). (b) Convex hull (pink) and bounding box (blue) approximations of the boundary of the unknown shape. (c) Bounding box of the points with k = 2 cut points along each segment of the boundary. Red lines show Voronoi cells. (d) Final boundary found by the erosion method.

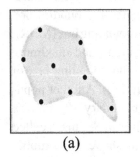 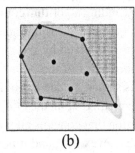 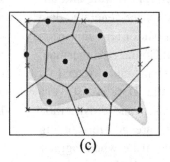 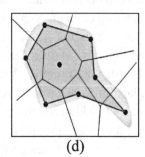

(a)   (b)   (c)   (d)

it. The point s becomes the new boundary point, and we update the boundary by joining s to the previous boundary point. We iterate this process by placing k new cut points along this segment to find the finer grained boundary. This erodes the bounding box to a polygon, which may be concave.

The erosion process is best explained by drawing the Voronoi diagram of S, shown in Figure 2(c). Each $s \in S$ serves as the center of a Voronoi cell, with all points within the cell closer to s than to any other point in S. The bounding box of S intersects boundary Voronoi cells. We place k cut points along each side of the bounding box and erode the boundary from the cut point to the center of the Voronoi cell it falls into. In practice, we do not need to create the Voronoi diagram explicitly, but simply implement the procedure that find the point $s \in S$ nearest to the cut point.

Algorithm 1 gives the pseudocode of the erosion algorithm. It has two parameters: the number of cut points k and the *maximum* number of iterations. As we increase these two parameters, we tend to get more accurate boundary; however, the trade-off is increasing computational time.

The eroded polygon may have zero-width spikes or crossed lines that will cause problems when using off-the-shelf packages to compute areas. The following post processing step will correct them. We start by adding a point from the original polygon (with spikes) to a new polygon

and check whether the new point causes a zero-width spike or a crossed line. If it does, we remove this point from the new polygon. The process terminates after all points on the original polygon have been checked. Because we got rid of these spikes, the new boundary will not contain all input data points; however, it still well approximates the boundary of the data points.

In contrast to alpha shape, erosion algorithm cannot learn boundary of a region that has holes inside it, such as a donut. In this case the erosion algorithm will simply produce a circle. However,

Algorithm 1. Erosion algorithm pseudocode

```
Input:
S: Set of points from a region
max: number of iterations
k: Number of cut points on boundary segment
Output: Polygon
Initialize: Create bounding box
for iter = 1 to max do
if converge then
return polygon
end if
for each boundary point i=1 to last do
line=createLine(i, i + 1))
place cut points (line, k)
for j = 1 to k do
s=nearestPoint(j, S)
if not polygon.contains(s) then
polygon.add(s)
end if
end for
end for
end for
```

in our domain interior holes are not desired. In fact, alpha shape may erroneously create holes in regions that have low density of points simply because these regions are less popular or less accessible.

Erosion algorithm depends heavily on the nearest-neighbor operation, whose average runtime is $O(n)$ in linear search, where $n = |S|$. However the average runtime can be improved to $O(log(n))$ (Panigrahy,2008) when data points are indexed by the Kd-tree (Bently, 1975). Since nearest neighbors have to be computed for q cut points on the boundary, the complexity of the erosion algorithm is $O(qn)$ in each iteration, but it will improve to $O(qlog(n))$ when data points are indexed. In comparison, the complexity of the convex hull and alpha shape methods is $O(nlog(n))$. In summary, the average runtime of erosion algorithm is $O(qlog(n))$ when data points are indexed by Kd-tree.

Evaluation

We evaluate the performance of the erosion algorithm on US zipcodes. For this study, we retrieved boundary files of 1000 random zipcodes from U.S. Census Bureau. We constructed two data sets by sampling points uniformly at random from each zipcode: a fine-grained data set consisting of 5000 points per zipcode, and a coarse-grained data set consisting of 500 points per zipcode. We measure the performance of the boundary finding algorithms in terms of precision and recall with respect to the ground truth, which is given by

zipcode boundary files. Let Z_i be the actual boundary of zipcode i, which defines a region with area $Area(Z_i)$. Let B_i be the boundary of i found by a boundary-finding algorithm, with area $Area(B_i)$. Precision and recall measure how well the learned boundary reconstructs the actual boundary. Precision is defined as: $P = Area(B_i \cap Z_i))/Area(B_i)$ where $Area(B_i \cap Z_i)$ is the area of the intersection between regions defined by boundaries B_i and Z_i. Recall is defined as: $Area(B_i \cap Z_i))/Area(Z_i)$. We use JTS library to compute intersections and areas. F-measure is the harmonic mean of precision and recall: $F = 2PR/(P + R)$ (see Table 1).

State-of-the-art alpha shape was computed using Hull (Clarkson, n. d.). We used "-A" parameter in Hull to automatically find the smallest alpha parameter to compute the alpha shape that contains all points. Thus for each zipcode, the alpha parameter was different depending on how the points were distributed.

The alpha shape achieves best precision and recall in both fine and coarse-grained data sets. The erosion method is only slightly worse. However, it is simpler to implement and computationally efficient. These advantages make it a competitive boundary finding method.

LEARNING ABOUT PLACES FROM SOCIAL METADATA

We have recently proposed a method that mines metadata users create in the course of organizing their own images on Flickr to learn about places and

Table 1. Comparison of performance of the erosion algorithm against other boundary finding methods

Method	Precision	Recall	F-score	Precision	Recall	F-score
	5000 points per zipcode			**500 points per zipcode**		
bounding box	0.524	0.999	0.679	0.552	0.992	0.701
convex	0.783	0.996	0.872	0.814	0.97	0.881
erosion (k=12,n=9)	0.972	0.945	0.958	0.969	0.845	0.902
alpha	0.993	0.958	0.975	0.982	0.871	0.922

relations between them (Intagorn et al., 2010). That method used convex hulls to represent boundaries of places, which led to inaccurate representation of places. This, in turn, decreased the quality of part-of relations we tried to learn. We solve this problem in the current work by using the erosion method. In this section we describe our method for aggregating geo-tagged metadata from Flickr to learn about places and relations between them. The method has two main steps: recognizing places and learning relations between them.

Recognizing Places

The first step is to identify places from social metadata and obtain a representative set of points. We define a place as an association between the name of that place and a collection of geographic points. We derive place names from textual metadata and obtain points from the geographic coordinates (latitude, longitude pairs) of the geo-tagged photos associated by all users with these names.

We use set names as definitions of a place. We assume that when the set refers to a place, points from that set belong to the same place. However, set names can refer to any concept, not necessarily a place; therefore, we have to filter out non-place names. We use GeoNames as a reference set for place names, although it may not be complete. We normalize data by lowercasing both Geonames and set names and use substring matching to check whether a set name contains a term from GeoNames. We plan to automatically recognize place names in future work.

Disambiguating Places: A given name can be ambiguous, mixing points from different places. For example, `victoria' can refer to a place in Canada or Australia, and many others. Similarly, `cambridge' can be found in United Kingdom and United States. A natural solution is to cluster points with the same name. Specifically, we assume that points associated with the same place are closer to each other than those associated with other places. To cluster points for a given geo-name, n,

we first obtain all points from all sets with name n. Places may contain non-contiguous subregions that are distant from each other, e.g., Hawaii and Alaska are part of United States, though far from the mainland. To link distant regions to a given place, we exploit constraints imposed by photo sets on Flickr. Specifically, we assume that points from the same set belong to the same place. If two clusters are found with points from the same set, we consider these clusters to belong to the same place.

We capture these constraints in a graph $G1_n = (V_n, E_n)$, an undirected graph of points associated with a geo-name n, and then analyze the graph to discover distinct places. Vertices V_n are points corresponding to photos in sets with name n, and E_n are the edges between them. Let s_{vi} be the set index of the vertex $v_i \in V_n$. An edge between two points is created if and only if $dist(v_i, v_j) < \tau$ or $s_{vi} = s_{vj}$. Here, $dist(v_i, v_j)$ is the Euclidean distance between points v_i and v_j and $\tau = 500$ km. Figure 3(a) shows the graph for points associated with 'cebu'. After creating the graph, we find its maximally connected components (Hopcroft & Tarjan, 1973), with each component corresponding to a different place. Thus, points associated with 'victoria' will be divided into two sub-graphs, one located in Canada and the other one in Australia.

After disambiguation, we cluster points again in each disambiguated place using the distance criterion only. This helps identify disjoint regions, for example, continental US, Hawaii, and Alaska in 'usa'. Sometimes, due to data sparseness, a contiguous place will end up in multiple clusters. In our data set, continental US is split into Eastern US and Western US. Our approach, therefore, has to represent places as non-contiguous regions.

Noise Filtering: As mentioned earlier, social metadata created by diverse users is noisy. This can significantly distort our representation of places and degrade the performance of the learning algorithm. For example, there are photos taken at Los Angeles International Airport (LAX) that appear in a set "Australia." Any representation of

Figure 3. Representing places and noise filtering. (a) A graph of points associated with place 'cebu', (b) Points are associated with sets containing name 'portugal'. Points in countries other than Portugal (in green) are filtered out.

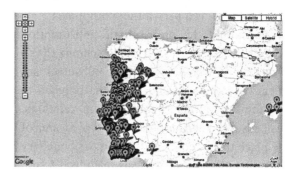

'australia' that includes parts of Los Angeles will lead to inaccurate relations between Australia and other places, and has to be filtered out.

As illustrated in the example above, noise can appear due to idiosyncratic tagging by individual users. This leads us to identify two characteristics of noise: (a) it is very different from other similar data (LAX points are very far from the other 'australia' points), and (b) it is created by a small number of users (it is highly unlikely that more than one user added points around LAX to a set named "Australia"). Let U_{ci} be number of users who geo-tagged photos in cluster i. We filter the noisy cluster out if $U_{ci} < \kappa$. In our experiments, we set $\kappa=2$. That is, if a given cluster contains points from only one user, it is very likely to be noise.

Noise can also lead to errors in estimating place boundary. For example, some of the points in sets called 'canada' are actually located in the United States, because people often include US border regions in their travel to Canada. The result is that the place 'canada' will include points in the United States. Most of them will occur as a single point or small group of points. We detect this type of noise by its locality. In our implementation, we average distance of a point to its K-nearest points and filter out $N\%$ of farthest points. In our experiments, N=5. Figure 3(b) shows the points associated with a place 'portugal'. Most of the points are located in Portugal, although a few others, shown in green, are in other countries, such

as Spain and Italy. The method described in this paper filters out these points.

Finally, we apply the erosion algorithm to disambiguated, de-noised points to identify the boundary of each region. Note that due to non-uniformity and sparseness, the coverage or recall of the boundary is not perfect. We hypothesize this problem will improve as users create and tag more data.

Discovering Relations Between Places

People express geospatial concepts at different levels of granularity, from continents and countries to cities, parks, and landmarks. In this section we discuss our scheme for reasoning about relations between places. Specifically, we use geospatial subsumption to learn containment, or part-of, relations, e.g., 'california' is part-of 'united states' and 'cebu' is part-of the 'philippines'. In addition to such well-known relations, we also learn examples of colloquial or "folk" knowledge, for example, where 'southern california' begins and ends, or that there is 'wild animal park' in 'san diego'. While not strictly correct according to a formal gazeteer, such relations are useful in that they allow us to learn how people conceptualize and talk about places.

Figure 4 shows boundaries constructed by erosion algorithm from points associated with 'cebu' and 'philippines'. Once overlaid, the larger 'philippines' region contains, or subsumes, the smaller 'cebu' region. We may reasonably conclude from this that 'cebu' is a part of the 'philippines'.

We extend the probabilistic subsumption method (Sanderson & Croft, 1999; Schmitz, 2006) to determine whether one place subsumes another. We use the boundary of the polygon to determine geographic subsumption relations. Basically, we determine the fraction of an area of one place that is contained within the boundary of another. Specifically, we say that place *A* subsumes place *B* if "most" of *B* is contained within the boundary of *A*, but not vice versa. Mathematically, *A* subsumes *B* if $p(A|B) >= t$ and $p(B|A) < t$, where *t* is a predefined threshold (Schmitz, 2006). In geographic subsumption, $p(B|A) = Area(A \cap B)/Area(A)$, and similarly for $p(A|B)$, where *Area(A)* is a function that returns the area of *A*, and *Area(A∩B)* returns the area of intersection of *A* and *B*.

Evaluation: We used the Flickr API to retrieve the names of members of seventeen public groups devoted to wildlife and nature photography. We then used a Web page scraping tool to retrieve sets created by these users. We retrieved a total of 166,526 sets from 7,618 pro users, and also the tags and geotags from images in these sets, which yields 1.3 millions of geographical points (photos) in total. We collected points associated with each place name by identifying photos that are contained in sets whose name matches a geoname in GeoNames.org. We used substring matching to match the geoname to set name. The geotags of these photos then become our points. We identified 1,774 geographic concepts in the data set and used associated points to create regions. Of these concepts, 610 are about the North America continent.

We evaluate relations learned by geographic subsumption by comparing them against existing relations in the Geonames hierarchy. Let *NL* be the number of learned relations, of which *C* also exist in the Geonames hierarchy. Recall measures

Figure 4. Polygons created from points representing 'cebu' (green) and 'philippines' (red). Note that 'cebu' is subsumed by the 'philippines'

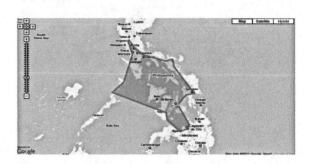

the coverage of the learned relations, i.e., fraction of the existing relations that the method actually learned. To measure recall, we first compute *NG*, the number of relations within the Geonames hierarchy that exist between places (place ids) that are linked to set names in our data set. Recall is then given by $R = C/NG$. Precision measures the quality of the learned relations, i.e., the fraction of the learned relations that are correct. We compute precision as $P = C/NL$.

Figure 5 shows recall and precision of the relations learned by geographic subsumption using the convex hull and erosion methods. Since our data set is very sparse, it turns out that convex hull is better able to generalize the boundary of a region, resulting in a higher number of learned relations. However, convex hull may overgeneralize, including irrelevant regions within a place's boundary, reducing its precision score. Compared to the convex hull, erosion algorithm leads to fewer errors.

We also compare our approach to tag-based probabilistic subsumption described in Schmitz (2006). This method computes $p(B|A)$ from the co-occurrence of tags *A* and *B*: i.e., $p(B|A)=Frequency(A,B)/Frequency(A)$, where *Frequency(A)* is the number of photos tagged with *A*, and *Frequency(A,B)* is the number of photos tagged with *A* and *B*. To collect data for this

Figure 5. Comparison of geographic subsumption on Flickr data using erosion and convex hull methods to represent place boundaries

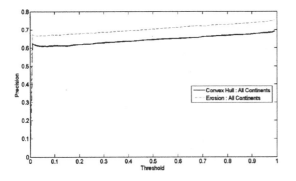 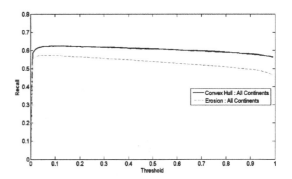

baseline, we queried Flickr to find the number of images that were tagged with keyword *A* and two keywords *A* and *B*. The keywords were geonames that matched set names in our data set. Unfortunately, since the baseline approach requires us to invoke Flickr's web service to obtain a co-occurrence count of each tag pair, it is infeasible for us to obtain all counts of the entire data set (which requires 1,774 choose 2 requests). Instead, we collect tag co-ocurrence statistics for photos on the North American continent (610 choose 2 requests).

Geographic subsumption significantly outperforms the baseline tag subsumption, as shown in Figure 6, for the following reasons. First, as observed by Schmitz (2006), users seldom annotate an image both with the most general and most specific tags. For example, using the baseline probabilistic subsumption method, p('university of south florida' | 'usa') = 0.0 and p('usa' | 'university of south florida') = 0.001. In other words, few users specify tags 'usa' and 'university of south florida' in the same photo. However, the geographic distribution of the tag 'usa' is likely to geographically subsume the distribution of the tag 'university of south florida'. Thus, geographic subsumption can solve the challenge of "general vs specific" concepts. Second, geographic subsumption can also solve the "popularity vs generality" challenge. For example, in baseline approach, p('california'

| 'usa') = 0.14 and p('usa' | 'california') = 0.12. The result is that 'california' will subsume 'usa'. However, this is simply because users specify the tag 'california' more frequently than the tag 'usa'. Finally, with proper parameters, we can solve the "ambiguity" challenge, for example, Victoria in Canada or Australia or Cambridge in United States or United Kingdom. In fact, this challenge can also lead to the "popularity vs generality" challenge, because when evidence for ambiguous tags is aggregated, the total frequency may become more than its parent's. For example, the tag 'victoria' has been used 847,467 times on Flickr and 'british columbia' 513,116 times. However, 'victoria' tag could include instances of Victoria in Australia, and other places named 'victoria', resulting in a higher tag count for this concept than its parent 'british columbia' concept. After our method disambiguates the term 'victoria', it correctly infers that 'british columbia' geographically subsumes 'victoria'.

Novel relations: Since Geonames is not complete, some of the relations that are judged incorrect because they do not exist in the Geonames hierarchy may actually be valid. Table 2 reports some of the novel relations learned from data. Some of these places, such as 'disneyland resort', do not exist in Geonames. In other cases, the place does exist but the relation does not. This is because Geonames is often to coarse-grained. For example,

Figure 6. An automatic comparison against the reference set on F-score between the proposed approach and the baseline at different values of the subsumption threshold

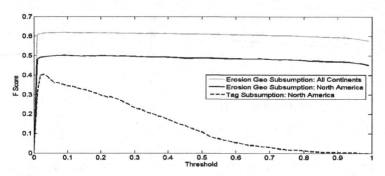

according to Geonames, 'bronx' is part of Kings County, rather than New York City. Similarly, 'disneyland' is part of Orange County, rather than the city of Anaheim. Our method finds correct finer-grained relations. At other times, learned relations appear to be patently incorrect. Neither 'la jolla' nor 'wild animal park' are in 'san diego' proper, but their proximity to San Diego makes many people associate them with the city. While an expert would judge this relation not to be correct, it still represents useful knowledge, since a visitor to San Diego should know about La Jolla and Wild Animal Park. Extracting such bits of folk knowledge is the distinguishing feature of our approach.

RELATED WORK

Several researchers have recently proposed approaches to learning conceptual hierarchies, or folksonomies, from social metadata. These approaches include graph-based (Mika, 2007), clustering (Brooks & Montanez, 2006) and hybrid methods that create similarity graph of tags (Heymann & Garcia-Molina, 2006). Schmitz (2006) has also applied a statistical subsumption model (Sanderson & Croft, 1999) to induce hierarchical relations of tags. All these methods use tags, and therefore, suffer from the "popularity vs generality" problem. Specifically, a certain tag may be used more frequently not only because it is more general, but because it is more popular among users. On Flickr, e.g., there are many more photos tagged with "Washington" than "United States". As was argued and demonstrated in the previous work (Plangprasopchok & Lerman, 2009), tag statistics alone may not be adequate for inducing relations.

In addition to tags, there are other types of user-generated metadata, such as set/collection hierarchies and geo-referencing tags (geo-tags) that are ubiquitous on the Social Web sites such. Geotags can potentially be used to resolve the "popularity vs. generality" problem and many others by providing an additional view on how one concept geospatially relates to others. Researchers have begun to exploit geo-tags to induce "place semantics" – an association between place and other features, such as textual and visual information. Rattenbury and Naaman (2009) proposed an automatic approach to determine whether a certain tag is used for representing place(s). This approach is based on the assumption that, in general, a place tag appears locally, rather than ubiquitously, within a certain area. Meanwhile, a couple of recent works proposed frameworks to find correlations among geotags, visual features and tags of photos, and then utilize them for tag recommendation from photos" location (Moxley et al., 2008) and visual features (Kleban et al., 2009), or conversely, estimating location from

Table 2. Novel relations learned from data

Child	Parent	Child	Parent
anaheim	la	golden gate bridge	san francisco bay
ballard	puget sound	greenfield	new york
brandywine park	wilmington	griffith park	la
bronx	new york city	griffith park	los angeles
bronx zoo	new york city	hollywood	la
bruce peninsula	georgian bay	la jolla	san diego
burbank	la	malibu	los angeles
burbank	los angeles	pasadena	la
cabo san lucas	los cabos	pearl harbor	oahu
cabrillo	san diego	queen anne	seattle
chinatown	los angeles	san diego wild animal park	san diego
coney island	new york city	san diego zoo	san diego
crescent beach	nova scotia	santa monica	la
dayton	new york	santa monica	los angeles
discovery park	seattle	sea world	orlando
disneyland resort	disneyland	times square	new york city
eastern market	detroit	union square	manhattan
elkhorn slough	monterey	university of south florida	tampa
eureka	victoria	university of washington	puget sound
georgia aquarium	atl	webster park	rochester
pasadena	los angeles	lake eola	orlando
disneyland	la	bainbridge island	puget

visual features and tags (Crandall et al., 2009). The aims of these works are different from ours: we further investigate the approach to induce hierarchical relations among geospatial concepts to construct and/or enrich geospatial ontologies.

Several works dealt with the problem of disambiguating places. Approaches proposed by Li et al. (2003) and Amitay et al. (2004) utilize gazetteers to identify places mentioned in some documents. In particular, when an ambiguous place is mentioned in a document, e.g., "Buffalo" can be one of 23 different cities in the United States, the rest of the document is scanned to obtain more clues, e.g. the term "NY". New clues in combination with the ambiguous name are then compared to some place names in the gazetteer. If there is one

exact match, the place is then identified and hence disambiguated. Our method does not assume prior knowledge, such as gazetteers, to disambiguate places, but locality of geographic coordinates and geographic subsumption relations. This way, our approach can enrich existing gazetteers, which, in turn, can be by other methods to achieve better performance.

Finding a boundary of a region from a set of points is a widely studied problem. Alpha shape is a popular algorithm in many shape reconstruction applications and has been generalized to 3D reconstruction. It was introduced by Edelsbrunner et al. (1994) and is a generalization of the convex hull (Cholewo, 1999). The algorithm is based on Delaunay triangulation. It produces a sub graph of Delaunay triangulation that is determined by a parameter alpha, which produces different level of details. When alpha is very large, the boundary produced by this method will be identical to the convex hull. However, alpha is a global parameter which is well suited only for uniformly distributed points (Cazals, 2005). In some shape reconstruction applications, data points are not uniformly distributed. For examples, if we would like to reconstruct Portugal boundary from the locations of photographs that people take, the points will be dense in the coastal areas but sparse in the inland regions.

Alani et al. (2001) present a method for approximating the shape of a region from a set of spatial points. Authors call their region approximation technique Dynamic Spatial Approximation Method. The algorithm creates a Voronoi diagram from different classes of geospatial concepts. The union of Voronoi cells of the same class approximates the geospatial concept. However, this method cannot apply to isolated regions that are not adjacent to other regions. Sharifzadeh et al. used machine learning techniques, such as k-nearest neighbors and support vector machines, to identify decision boundary between points belonging to different geospatial concepts and used the decision boundary as region boundary (Sharifzadeh, 2003). These techniques assume that the regions whose boundaries are being learned are disjoint.

However, both of these approaches make assumptions that are not suitable for our use. First, they need points from more than one class to limit Voronoi cells and create the decision boundary. On the other hand, we cannot assume complete data from a specific area, which includes all classes required to learn the decision boundary. Second, they assume disjoint regions. While this assumption holds for the zipcodes data set, it does not apply to social metadata, where we often find related classes. For example points belonging to Los Angeles and California do not form disjoint regions with an explicit boundary between them.

They solve this problem by using a part-of database to resolve relations between classes. In our work, on the other hand, we learn these relations.

CONCLUSION

User-generated content and metadata may be mined to discover geospatial knowledge that can enrich and supplement existing geospatial directories and gazeteers. We have explored a method to extract knowledge about places and relations between them from geo-tagged metadata on the social photo-sharing site Flickr. Our approach aggregates metadata across many individuals to discover points associated with places, constructs a boundary of these places and then uses geographic subsumption to discover relations between places. Key to this is the ability to learn accurate place boundaries. In this paper, we introduced a simple algorithm that can reconstruct a non-convex boundary of a set of points. Starting with a bounding box of the set of points, the algorithm erodes the boundary to the point nearest the boundary. We compared the performance of the erosion algorithm to other boundary finding algorithms on a synthetic data set consisting of points sampled from US

Zipcodes: While erosion algorithm's performance is slightly worse than that of the existing state-of-the-art approach, its speed and simplicity make it a competitive boundary finding algorithm.

ACKNOWLEDGMENT

This work is supported, in part, by National Science Foundation under award CMMI-0753124. Suradej Intagorn gratefully acknowledges the support of the Thai Government.

REFERENCES

Alani, H., Jones, C. B., & Tudhope, D. (2001). Voronoi-based region approximation for geographical information retrieval with gazetteers. *International Journal of Geographical Information Science*, *15*(4), 287–306. doi:10.1080/13658810110038942

Amitay, E. Har'EI, N., Sivan, R., & Soffer, A. (2004). Web-a-where: Geotagging web content. In *Proceedings of the 27th Annual International ACM SIGIR Conference on Research and Development in Information Retrieval.*

Bentley, J. L. (1975). Multidimensional binary search trees used for associative searching. *Communications of the ACM*, *18*(9), 509–517. doi:10.1145/361002.361007

Brooks, C. H., & Montanez, N. (2006). Improved annotation of the blogosphere via autotagging and hierarchical clustering. In *Proceedings of the 15th International World Wide Web Conference.*

Cazals, F., Giesen, J., Pauly, M., & Zomorodian, A. (2005). Conformal alpha shapes. In *Proceedings of the Second Symposium on Point Based Graphics* (pp. 55-61).

Cholewo, T. J., & Love, S. (1999). Gamut boundary determination using alpha-shapes. In *Proceedings of the IS&T and SIDs Seventh Color Imaging Conference* (pp. 200-204).

Clarkson, K. (n. d.). *A program for convex hulls.* Retrieved from http://www.netlib.org/voronoi/hull.html

Crandall, D. J., Backstrom, L., Huttenlocher, D. P., & Kleinberg, J. M. (2009). Mapping the world's photos. In *Proceedings of the 18th International World Wide Web Conference.*

Djurcilov, S., & Pang, A. (1999). Visualizing gridded datasets with large number of missing values. In *Proceedings of the Visualization Conference* (pp. 405-408).

Edelsbrunner, H., & Mucke, E. P. (1994). Three-dimensional alpha shapes. *ACM Transactions on Graphics*, *13*(1), 43–72. doi:10.1145/174462.156635

Golder, S., & Huberman, B. A. (2006). The structure of collaborative tagging systems. *Journal of Information Science*, *32*(2), 198–208. doi:10.1177/0165551506062337

Graham, R. L. (1972). An efficient algorithm for determining the convex hull of a finite planar set. *Information Processing Letters*, *1*, 132–133. doi:10.1016/0020-0190(72)90045-2

Heymann, P., & Garcia-Molina, H. (2006). *Collaborative creation of communal hierarchical taxonomies in social tagging systems* (Tech. Rep. No. 2006-10). Stanford, CA: Stanford University.

Hopcroft, J., & Tarjan, R. (1973). Algorithm 447: Efficient algorithms for graph manipulation. *Communications of the ACM*, *16*, 372–378. doi:10.1145/362248.362272

Intagorn, S., Plangprasopchok, A., & Lerman, K. (2010). Harvesting geospatial knowledge from social metadata. In *Proceedings of the 7th International Conference on Information Systems for Crisis Response and Management.*

Kavouras, M., Kokla, M., & Tomai, E. (2006). Semantically-aware systems: Extraction of geosemantics, ontology engineering, and ontology integration. In Stefanakis, E., Peterson, M. P., Armenakis, C., & Delis, V. (Eds.), *Geographic hypermedia* (*Vol. 3*, pp. 257–273). Berlin, Germany: Springer-Verlag. doi:10.1007/978-3-540-34238-0_14

Keating, T., & Montoya, A. (2005). *Folksonomy extends geospatial taxonomy*. Directions Magazine.

Kleban, J., Moxley, M., Xu, J., & Manjunath, B. S. (2009). Global annotation on georeferenced photographs. In *Proceedings of the Conference on Image and Video Retrieval* (p. 12).

Li, H., Srihari, R., Niu, C., & Li, W. (2003). InfoXtract location normalization: A hybrid approach to geographic references in information extraction. In *Proceedings of the Workshop on the Analysis of Geographic References NAACL-HLT*.

Mathes, A. (2004). Folksonomies: Cooperative classification and communication through shared metadata. [University of Illinois Urbana-Champaign.]. *Urbana (Caracas, Venezuela)*, IL.

Mika, P. (2007). Ontologies are us: A unified model of social networks and semantics. *Web Semantics*, *5*(1), 5–15. doi:10.1016/j.websem.2006.11.002

Moxley, E., Kleban, J., & Manjunath, B. S. (2008). Spirittagger: A geo-aware tag suggestion tool mined from flickr. In *Proceedings of the 1st ACM International Conference on Multimedia Information Retrieval* (pp. 24-30).

Newsam, S., & Yang, Y. (2008). Integrating gazeteers and remote sensed imagery. *GIS, 26*.

Panigrahy, R. (2008). An improved algorithm finding nearest neighbor using kd-trees. In *Proceedings of the 8th Latin American Symposium on Theoretical Informatics* (pp. 387-398).

Plangprasopchok, A., & Lerman, K. (2009). Constructing folksonomies from user-specified relations on flickr. In *Proceedings of the International World Wide Web Conference*.

Rashmi, S. (2005). *A cognitive analysis of tagging*. Retrieved from http://rashmisinha.com/2005/09/27/a-cognitive-analysis-of-tagging/

Rattenbury, T., & Naaman, M. (2009). Methods for extracting place semantics from Flickr tags. *ACM Transactions on the Web*, *3*(1), 1–30. doi:10.1145/1462148.1462149

Sanderson, M., & Croft, B. (1999). Deriving concept hierarchies from text. In *Proceedings of the 22nd Annual International ACM SIGIR Conference on Research and Development in Information Retrieval* (pp. 206-213).

Schmitz, P. (2006) Inducing ontology from flickr tags. In *Proceedings of the Collaborative Web Tagging Workshop*.

Sharifzadeh, M., Shahabi, C., & Knoblock, C. A. (2003). Approximate thematic maps from labeled geospatial data. In *Proceedings of International Workshop on Next Generation Geospatial Information*.

ENDNOTES

[1] We distinguish between two types of community-generated knowledge. Active knowledge, such as Volunteered Geographic Information (VGI), is created through collaborative efforts, e.g., to map Haiti. Passive knowledge emerges from actions people do to fulfill personal information needs, e.g., categorize their photos.

[2] The collection feature is limited to paid "pro" users. Pro users can also create unlimited number of photo sets, while free membership limits a user to three sets.

This work was previously published in the International Journal of Information Systems for Crisis Response and Management, Volume 3, Issue 2, edited by Murray E. Jennex and Bartel A. Van de Walle, pp. 33-47, copyright 2011 by IGI Publishing (an imprint of IGI Global).

Chapter 8
Interaction Design Principles for Web Emergency Management Information Systems

Teresa Onorati
Universidad Carlos III de Madrid, Spain

Alessio Malizia
Universidad Carlos III de Madrid, Spain

Paloma Díaz
Universidad Carlos III de Madrid, Spain

Ignacio Aedo
Universidad Carlos III de Madrid, Spain

ABSTRACT

The interaction design for web emergency management information systems (WEMIS) is an important aspect to keep in mind due to the criticality of the domain: decision making, updating available resources, defining a task list, and trusting in proposed information. A common interaction design strategy for WEMIS seems to be needed, but currently there are few references in literature. The aim of this study is to contribute to this lack with a set of interactive principles for WEMIS. From the emergency point of view, existing WEMIS have been analyzed to extract common features and to design interactive principles for emergency. Furthermore, the authors studied design principles extracted from a well-known (DERMIS) model relating them to emergency phases and features. The result proposed here is a set of design principles for supporting interactive properties for WEMIS. Finally, two case studies have been considered as applications of proposed design principles.

DOI: 10.4018/978-1-4666-2788-8.ch008

INTRODUCTION

The management of information and resources is a crucial activity for the agencies and organizations managing emergency situations. The scope of Emergency Management Information Systems (EMIS) is to support the activities usually performed by emergency workers: they are focused on the organization of available information and resources (Van de Walle & Turoff, 2007). Emergency management is a collaborative and multi-organizational endeavor (Waugh & Streib, 2006) and, therefore, EMIS have to provide communication channels and collaborative tools for teams that are not only geographically distributed but also functionally independent. By functionally independent we mean that different teams might have different goals, perspectives and operation protocols. Nowadays, many agencies and organizations for emergency management use web-based EMIS (WEMIS) as a support tool for the cooperation and the coordination in different emergency phases. WEMIS employ Internet protocols and facilities for the communication during coordination activities. Eventually, they do not require an installation procedure and they have a high portability: all devices with an Internet connection can access to them. In this way, also people that work in direct contact with the emergency area, as firemen or policemen, can use them with mobile devices and communicate with coordination offices.

In this paper, we focus on WEMIS that make it possible to efficiently communicate and share information. More specifically, our interest lies on the design of the interaction with WEMIS. Emergency workers have to manage complex situations where short time decision making is fundamental: victims and damages depend on the emergency solution strategy. For this reason, users need a quick and trustworthy interaction with the system: they have to know exactly the next task to perform, which kind of results and consequences to expect, and which information

and status have to be updated. In other words, the WEMIS has to support at some level situation awareness (Endsley, 2000) by providing accurate, timely and appropriate information at each stage so that each user can understand the situation (situation assessment) and decide how to react properly. Considering emergency management as a team work, we should transcend the concept of situation awareness to move onto activity awareness which is the "awareness of project work that supports group performance on complex tasks" (Carroll et al., 2003). Even though no system can guarantee situational or activity awareness, a good interaction design can help to support it by ensuring that users will have the right information at the right time and in the right form both concerning the situation as well as the performance of other cooperating teams.

The design of the interaction with the WEMIS is therefore a cornerstone as systems should support users in developing their tasks without interfering with their usual protocols or imposing any kind of burden (Carver & Turoff, 2007). Moreover, the adoption of this kind of systems is not straightforward as many factors influence their real usage in emergency situations (Aedo, Diaz, Carroll, Convertino, & Rosson, 2009). For instance, the subjective impressions of users about their personal capability to use the system, the degree of control, the support from other colleagues or from the organization, have a strong influence in the final decision to use a system (Mathieson, Peacok, & Chin, 2001). The interaction design process for WEMIS needs to address all these issues properly: a possible approach could be the participatory design where users and stakeholders are actively involved.

There are some design principles like those defined in (Turoff, Chumer, Van de Walle, & Yao, 2004) but until now, interactive design models have not been specifically defined for the emergency domain and in particular for WEMIS. The aim of our work is to contribute to this lack to guarantee a quick and trustworthy interaction among users and systems improving the entire emergency management. Our proposal is a set of formal interactive principles that designers could apply during the design process of WEMIS.

In the first section, we present a survey about existing WEMIS to find out common characteristics and design aspects. In the second section, we categorized WEMIS features into twenty-three different classes relating them with emergency supported by the framework proposed by Chen, Sharman, Rao, and Upadhyaya (2008). In the third section we present our main contribution, a set of interactive properties for WEMIS. We relate features' classes, as previously defined, to a set of design principles extracted from a specific design model for the emergency management domain: DERMIS (Turoff, Chumer, Van de Walle, & Yao, 2004). In the fourth section these principles have been formalized using the PIE model (Dix, Finlay, Abowd, & Beale, 2004), obtaining a number of formal interactive properties fulfilled if our design principles are employed. Finally, two case studies have been considered as application of proposed interaction design principles.

A STUDY OF EMERGENCY MANAGEMENT INFORMATION SYSTEMS IN THE WEB

The main basis for the design process of existing WEMIS is the analysis of data collected from real experiences in emergency management. First of all, we have analyzed the literature about functionalities and requirements an emergency management information system has to provide. Successively, we have collected several systems used during real emergency situations.

In literature, there are several studies about factors that can influence positively or negatively the management process and consequently the usage of support systems (Jennex, 2004). A negative factor is represented by the stress as influence for the decision making process: Patton and Flin

(1999) and Turoff (2002) have pointed out possible solutions, such as testing operational demands, allocating resources, improving the confidence in available information and providing modifiable templates for actions. Another factor is the communication: delays in communicating messages, such as notifications or commands, could cause misunderstood among involved emergency workers (Andersen et al., 1998). A positive factor could be the application of new technologies to emergency phases: in this way it could be possible to solve problems related to information, such as overload, loss or outdated (Fisher, 1998). In particular, Gheorghe and Vamanu (2001) have proposed new features for integrating Geographical Information Systems with WEMIS.

For instance, the Sahana web system has been developed from the observation of the emergency management process applied during the tsunami of 2004 in Sri Lanka (De Silva, De Silva, Careem, Raschid, & Weerawarana, 2006); similarly SIGAME was based on the lessons learned from the massive wild fires in Galicia in 2007 (Aedo, Díaz, & Díez, 2009). However, apart from individual experiences there is a lack of clear definition of the list of interactive tasks to perform for emergency planning and solving. Moreover, until now a common and formal design strategy has not been defined yet to develop WEMIS. In this section we analyze a set of WEMIS to extract common and frequent characteristics.

The set of WEMIS that we considered has been extracted from the W3C Working Group, called the Emergency Information Interoperability Framework (EIIF), which aim is to define standards for emergency management domain. In Table 1, there are the eight WEMIS that we have considered from the W3C classification. For each of them we have collected a detailed description about the management process, available features, used standards and real use cases. All these data have been extracted both from the W3C classification and by studying system's references. Table 1 summarizes the main features of these systems.

Table 1. Considered WEMIS for the analysis

WEMIS	Main Features	Reference
Sahana	Tracing // Inventory Management // Aid Distribution // Situation Mapping // Shelter tracking // Responder Management	www.sahana.lk
Fema	Traditional shelters // Household Pet // Medical Special Needs // Kitchens // Points of Distribution // Food warehouses // GIS mapping	www.fema.gov
ReliefWeb	Mapping // Funding // On-line library // Professional Resources // Web feed service	www.reliefweb.int
UN OCHA	Mapping // Flexible architecture // Synchronization	3w.unocha.org
DesInventar	Disaster datacard // Database Support// GIS systems // Google maps and Google Earth tools // Excel and XML formats	www.desinventar.net/
FEMIS	Resource tracking // Task lists // Contact lists // Event logs // Status boards // Hazard modeling // Evacuation modeling	www.pnl.gov/FEMIS
UN GDASC	Disaster alerts // Situation overview // List and status of emergency teams // Satellite based maps	www.gdacs.org
SIGAME	Coordination between communities // Coordination of the supra-communitarian aids // Monitoring of resources	www.sigame.es

From our study we can conclude that there are some common features, but the design and the implementation is usually completely different. For example, in Sahana the mapping of available resources and organizations is a textual list with the identification of the data and the location whilst in DesInventar it is represented as an interactive map. Another example is the notification of disaster information: in GDACS it is shown directly on the homepage, instead in FEMA it is in a different page accessible from the main menu.

In order to obtain a common list of features for WEMIS that could help to define a common interaction model, we have combined characteristics obtained from the two analyses we have performed: the first one about the literature and the second one about real WEMIS. In the next section we are going to group these characteristics into general classes of functionalities.

A CATEGORIZATION OF EMERGENCY CHARACTERISTICS BY PHASES

The emergency management process involves a complex set of activities, each of which has

a specific scope and requires different kinds of resources, information, cooperation degree and worker expertise. Moreover, an important dimension to consider is timing: the emergency process can be seen as a sequence of stages where the tasks to be performed strongly depend on specific requirements of the emergency phase.

In literature, emergency phases have been identified in different ways. In 2009, Harrald has presented an interesting study about the relation between agility and organizational systems for the management of emergencies. In this work, the author has considered four emergency phases related to two dimensions: the organizational size and the timing (Harrald, 2009). During the first phase (*reaction and mobilization*), involved people work on the preparation and the prevention. When an extreme event occurs, the organizational size grows up in order to individuate the response teams (second phase, *integration*) and successively tasks to perform (third phase, *production*). Finally, during the fourth phase (*demobilization*) the large presence of response teams decreases. A similar approach has been adopted by Jennex and Raman (2009). Authors have individuated four emergency phases related to organizational and timing dimensions: *situational analysis, initial response, emergency response* and *recovery response*. Along these phases, several decision points have been defined depending on the progress of the emergency through different events: *initiating, control, restoration, normalizing* and *terminating*.

Other interesting studies have been published considering a collaborative approach. Waugh and Streib (2006) identified four phases depending on role and tasks to perform: (1) prevention and mitigation; (2) planning and training; (3) response; (4) recuperation. Another framework based on a lifecycle approach is the emergency response coordination lifecycle presented in Chen et al. (2008). Compared to Waugh et al. (2006), this framework introduces a distinction between activities related to front-end situations and those concerning back-end organizations. In this way it is possible to specify a task list depending on the role and responsibility of workers. For example, front-end workers use different kinds of devices compared to back-end workers: applying the Chen's framework it is possible to face this difference with an appropriate design strategy. For instance, in Catarci et al. (2007), authors presents the European project Workpad: a 2-level framework for the collaboration among emergency teams that confirms the current trend to manage emergencies distinguishing between back-end and front-end communities. The scope is to provide specific services for each community. For example, the back-end community needs an efficient way to organize information collected about the disaster, like maps, resources and localization of emergency points. The front-end community, on the other end, needs a fast communication channel to receive back-end information and organize the emergency response.

In Chen et al. (2008), the life cycle is divided into three stages: pre-incident, during-incident, post-incident. The first one concerns with activities related to planning, training and organizing involved teams. The response management of occurred crisis and the coordination of workers are part of the during-incident phase. Furthermore, the during-incident stage is viewed as the composition of two sub-cycles: mini-second and many-second. The mini-second coordination cycle is the onsite organization of teams and front-end emergency workers for a rapid intervention and prevention. The many-second coordination cycle concerns the emergency operations center (EOP) and all supervisor structures for an efficient communication with the onsite response. The post-incident or recovery phase focuses on the return to a normal situation and the analysis of collected information in order to improve future planning. Moreover, the framework defines five basic elements and it applies them to each stage of the coordination phase: (1) *task flow* with tasks to perform and relations among them; (2) *resources management*; (3) *information management*; (3) *decision making* about organization and structure; and (5) *responders* involved to complete the task flow.

We categorized the set of features presented in Table 1 to fit general classes of features, and thus, summarizing a set of common features into a general category; for example: *mapping, situation mapping, Google maps and Google Earth tools* taken from Table 1 all fit into the *Spatial Registry* feature class shown in Table 2.

By relating features classes and different emergency phases within the Chen's lifecycle framework we propose an organization of information that before was spread among different aspects of WEMIS. The result is shown in Table 2: we have defined eighteen classes structured into three main emergency phases.

For pre-incident response, we have individuated five classes for planning, training and modeling information. The first class is the *Organization Registry* for the management of organizations and teams working in the disaster area, including personnel information about each worker and its responsibilities and capabilities, so that the WEMIS can have data to decide which are the tasks that can be assigned to each worker and the kind of information he would need to improve his situation assessment. The second one is the

Table 2. Emergency coordination life cycle and classes of features

Stage		General Features
Pre-Incident Response		Organization Registry: collect information about involved organizations in the disaster region.
		Spatial Registry: associate emergency information with geographic information system
		Emergency Message Center for training and meetings
		Hazard and vulnerability modeling
		Demographic Registry
During Incident Response	Mini-Second Coordination Cycle	Task Tracking for front-end teams
		Victims Registry
		Emergency Message Center for front-end teams
		Emergency Teams Registry for front-end teams
	Many-Second Coordination Cycle	Emergency Message Center for back-end workers to coordinate all involved organizations
		Task Tracking for operations and decisions
		Collaborative Suggestions Collection about the task list
		Requests Registry
		Shelter and Resource Registry
		Victims Registry
		Volunteer Registry
		Situation Mapping
Post-Incident Response		Task tracking with final results and real processes
		Discussion forum among involved organizations and emergency workers
		Photo library
		Analysis and Improvement for task tracking
		Reports Registry
		Estimation of population at risk for future pre-incident planning

Spatial Registry to establish a relation between emergency information and geographic localizations, so distributed teams as well as resources can be organized in an efficient way. The third class is the *Emergency Message Center* to define a communication channel for planning meetings and trainings. Next, the *Hazard and Vulnerability Modeling* concerns the modeling of information collected during past disasters to evaluate the risks and vulnerabilities. The last class is the *Demographic Registry* that manages people living or working in disaster areas.

For the mini-second coordination cycle during the incident response, we identified four classes and main features about tracking tasks and man-

aging communications. The *Task Tracking* class keeps track of performed tasks, so that support for activity awareness can be provided, collecting information from front-end teams. Next, the *Victims Registry* manages affected people, updating lost, dead and injured. The third class is the *Emergency Teams Registry* to follow activities of each team. The last one is the *Emergency Message Center*, the same one we have defined for the pre-incident stage, but in this case the communication channel is established among front-end teams to share information and resources.

Activities for the many-second coordination cycle during the incident response are grouped into eight classes related to a higher level of management. The *Task Tracking* is for main operations and decision-making, where an operation is a set of simple tasks. The *Victims Registry* is defined crossing different affected areas. In the *Emergency Message Center*, the communication channel is established among back-end workers to coordinate operations and organizations. The *Collaborative Suggestions Collection* collects information and suggestions from emergency teams in order to choose the best-fitting plan and task list. Next, several registries are defined about *Requests*, *Shelter* and *Volunteer* to keep updated available information. The last class is the *Situation Mapping* to monitor the general response process, collecting information from all involved teams and performed operations.

Finally, in the post-incident response we have defined six classes related to the improvement of future emergency response. In this case the Task Tracking class presents final results for performed tasks: this information can be employed for next analysis, as Improvement, Estimation of population at risk and Report Registry. Moreover, a Discussion Forum supported by a Photo Library is created among organizations and emergency workers for a collaborative exchange of opinions and best practices with a view to elicit knowledge sharing.

In the next section, we derive from this categorization a set of design principles and formalized interactive properties.

FROM DERMIS TO PIE MODEL: INTERACTIVE DESIGN PRINCIPLES FOR WEMIS

This section presents our proposal to structure the interaction in the WEMIS domain: from the emergency life cycle to formal properties for interaction design principles. The definition of our interaction design principles has two different stages. During the first one, the relation between emergency phases, features and design principles is established to derive design principles from existing practices. In this case, we are going to merge the results obtained in Table 2 and the Dynamic Emergency Response Management Information System (DERMIS) model presented in (Turoff et al., 2004). During the second stage, emergency design principles are formalized through the PIE (Program, Interpretation, Effect) formalism (Dix et al., 2004) to obtain a set of interactive properties related to phases and features. We will briefly describe here the DERMIS and PIE model; for an extended description see the Appendix.

In Turoff et al. (2004), authors present a framework for the design of EMIS, called DERMIS model. From the analysis of organization and planning of real emergency management agencies, the DERMIS model has been structured into four different sections: *Design Premises*, *Conceptual Design*, *General Design Principles and Specifications*, and *Supporting Design Considerations and Specifications*. We have formalized and specialized for interaction the set of general design principles by using the PIE model. The scope of the PIE model is to structure the interaction between users and systems (Dix et al., 2004), through a formal language. The interaction starts with a set of user's actions as input: each action is called *command* (C) and a sequence of commands is represented as a *program* (P). The output of the system is the result obtained from the execution of the program and it is called *effect* (E). The sequence of system functions for generating the output is represented by an interpretation function I that relates programs with effects. For example, if a user wants to print a file, the print function is the program and the printed document is the effect. Furthermore, the effect generates a set of *results* (R) and a set of *observable effects* (O). Results represent the system output (e.g. the printing of the document); instead, the observable effect is what the screen visualizes (e.g. the editor and the open document). At last, if the screen cannot visualize the entry observable effect, but just a part of it, this part is the *display* (D) (e.g. the display is the part of a document currently visualized in the screen and the observable effect is the entire document that can be visualized with the scrolling). Programs, effects, results, observable effects and displays are the main elements of the PIE.

THE ENRICHED EMERGENCY LIFE CYCLE: INTERACTION DESIGN PRINCIPLES FOR WEMIS

We have enriched the emergency life cycle with interactive principles defined through two different phases. During the first one, we have related principles from the DERMIS model to emergency features and phases. In particular, for each stage of the coordination life cycle, we have considered individuated classes of features (see Table 2) in order to recognize appropriate DERMIS principles. For example, let us consider the during-incident response stage and the mini-second coordination cycle (see Table 3). *Task Tracking* and *Victims Registry* features can be related with *Information Source and Timeliness* and *Up-to-date Information and Data* principles. All of them are about the visualization and the updating of information: in order to perform an efficient task and victims tracking it is important to provide appropriate data in a well-structured way to facilitate the access. The second class of features is *Emergency Message Center for front-end teams*. In this case, related principle is *Open Multi-Directional Communication* to structure the communication of updated information. The last class is *Emergency Teams Registry*. Here, we need *Authority, Responsibility, and Accountability* to support the definition of different levels of responsibility and the social interaction among workers.

During the second phase for the definition of the enriched life cycle, we have formalized DERMIS principles through the PIE model properties. In particular, each principle has been formalized in order to obtain the proper interactive property among four possibilities: *observability*, *reach-ability*, *predictability*, *transparency* and *meta-communication*. While *observability*, *reachability*, *predictability* and *transparency* are already defined within the PIE model (Appendix), our contribution also consists of introducing *meta-communication* (de Souza, 2004) to represent rules and semantics for the communication between users and designers, so that participatory design can be supported. A high *meta-communication* corresponds to a clear understanding of the designer's strategy by users: in this way, users can access to available information through an efficient interaction. The *reachability* is implicitly defined into the *observability* and *predictability* as the possibility to reach a specific state of the system from any other through a sequence of actions. Moreover, *observability* and *predictability* represent the basis to define *transparency* and *meta-communication*. The *observability* is referred to visualized information on the screen: a system is observable if users can understand completely the output of the system observing the display. The *predictability* is similar to the *observability*, but for future states: a system is predictable if observing the display users are able to determine which states will be reached by the system through future actions. Users of an observable and predictable system know exactly the current and future results they will obtain from running functionalities. In the Appendix, there is a detailed description of each interactive property. From these definitions, now we can derive the *transparency* as a combination of *predictability* and *observability*: a transparent system visualizes clearly on the screen all information and data about performed operations, events and effects. In this way, users can know exactly current and future states, avoiding possible mistakes. As confirmed by the formalization process, in the emergency management domain, the *transparency* represents the solution to ensure that users get updated and relevant information and an efficient exchange of messages, so that it might contribute to develop an appropriate situational awareness.

Table 3. The enriched emergency life cycle for mini-second coordination

Interactive Principles	**Information Source and Timeliness** → a complete representation of information → considering information as a program P and its visualization as an effect E, we visualize the complete status of available information → transparency $\forall e \in E.predict(observe(e)) = (e)$, where $\forall e \in E.predict(observe(e)) = result(e)$	**Up-to-date Information and Data** → "gone away for a cup of tea" metaphor → predictability $\forall e \in E.predict(observe(e)) = result(e)$	**Open Multi - Directional Communication** → all users have to be aware of the current state of the system, from information to communication → considering the information as a program (P) and its visualization as an effect (E), we want to communicate P in order to avoid misunderstanding about E → transparency and meta-communication $\forall e \in E.predict(observe(e)) = (e)$ & $\forall e \in E.I_D^{-1}(e) = I_U^{-1}(e)$	**Authority, Responsibility, and Accountability** → we have to adapt the system to different roles and users that interact with it → improve the meta-communication between designer and user → meta-communication $\forall e \in E.I_D^{-1}(e) = I_U^{-1}(e)$
DERMIS Design Model (Turoff) Design Principles	**Information Source and Timeliness** – In an emergency it is critical that every bit of quantitative or qualitative data brought into the system dealing with the ongoing emergency be identified by its human or database source, by its time of occurrence, and by its status. Also, where appropriate, by its location and by links to whatever it is referring to that already exists within the system	**Up-to-date Information and Data** – Data that reaches a user and/or his/her interface device must be updated whenever it is viewed on the screen or presented verbally to the user	**Open Multi - Directional Communication** – A system such as this must be viewed as an open and flat communication process among all those involved in reacting to the disaster	**Authority, Responsibility, and Accountability** – Authority in an emergency flows down to where the actions are taking place
General Features	Task Tracking for front-end teams & Victims Registry Emergency Message Center for front-end teams Emergency Teams Registry for front-end teams			
Stage	Mini-Second Coordination Cycle			
	During Incident Response			

In Figure 1, there is a graphical representation of the extended PIE model. There are two actors: the designer and the user. Each one defines an interpretation function (respectively, I_D and I_U) between sequences of actions (respectively, P_D and P_U) and expected system states (the effect E). Other involved elements are the available information (the observable effect O), what is now visible in the screen (the display D) and the system's output (the result R). Furthermore, the defined interactive properties are represented as functions among these elements: *observe* between E and O for observability, result between E and R for predictability, *transparent* among E, O and R for transparency and *m-communication* between designer and user for meta-communication. In the Appendix there is a formal definition of the elements and properties introduced here.

In Table 3 there is the life cycle we have obtained at the end of the formalization for the mini-second coordination cycle in during-incident response stage. The first principle we have analyzed is *Information Source and Timeliness*. Representing the information as a program (P) and its current state as an effect (E), the principle can be traduced displaying all available data of the state (the result R of the effect E): this corresponds to the *transparency* property.

The second principle is *Up-to-Date Information and Data*. In this case we can recognize the

gone away for a cup of tea metaphor: the requirement is that users could understand and easily access information through the interface, even if they are distracted by external factors. Employing the same annotation of the previous principle, this means that observing the state of the information (E) from the screen (O), all available data about the information (R) are clearly visualized: this corresponds to the *predictability*.

The third principle we are going to present here is Open Multi-Directional Communication about the updating of information and the communication process to exchange this information among people. In this case, we have formalized two different properties: the transparency and the meta-communication. The transparency is obtained in the same way we have shown for Information Source and Timeliness principle. For the communication channel, users have to receive messages without misunderstanding problem. Considering the message to send as a program (PD), the received message by the users as another program (PU), the scope of the communication as an effect (E), the principle can be formalized requiring that given E, PD and PU have to be the same, as in the meta-communication property.

Finally, the last principle for the mini-second coordination cycle is Communication Authority, Responsibility, and Accountability. The idea is to adapt the system to different roles and respon-

Figure 1. Extended PIE model with interaction properties

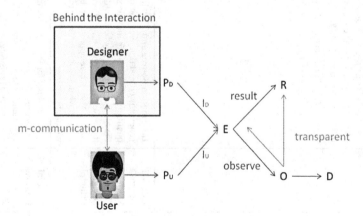

sibilities users might have. For this scope, it is important to understand available information and data in the screen and eventually adapt them to users' needs. Interpreting the principle in this way, we can formalize it with the meta-communication process as in the previous case. In this case, the designer has to communicate efficiently which programs and effects are accessible, depending on roles and responsibilities.

The Table 3 shows a view of the enriched emergency life cycle obtained as the result of the formalization process defined here. There are three columns: stages, general features and design principles with PIE formalism. The stages column contains emergency phases as they have been defined in Chen et al. (2008). The general features column has the classes we have defined in Table 2. The last column concerns the design principles that we have extracted from the DERMIS model and the formalization process applied to define interactive properties. While, the DERMIS model identifies important characteristics for emergency management systems and gives to designers guidelines to implement them, our properties are related to interactive aspects of the relation between users and systems, designers and systems and users and designers. As we have shown in Table 3, we have extracted a set of interactive properties with a bottom-up process, where design principles from DERMIS model represent the first step for the formalization of emergency features. The PIE model, instead, has been used as formalism for the final step to obtain the properties. Moreover, we have applied an extended PIE model, where we have added the meta-communication as interactive property between users and designers.

CASE STUDIES

During the analysis of considered web-based emergency management information systems, not only main features but also several case studies

have been collected. In this section, we are going to show how identified interactive principles can be applied to selected case studies. The first case study we are going to consider is a simulation for the Lost Person Finder module of Sahana (see Table 1). The second case study is an evaluation performed by ReliefWeb (see Table 1) about the Earthquake of 2005 in Pakistan.

Lost Person Finder (LPF) by Sahana: In 2009, a joint civilian and military simulation of a bomb explosion was performed to test management activities by first responders and medical personnel in a hospital. In particular, the aim was to evaluate the usage of the module *Lost Person Finder(LPF)* of Sahana. The LPF module collects information about victims and tries to identify them looking for additional information from organizations and local communities. Moreover, it provides an update mechanism to communicate any changes about the status of victims. In this particular case study, a group of volunteers was in charge of collecting photos and data about victims and publishing them for local communities. At the end of the simulation, involved people were interviewed to collect information about the usage of the LPF module.

The analysis of obtained results has pointed out several interaction problems that we can solve applying defined interactive principles. The LPF module is used during the incident response by people involved in the mini-second coordination cycle, such as hospital volunteers and first responders. During the simulation, volunteers have retrieved several problems interacting with systems: for example, they could not update easily information due to a great amount of data. We have organized pointed out problems into two groups. The first one is about difficulties in updating available information and resources for teams working in front end: we can apply here the predictability and the *observability* (associated to *Up-to-Date Information* and *Link Relevant Information and Data*) to guarantee an improved integration and synchronization of available data. The second group of retrieved problems are about the user interface of the module: users had difficulties in using available functionalities and organizing resources. In this case, we can apply the *transparency* and the *meta-communication* principles (associated to *Open Multi–Directional Communication*) to ensure an efficient communication and visualization of available functionalities, depending on the role and expertise of users.

Evaluation of the Pakistan Earthquake 2005 by ReliefWeb: In 2005, a violent earthquake destroyed the Pakistan with the death of some 75,000 people and damage for 5.8 billion of dollars. The ReliefWeb system has collected useful information about the management process of this emergency. Analysing this information, we can plan to apply identified principles to improve the emergency management process. In particular, the evaluation is about the many-second coordination cycle during the incident response.

The evaluation points out factors related to organizations and humanitarian agencies that come from outside. Volunteers and emergency workers had difficulties in understanding the social and political situation and consequently in offering an adequate help to victims. They need to be aware about social and political interaction within local communities and government in order to work in the most efficient and effective way. In particular, we can relate problems with social and political mechanisms to the management and the visualization of useful data: in this way, we can guarantee updated and specific information for emergency workers and people involved in the management process. As in the previous case, also here we can apply proposed interactive principles as possible solutions: *observability* (associated to *Link relevant Information and Data*), *transparency* (associated to *Information Source and Timeliness*), and *meta-communication* (associated to *Psychological and Social Needs*). Other factors we are going to consider here are about the organization of involved emergency workers in clusters and the coordination among clusters. An efficient organization could improve the in-

teraction among workers, volunteers and victims, solving possible misunderstood and delays. In this case, it is necessary to establish an efficient communication system for sharing information and data. For this reason, we can apply principles such as *transparency* and *meta-communication* (associated to *Open Multi-Directional Communication*). A transparent communication channel allows users to receive updated information and data during the entire management process.

CONCLUSION AND FUTURE WORKS

In this paper, we have presented a first contribution in the definition of interaction design principles for WEMIS. From the analysis of existing WEMIS, we have found out that there is no unique definition of interaction design for the emergency management process. For this reason, as a first contribution in Table 2 we have defined the relation between emergency phases and WEMIS features' classes. Successively, we have related them to a set of design principles extracted from the DERMIS model. Finally, through collected information about emergency phases and WEMIS features, we have formalized these principles using the PIE model and we have obtained a set of interaction design principles for WEMIS guaranteeing such systems properties: observability, predictability, reachability, transparency and meta-communication.

The next step will be the definition of formal patterns for the interaction design of WEMIS. The idea is to solve critical design aspects collected during the analysis of existing WEMIS and literature with the application of interactive properties presented in this paper. Designers might use these patterns for designing web based systems in the emergency domain in order to guarantee an efficient interaction (by supporting the properties obtained by the application of such design principles). Not only design, but also redesign activities could be an application of desired patterns. Moreover, we could consider case studies presented in the previous section to evaluate the desired contribution. In particular, we could apply the interactive principles to redesign systems used in such cases. The redesigned systems do not present the same problems pointed out by users previously: the interaction with emergency workers is improved such as the entire management process. Finally, we could evaluate considered systems performing a set of tasks with emergency workers and experts. From the comparison between results obtained in this evaluation and interaction problems extracted from the literature about WEMIS, we could modify, if necessary, identified interaction principles to improve their efficiency.

ACKNOWLEDGMENT

This work is been supported by Spanish Ministry of Science and Innovation (UrThey project, TIN2009-09687).

REFERENCES

Aedo, I., Díaz, P., Carroll, J. M., Convertino, G., & Rosson, M. B. (2010). End-user oriented strategies to facilitate multi-organizational adoption of emergency management information systems. *Information Processing & Management, 46*(1), 11–21. doi:10.1016/j.ipm.2009.07.002

Aedo, I., Díaz, P., & Díez, D. (2009). Cooperation amongst autonomous governmental agencies using the SIGAME web information system. In *Proceedings of the EGOV Conference.*

Andersen, H. B., Garde, H., & Andersen, V. (1998). MMS: An electronic message management system for emergency response. *IEEE Transactions on Engineering Management, 45*(2), 132–140. doi:10.1109/17.669758

Carroll, J. M., Neale, D. C., Isenhour, P. L., Rosson, M. B., & McCrickard, D. S. (2003). Notification and awareness: Synchronizing task-oriented collaborative activity. *International Journal of Human-Computer Studies, 58*, 605–632. doi:10.1016/S1071-5819(03)00024-7

Carver, L., & Turoff, M. (2007). Human computer interaction: The human and computer as a team in emergency management information systems. *Communications of the ACM, 50*(3), 33–38. doi:10.1145/1226736.1226761

Catarci, T., Leoni, M. D., Rosa, F. D., Mecella, M., Poggi, A., Dustdar, S., et al. (2007). The WORKPAD P2P service-oriented infrastructure for emergency management enabling technologies: Infrastructure for collaborative enterprises. In *Proceedings of the16th IEEE International Workshops on Enabling Technologies: Infrastructure for Collaborative Enterprises* (pp. 147-152).

Chen, R., Sharman, R., Rao, H. R., & Upadhyaya, S. J. (2008). Coordination in emergency response management. *Communications of the ACM, 51*(5), 66–73. doi:10.1145/1342327.1342340

De Silva, C., De Silva, R., Careem, M., Raschid, L., & Weerawarana, S. (2006). Sahana: Overview of a disaster management system. In *Proceedings of the IEEE International Conference on Information and Automation* (pp. 361-366).

de Souza, C. (2004). *The semiotic engineering of human-computer interaction.* Cambridge, MA: MIT Press.

Dix, A. J., Finlay, J. E., Abowd, G. D., & Beale, R. (2004). *Human-computer interaction* (3rd ed.). London, UK: Prentice Hall.

Endsley, M. R. (2000). Theoretical underpinnings of situation awareness: A critical review. In Endsley, M. R., & Garland, D. J. (Eds.), *Situation awareness analysis and measurement.* Mahwah, NJ: Lawrence Erlbaum.

Fischer, H. W. (1998). The role of the new information technologies in emergency mitigation, planning, response, and recovery. *Disaster Prevention and Management, 7*(1), 28–37. doi:10.1108/09653569810206262

Gheorghe, A. V., & Vamanu, D. V. (2001). Adapting to new challenges: IDSS for emergency preparedness and management. *International Journal of Risk Assessment and Management, 2*(3-4), 211–223. doi:10.1504/IJRAM.2001.001506

Harrald, J. R. (2009). Achieving agility in disaster management. *International Journal of Information Systems for Crisis Response and Management, 1*(1), 1–11. doi:10.4018/jiscrm.2009010101

Jennex, M. E. (2004). Emergency response systems: The utility Y2K experience. *Journal of Information Technology Theory and Application, 6*(3), 85–102.

Jennex, M. E., & Raman, M. (2009). Knowledge management in support of crisis response. *International Journal of Information Systems for Crisis Response and Management, 1*(3), 69–83. doi:10.4018/jiscrm.2009070104

Patton, D., & Flin, R. (1999). Disaster stress: An emergency management perspective. *Disaster Prevention and Management, 8*(4), 261–267. doi:10.1108/09653569910283897

Turoff, M. (2002). Past and future emergency response information systems. *Communications of the ACM, 45*(4), 29–32. doi:10.1145/505248.505265

Turoff, M., Chumer, M., Van de Walle, B., & Yao, X. (2004). The design of a dynamic emergency response management information system (Dermis). *Journal of Information Technology Theory and Application, 5*(4), 1–36.

Van de Walle, B., & Turoff, M. (2007). Emergency response information systems: Emerging trends and technologies. *Communications of the ACM, 50*, 3.

APPENDIX

The DERMIS Design Model

In Turoff et al., (2004), authors present a framework for the design of DERMIS systems, called DERMIS model. It has been developed collecting data from real emergency management experiences in USA. In particular, authors have considered the Office of Emergency Preparedness (OEP) and the Federal Emergency Management Agency (FEMA).

The first section is a collection of nine premises that designers have to accomplish in an initial phase of the design process: for example, the first one is titled *System Training and Simulation* and underlines the importance of training during the pre-incident phase.

The second section, *Conceptual Design*, presents five criteria to design the interface: *Metaphors*, *Human Roles*, *Notifications*, *Context Visibility* and *Hypertext*. They adapt human-computer interaction concepts to the emergency domain. For example, *Human Roles* represent groups of users that perform same activities: roles are defined depending on the level of responsibility, required tasks and emergency phases.

The third section is the *General Design Principles and Specifications*. There are eight principles: *System Directory*; *Information Source and Timeliness*; *Open Multi-Directional Communication*; *Content as Address*; *Up-to-date Information and Data*; *Link Relevant Information and Data*; *Authority, Responsibility, and Accountability*; *Psychological and Social Needs*. These principles are general guidelines for designers: they are more specific than other sections of the model and for this reason they will be applied during the first step of our definition of interactive properties. The first principle is *System Directory* that guides the designer in structuring data and information with a hierarchical tree for a useful text search. Other principles related to the management of available information are the *Information Source and Timeliness*, the *Content as Address*, the *Up-to-date Information and Data* and the *Link Relevant Information and Data* that are about the usage of database, the definition of semantic links and addresses, and the visualization of last updating in the interface. Moreover, the *Open Multi-Directional Communication* defines the entire emergency response process as an open communication among involved people and the *Authority, Responsibility, and Accountability* assigns responsibilities. The last guideline is the *Psychological and Social Needs* of emergency workers and affected people during a disaster.

The last section is *Supporting Design Considerations and Specifications* with general requirements and support tools: databases, collective memory to manage relations and dependencies among events, and online communities of domain experts.

The PIE Model

The PIE model is an interaction model presented in Dix et al. (2004): it presents a set of interactive properties through a formal language based on the fist-order-logic. This model defines five elements: a *program* (P), an *effect* (E), a *result* (R), an *observable effect* (O) and a *display* (D). The interaction among these elements is represented through five properties: *observability*, *predictability*, *reachability*, *transparency* and meta-*communication*. While *observability*, *predictability*, *reachability* and *transparency* are already defined in the PIE model, the *meta-communication* is an additional property that we have introduced in the model. Next, we are going to present a formal description of these interactive properties.

The *observability* represents what is displayed on the screen about the effect: this means that here we are measuring how many data users can know about the effect just observing the screen. In particular, a system is observable if users can understand completely the output of the system (effect) from information visualized on the screen (observable effect). The formalization of the property is:

$$\exists\ observe : O \rightarrow R.\forall e \in \text{E}.observe(e) = e,$$

where observe is a function from the set of observable effects O to the set of results R and it is defined for each effect in E.

The *predictability* is related to the *observability*, but for future states: observing the screen, users must be able to determine future states of the system. Another way to explain this property is through the metaphor of *gone away for a cup of tea problem*, also presented in Dix et al. (2004). To illustrate this metaphor, consider a user that is working on a program and decides to leave her tasks to get a cup of tea. When she comes back, in order to continue efficiently her tasks she needs to remember the command history and which action will be the next to perform. A predictable system displays useful information on the screen about past and future commands to help the user in resuming her work. Formally, the predictability is expressed as:

$$\exists\ observe : O \rightarrow R.\forall e \in \text{E}.predict(observe(e)) = result(e),$$

where predict is a function from the set of observable effects O to the set of results R and it is defined for each effect in E through the *observability* property (observe(e)).

The *reachability* represents the possibility to reach any states of the system from one of them. In this way, users should have undo facilities in order to come back and eventually correct mistakes. The formal definition of this property uses the *doit* function for the transaction from a state to another one through a sequence of commands. The formalization is:

$$\forall e, e^{'} \in \text{E}.\exists p \in P.doit(e, p) = e^{'},$$

where e and e' are effects in E, P is the set of programs and the transaction function doit is defined from couple (effect, program) to a new effect.

The *transparency* corresponds to a stronger kind of *predictability*: if the system is fully predictable, it is transparent too. The screen displays all information and data about the state of the system and users can know exactly current effects. Moreover, the *transparency* allows predicting future state and avoiding possible mistakes. The formal form is:

$$\exists transparent : 0 \rightarrow \text{E}.\forall e \in \text{E}.predict(observe(e)) = (e),$$

where transparent is a function from the set of observable effects O to the set of effects E, through the *observability* and the *predictability* (predict(observe(e)) properties.

The *meta-communication* represents rules and semantics for the communication between users and designers. In order to introduce this additional property, we have extended the PIE model with two new actors, the designer and the user. Next, we have substituted the set of programs P with two new elements: the program defined by the designer P_D and the program defined by the user P_U. Both of them are related to the effect E through two different interpretation functions, I_D and I_U. The *meta-communication* is defined between these two elements. Given an effect E, P_D and P_U are the user and designer's programs to reach it. If they are composed by the same sequence of commands, the designer has communicated correctly her design strategy to the user and the meta-communication is high. Otherwise, if P_D differs from P_U, the communication between designer and user does not work properly and the level of meta-communication is low. The formal form is:

$$\exists m - communication : E \rightarrow P \forall e \in E.I_D^{-1}(e) = I_U^{-1}(e),$$

where m-communication is a function defined from the set of effects E to the set of programs P, through the inverse function of the designer interpretation I_D^{-1} and the inverse function of the user interpretation I_U^{-1}.

The Enriched Life Cycle

The enriched life cycle is based on the coordination life cycle for the management of emergencies presented by Chen et al in (2008). It is divided into three stages: pre-incident, during-incident, post-incident. For each stage, we have identified a set of classes of features, extracted from the analysis of existing WEMIS, and interactive principles form the PIE Model. While in the first table (see Table 4) we show the formalization of DERMIS principles through the interactive principles from the PIE model, the second table (see Table 5) presents the entire enriched life cycle.

Table 4. The formalization of DERMIS Principles through PIE Model

DERMIS Principles	Formalization by PIE Model
System Directory: The system should provide a hierarchical structure for all the data and information currently in the system and provide a complete text search to all or selected subsets of the material.	Easy access to available information → considering information as a program P and its visualization as an effect E, we can reach an effect from any other effect → reachability $$\forall e, e' \in E. \exists p \in P.doit(e, p) = e'$$
Information Source and Timeliness: In an emergency it is critical that every bit of quantitative or qualitative data brought into the system dealing with the ongoing emergency be identified by its human or database source, by its time of occurrence, and by its status. Also, where appropriate, by its location and by links to whatever it is referring to that already exists within the system.	A complete representation of information: considering information as a program P and its visualization as an effect E, we visualize the complete status of available information → transparency $$\forall e \in E.predict(observe(e)) = (e)$$, where $$\forall e \in E.predict(observe(e)) = result(e)$$
Open Multi - Directional Communication: A system such as this must be viewed as an open and flat communication process among all those involved in reacting to the disaster.	All users have to be aware of the current state of the system, from information to communication → considering the information as a program (P) and its visualization as an effect (E), we want to communicate P in order to avoid misunderstanding about E → transparency and meta-communication $$\forall e \in E.predict(observe(e)) = (e) \,\&\, \forall e \in E.I_D^{-1}(e) = I_U^{-1}(e)$$
Content as Address: The content of a piece of information is what determines the address.	Through the address in the information, it is possible to create group of interest and discuss with members just about this subset of information → just visualize a piece of information → considering information as a program P and its visualization as an effect E, we visualize just a piece of information (O) and not all (R) → observability $$\exists observe : O \to R. \forall e \in E.observe(e) = e$$
Up-to-date Information and Data: Data that reaches a user and/or his/her interface device must be updated whenever it is viewed on the screen or presented verbally to the user.	"Gone away for a cup of tea" metaphor → predictability $$\forall e \in E.predict(observe(e)) = result(e)$$
Link Relevant Information and Data: An item of data and its semantic links to other data are treated as one unit of information that is simultaneously created or updated.	The set of available information has to be consistent depending on considered data and their visualization → considering the effect E as available data and the observable effect O as their visualization, we want to ensure that what users observe is a consistent representation of available data → observability $$\exists observe : O \to R. \forall e \in E.observe(e) = e$$
Authority, Responsibility, and Accountability: Authority in an emergency flows down to where the actions are taking place	We have to adapt the system to different roles and users that interact with it → improve the meta-communication between designer and user → meta-communication $$\forall e \in E.I_D^{-1}(e) = I_U^{-1}(e)$$
Psychological and Social Needs: Encourage and support the psychological and social needs of the crisis response team	The psychological and social support for the crisis response team can be seen as an improvement for misunderstanding during the interaction with the system → the interpretation of tasks and activities by users has to follow previously defined action plan → meta-communication $$\forall e \in E.I_D^{-1}(e) = I_U^{-1}(e)$$

Table 5. The enriched emergency life cycle

Stage		General Features	DERMIS Principles	Interactive Principles
Pre-Incident Response		• Organization Registry: collect information about involved organizations in the disaster region. • Spatial Registry: associate emergency information with geographic information system • Emergency Message Center: exchange of messages for training and meetings • Hazard and vulnerability analysis modeling • Demographic Registry	System Directory	Reachability $$\forall e, e^{'} \in E. \exists p \in P.doit(e,p) = e^{'}$$
			Content as Address	Observability $$\exists observe : O \rightarrow R.\forall e \in E.observe(e) = e$$
			Link relevant Information and Data	Observability $$\exists observe : O \rightarrow R.\forall e \in E.observe(e) = e$$
During Incident Response	Mini-Second Coordination Cycle	• Task Tracking for front-end teams & Victims Registry • Emergency Message Center: exchange of messages among front-end teams • Emergency Teams Registry: manage data about front-end teams	Information Source and Timeliness	Transparency $$\forall e \in E.predict(observe(e)) = (e) \text{ where}$$ $$\forall e \in E.predict(observe(e))$$ $$= result(e)$$
			Up-to-date Information and Data	Predictability
			Open Multi-Directional Communication	Transparency $$\forall e \in E.predict(observe(e)) = (e) \text{, where}$$ $$\forall e \in E.predict(observe(e))$$ $$= result(e)$$ Meta-communication $$\forall e \in E.I_{D}^{-1}(e) = I_{U}^{-1}(e)$$
			Authority, Responsibility, and Accountability	Meta-communication $$\forall e \in E.I_{D}^{-1}(e) = I_{U}^{-1}(e)$$
	Many-Second Coordination Cycle	• Emergency Message Center: exchange of messages to coordinate all involved organizations • Task Tracking: track all operations and decisions • Collaborative Suggestions Collection: collect suggestions about the task list • Requests Registry • Shelter and Resource Registry • Victims Registry: collect information about victims and missing people • Volunteer Registry • Situation Mapping	Information Source and Timeliness	Transparency $$\forall e \in E.predict(observe(e)) = (e) \text{, where}$$ $$\forall e \in E.predict(observe(e))$$ $$= result(e)$$
			Open Multi-Directional Communication	Transparency $$\forall e \in E.predict(observe(e)) = (e) \text{, where}$$ $$\forall e \in E.predict(observe(e))$$ $$= result(e)$$ Meta-communication $$\forall e \in E.I_{D}^{-1}(e) = I_{U}^{-1}(e)$$
			Content as Address	Observability $$\exists observe : 0 \rightarrow R.\forall e \in E.observe(e)$$ $$= e$$
			Link Relevant Information and Data	Observability $$\exists observe : 0 \rightarrow R.\forall e \in E.observe(e)$$ $$= e$$
			Psychological and Social Needs	Meta-communication $$\forall e \in E.I_{D}^{-1}(e) = I_{U}^{-1}(e)$$
Post-Incident Response		• Task tracking: collect final results and real processes for possible improvement • Discussion forum: establish a constructive discussion among involved organizations and emergency workers • Photo library • Reports Registry • Estimation: estimate population at risk for future pre-incident planning	Content as Address	Observability $$\exists observe : 0 \rightarrow R.\forall e \in E.observe(e)$$ $$= e$$
			Link Relevant Information and Data	Observability $$\exists observe : 0 \rightarrow R.\forall e \in E.observe(e)$$ $$= e$$

Chapter 9
RimSim Response
Hospital Evacuation:
Improving Situation Awareness and Insight through Serious Games Play and Analysis

Bruce Campbell
Rhode Island School of Design, USA

Chris Weaver
University of Oklahoma, USA

ABSTRACT

To aid emergency response teams in training and planning for potential community-wide emergency crises, two coordinated research teams centered in King County, Washington have developed software-based tools to provide cognitive aids for improved planning and training for emergency response scenarios. After reporting the results previously of using the tools in pilot studies of increasing complexity, the implementation teams have been searching out community-wide emergency response teams working on emergency response plans that might benefit from use of the tools. In this paper, the authors describe the tools, the application of them to a countywide hospital evacuation scenario, and the evaluation of their value to emergency responders for improving situation awareness and insight generation.

INTRODUCTION

Communities are preparing diligently for potential community-wide crises arising from natural and man-made causes. First responders are those people who train to fulfill emergency response roles on behalf of community residents, seeking to limit loss of life, protect property, and reduce the cost of long-term recovery periods associated with crisis scenarios. The cost of providing physical drills to train for participation in community-wide crises is exorbitant and the 24/7 demands for first responders can preclude participation in training even if a physical drill is made available. As a result, research teams are exploring the use of software-based simulation environments to

DOI: 10.4018/978-1-4666-2788-8.ch009

help extend training and planning opportunities to synchronous and asynchronous activities using role-play interfaces to simulate the performance of activities independently as well as with other role-players. This paper reviews the activities and results of one research team attempting to evaluate the use of software-based simulation environments as serious games for first responder training and planning purposes.

As part of the research and development agenda for visual analytics (Thomas & Cook, 2005), researchers have been developing integrated tools for improving analytic capabilities that facilitate application of human judgment to evaluate complex data associated with emergency response efforts. As coordinating artifacts, geospatial visualization assists in knowledge construction and decision support (MacEachren & Gahegan, 2004). In 2006 the Pacific Area Regional Visual Analytics Center (PARVAC) team at the University of Washington began working with regional emergency operation centers to explore the use of geospatial visualizations as a key component in an emergency response crisis scenario simulator, being built to allow first responders to plan and train for community-wide potential emergency scenarios. The team built a series of software components to support a modular

architecture they call RimSim, seen in Figure 1, which advises development of simulation environments for first responder planning and training using emergency response scenarios identified by interested partners in the Pacific Northwest region of the United States.

The PARVAC team assesses shared use of geospatial visualization over time through iterations of scenario development tasks combined with simulated scenario game role-play (Campbell & Mete, 2008). The team builds supporting tools to help first response coordination teams explore four concepts in a community-wide emergency response effort: recognition-primed decision-making (Klein, 1998), situation awareness (Adams & Tenney, 1995), distributed cognition (Hutchins, 1996), and distributed intelligence (Pea, 1993). All four suggest models relevant to the use of interactive visual artifacts in the coordination of complex team activities under time-bounded conditions. By working closely with the emergency response community, the team has explored expedient methods for improving emergency response activities effectiveness, which can be attained by improving response behavior in association with any or all of the four models.

Figure 1. RimSim modular architecture

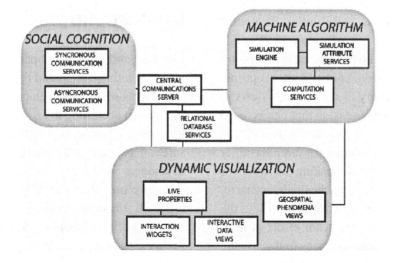

Through a series of pilot studies with medical logistics teams that contained a handful of participants, the PARVAC team observed that first response teams improved their emergency response activities performance by using two different methods of considering an emergency response scenario. Through software-supported simulators that used software-based agents to simulate first responder behaviors, first responders participated in simulation sessions that could provide experience with a scenario through repeated game role-play (Campbell, 2010). Simulator-support visualization products enabled a team of responders to play and replay a response effort sequentially in order to review their decisions and the ramifications of decisions made in geospatial and temporal space with constraints that would be typical of an emergency response effort. Through a separate tool development process, software-based sense-making and probing tools provided an emergency response team the opportunity to review an emergency response scenario effort as an information space that need not be queried or investigated sequentially, but instead selectively across time and space to look for interesting patterns that provide insights into the nature of the scenario.

To support the two observed methods of exploring a crisis-wide emergency response scenario in emergency operation centers, the PARVAC team iterated upon the development of two separate software tools they now call RimSim Response! (RSR) and RimSim Visualization (RSV). Although the RSR tool allows temporal investigation game play sessions, it does not provide an optimal interactive visual query of coordinated views in an emergency response session that can be used for insight generation – the focus is on role play to build a better real-time situation awareness rather than an after-game sense-making analysis for suggesting alternative approaches to the scenario. The RSV software, on the other hand, allows the team to visually query the team effort in non-sequential temporal investigations to discuss the ramifications of actions made by all participants.

The PARVAC team spent six months working closely with the King County Health Coalition (KCHC), in King County, Washington, to adapt the RSR tool for use in a countywide hospital evacuation scenario that the KCHC used as part of a five-part planning and training process that spanned a year of activities. The process involved an all-member brainstorming scenario-building exercise, an all-member paper-based tabletop exercise, a role-playing computer-mediated simulation drill, a role-playing RSR tool use drill, and a role-playing physical evacuation drill. Each of the five activities involved KCHC staff (the latter three with a subset of the KCHC membership) and allowed the scenario to evolve to better support future planning and training for hospital evacuations. By participating in the scenario development process, the PARVAC team gained the trust of the KCHC that led to KCHC participant willingness to participate in both an RSR and RSV tool assessment process. The resultant RSR and RSV tools were then brought to a new emergency evacuation team in Hartford County, Connecticut to make a second assessment using role-players who had not gone through the scenario development process. Two metrics were chosen to assess the usefulness of the RSR and RSV tools. The team assessed situation awareness quality for the RSR tool and insights generated quality for the RSV tool.

BACKGROUND

The RSR and RSV developers have collaborated via an iterative design and development process that invites software developers to participate as they see fit according to an open source development model. Decisions to include code in the base software packages are made by committee. Some developers work only with the RSR code base and some only with the RSV code base. Others, especially those who communicate often with the KCHC, work with both of the tools as described in this section.

The RSR Tool

The RSR tool was built as a series of software components that could support role-playing games that took place within a simulated emergency response scenario. For the scenario designer, The RSR software affords the opportunity to design and implement a basic emergency response scenario without any necessary programming to extend the base RSR software. To seed a scenario's incidents and resources, the scenario designer uses the editor tool shown in Figure 2.

As the designer can place visual scenario details anywhere on a Java World Wind-based virtual globe that appears in the tool, the scenario can take place anywhere on the surface of Earth and incorporate spatial scales varying anywhere from a small urban neighborhood to a full global reach. The designer draws responsibility jurisdictions as N-sided polygons, as well as N-sided polygons for out of bounds regions in which no simulation activity can take place. The designer drags and drops emergency response incidents geospatially and sets a begin time for each. The designer places the emergency response resources geospatially that a simulation role-player can use to meet incident demand, individually by hand or as in bulk by using one of many menu-driven statistical distributions built into the soft-

ware. For more specific scenarios, a developer can extend the software before providing it to the emergency response team first responder. A community of developers builds the base RSR tool, including the scenario editor that we describe in (Campbell & Schroder, 2009), and makes it available for download and exploration.

After the designer lays out the basic scenario, the RSR configuration file provides an opportunity to substitute a software-based agent into the simulation for each role identified for game play in the scenario. Those roles that aren't assigned a software-based agent are expected to be played by a live participant who participates in an active role-play simulation session over the Web by taking over the available simulation interaction controls at any time during the session. To gage the progress of scenario development and tool reliability, the RSR development community runs many simulated sessions with software-based agents playing jurisdictional roles in order to debug agent behavior and to experience the visualization as the game players will. To be able to test often, beta-testing play sessions are announced via e-mail and are open for anyone to play and require only a Java runtime and broadband Internet connection.

During each session, the RSR software logs all key game session variables necessary to replay an emergency response simulation session. As a

Figure 2. The RSV scenario editor

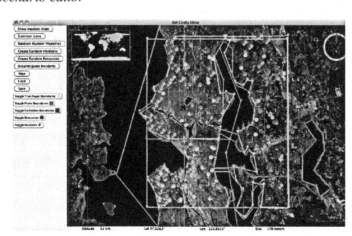

result, The RSR development team, beta-testing play team, or emergency response planning and training teams can review game play in any session that has been run if provided the session's log file for loading into the RSR tool. Playing the role-play simulation allows first-responders the opportunity to improve their situation awareness of a potential real-world scenario – especially in the area of distributed team activity. Replaying a wide variety of simulated emergency response scenario sessions enables a participant to observe patterns of resource allocation across the whole emergency crisis time and place extent. Klein's recognition-primed behavior theory predicts that repeated exposure to emergency response patterns should help with situation awareness, should a community-wide emergency ever develop in which the participant has response responsibilities (Klein, 1998).

The RSV Tool

The RSV tool was built to enable interactive visual querying through a series of highly coordinated views. Data logged during an RSR tool use session can be loaded into the RSV to explore the data in a highly interactive, non-sequential manner. Transcribed verbal communications between role-playing game participants are also loaded to provide an analysis of geospatial game state in conjunction with participant response.

The underlying Improvise software platform (Weaver, 2004) upon which the RSV tool exists enables rapid control widget and views construction. In the RSV the views exist within an interface that provides two tabs to access the two pages of views seen in Figures 3 and 4. The RSV development team allocate widgets such that one visual page provides an emphasis on geospatial analysis of responders, resources, and incidents, while the other provides emphasis on simulation participant communications and actions. The two pages are coordinated such that both sheets update dynamically when the user interacts through the interface in order to refresh all views immediately independent of which page initiates changes.

The view in the upper-left of Figure 3 identifies the same geospatial extent used in the RSR simulation tool session, embedded within a view container in order to be able to communicate with all other RSV tool controls and views. The inset map view in the bottom center of Figure 3 allows an analyst to drill down on a map view of the community with visualization layers that show many community objects relevant to emergency response (fire hydrants, buildings, waterways,

Figure 3. The RSV geospatial page

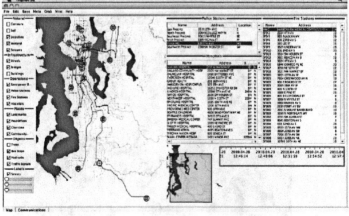

Figure 4. The RSV communications page

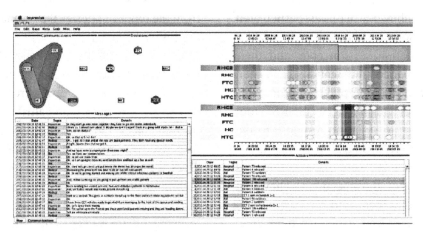

roads and highways, police stations, hospitals, etc.). The analyst can toggle the layers using the checkboxes that run down the left-hand margin of the page.

The inset view in Figure 3 lets the analyst consider the context of the scenario's geographical coverage. The analyst can drag the green rectangle to move to a different location or can drag its corners to grow or shrink the extent (effectively zooming in or out while at the same time allowing distortion should the analyst so desire). The rectangle updates itself whenever action in another view changes the map in the lower-left (they are coordinated in both directions). Zooming and panning in the upper-left map view retains the same interaction behaviors the analyst is familiar with from using the RSR.

Tabular lists of key response resources appear on page one. Upon selecting a named object, the maps pan to orient with that object in the center (while maintaining the current zoom level). These short-cut navigation aids are very useful for an analyst. For example, if an analyst is curious about activity at a specific hospital, he or she can select the hospital by name and then drag and paint upon the time sliders in the upper-right to see events that occur in the area. Often-used and trained-upon driving emergency routes to hospitals are emphasized with a salmon colored road network

of just those roads used whenever possible during a community crisis event.

An analyst can track resource and incident properties over time, and can track player communications and actions using the three timelines in the upper-right of Figure 4. The uppermost timeline shows details about the current incident load as incident requirements wax and wane. The middle timeline shows each player's communications behavior over time. The lower timeline shows each action made by the players over time. Each timeline accepts typical improvise-provided panning, zooming, rubber banding, and brushing behaviors in order to adjust start and stop times, filter the visual time span, and choose a specific time step for detailed analysis in coordinated views.

The upper views visually show the connections between players for messages sent and actions taken. By scrolling the timeline for messages and zooming in and out of the time range represented, an analyst can get a visual sense of how much players communicate with other players. The analyst can also ascertain how often players work in unison to perform actions (such as resource allocation tasks, for example). The views automatically change the position of player names in order to best show highest connectivity based on the range of messages or actions currently selected for viewing. The thickness of lines surrounding sets of players identifies the relative number of occurrences that combination of players participated in a communication event or other game action.

After a simulation session ends and their interpersonal verbal communications have been transcribed to text, the role-playing participants are invited to sit as a team and use the RSV tool to investigate their role-play in response to the scenario goals. Players can ask each other questions about why actions were made and discuss the ramifications of those actions. They can investigate recurrent themes to see how often a specific point-of-view leads to player actions. Over time, the assessment team developed a process whereby the players explicitly point out scenario specific insights that were made during game play and reach a consensus as to the value of that insight on a scale of 1 to 100. The scoring process enables a quantitative analysis across simulation role-playing sessions.

SHARED DATA MODEL

A shared data model helps coordinate RSR and RSV tool development as the same data model is incorporated within both tools. Working backwards from the types of queries they wanted to support, the PARVAC team generated the data model in Figure 5 that could support the RSV tool with three types of data: communications between first response coordinators, actions requested by first response coordinators, and attributes of physical phenomena in the field (specifically incidents, response resources, and responders). The three types are connected in a causal manner such that communications between first responders can lead to actions being taken which can then lead to changes to tracked attribute values of objects in the field. Alternatively, actions taken by a member of the response team can lead to updating key attributes in the field and communications with or between other key first responders.

Whenever analysis desired by a first responder team cannot be performed because the software does not have needed data attributes available, the software development team revisits the shared data model and, if warranted, updates the data model to incorporate newly identified attributes. Then, the model drives software changes to both the RSR and RSV tools.

Figure 5. RSR and RSV shared data model

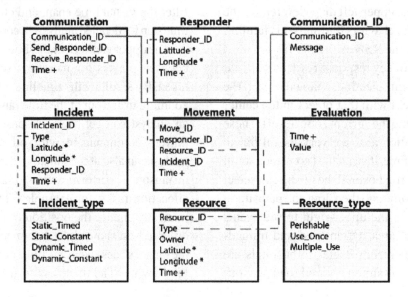

In Figure 5, attributes with asterisks can be used to analyze data geospatially (Responder.Latitude, Responder. Longitude, Incident.Latitude, Incident.Longitude, Resource.Latitude, Resource. Longitude). Attributes identified with plus signs enable temporal analysis (Communication.Time, Responder.Time, Incident.Time, Move.Time, Resource.Time, Evaluation.Time). Incident types are connected with a lookup table that can grow to manage scenarios of increasing complexity. Other attributes connected with dashed lines have a primary key – foreign key relationship associated with geospatial relationships. Attributes connected with a solid line are related via temporal relationships that lack concrete geospatial location.

The shared RSR and RSV data model supports evaluation of individual and team first responder behavior upon analysis. For example, by looking at an RSR session's data that complies with the data model for a response activity, an analyst can ascertain that at 6:10pm two responders communicated about downed power lines. At 6:12pm a medical unit resource was dispatched to head to that location. The medical unit then made its way from its base at the hospital to the downed power line incident location, affecting the incident and the resource entities.

To test out the data model's effectiveness, representative entities and relationships were chosen from multiple pilot scenario first response efforts to visualize in the RSV tool. Scenario specific response resources such as police, fire, and medical units were tracked geospatially through the responder entity over time. Vehicles carrying medical supplies or medical patients were tracked as resource entities and supply levels for resources are tracked by separate resource entity tuples for each location. The data model turned out to be sufficient for the development team and pilot subjects for all pilot scenario purposes.

In the case of a community-wide first response effort, first responders become critical attributes to track geospatially. Fire, police, medical, and other specialists are deployed by an event com-

mander or report in from the field. The data model provides an analyst the necessary data to track the movement of first responder personnel through geospatial visualization that includes visual layers to identify location and the resources each first responder manages throughout the first response effort. By visualizing the results of actions taken in the field, first responders can review their actions and the communications that led to those actions.

HOSPITAL EVACUATION SCENARIO

After presenting results from various pilot efforts at ISCRAM 2010 (Campbell & Weaver, 2010), the RSR tool development team teamed up with a University of Washington emergency response coordinator to recruit the King County Hospital Coalition (KCHC) in Seattle and Bellevue, Washington, USA to help build a game play scenario for a county-wide emergency hospital evacuation event that had been proposed as future work at the ISCRAM 2010 conference in Seattle. The scenario became the main scenario used for a full year to plan and train responders for the scenario and is described fully in Campbell (2010). The evacuation of patients in a single hospital that housed 200 to 250 patients to the other 20 hospitals participating in a community of mutual aid agreement became the focus of multiple tabletop session and live emergency response drills during 2010. The RSR tool development team attended the activities and participated in the activity preparation, activity performance, and post-activity discussions to iterate upon the RSR scenario that selected KCHC participants would play within the simulation tool. Tool developers observed the scenario to be well developed for inclusion in the RSR tool as a planning aid that could also help train emergency responders who participated in role-playing sessions.

Throughout the year of focus on hospital evacuation scenarios, the RSR and RSV tools were enhanced for any interested KCHC member's use. As the scenario evolved to identify a specific hospital to be evacuated, The RSR tool developers included an interactive hospital patient floor plan view as seen in Figure 6. Through mouse-based interaction with the floor plan view, role players could select patients, interrogate their current state, prepare each patient for evacuation, and suggest a route to take during physical transport. Upon releasing a patient, the view presented an animation of simulated movement of the patient for all role-players to consider and possibly incorporate into their situation awareness.

Two RSR-based game play sessions took place on April 26 and 28, 2010 with selected KCHC participants who had been studying their roles for more than six months and RSV tool use took place the following week in order to evaluate play performance by the team. The play session implementation team was surprised to see the level to which combined RSR and RSV tool emergency response users continued to point out improvements to the scenario roles and scenario event task orchestration. To improve scenario task coordination, the role-playing participants created a task list of eleven additional software changes the RSR tool development team was asked to make in order for the RSR to become more useful to planning and training participants who role-

played the scenario. As a result, a final evaluation of tools could not be made effectively without additional refinement.

The RSR tool team's negotiated time frame for working with the KCHC emergency response medical logistics team ended before they had incorporated all the RSR tool software changes suggested by KCHC scenario developers. The RSV tool development and assessment teams had met the objective of developing a workable process to obtain insight generation metrics after a scenario play session. To continue scenario development within the RSR tool, the development team turned their attention to a new team of role-players who had not had the experience of participating in the scenario development, but who would likely have to perform the roles contained in the simulator should an evacuation event occur at their hospital. None of the participants in the second role-playing team had experience with the geography of King County, Washington as they were based in Hartford County, Connecticut (2500 miles to the east).

The assessment team felt the additional requirement of learning new geography would provide interesting results when running two more RSR-based game play sessions with the Hartford county logistics staff. The assessment team hoped the second team's inclusion in RSR and RSV tool

Figure 6. RSR during hospital evacuation scenario

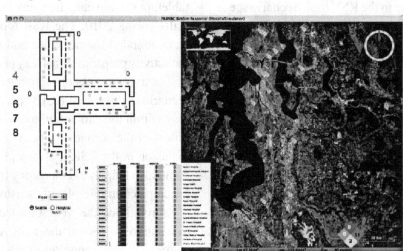

use trials would provide a focused perspective on using both tools together for training purposes. The second subject pool was not key emergency response personnel who would be difficult to schedule meeting times with outside of the normal activities of their jobs. The assessment team scheduled the tool use sessions that took place on June 8 and 10, 2010. Similar to the relationship developed between the PARVAC and KCHC, the relationship with the Hartford County team has built mutual trust that can continue to motivate all participants as developers iterate design further on the RSR and RSV tools.

TOOL ASSESSMENT

Situation Awareness

In the June trials, game role-players played the simulated hospital evacuation using the RSR tool while the assessment team recorded their efforts. To gauge situation awareness, the assessment included freeze-probe questionnaires that were asked at ten random points in time during the game session. The five questions asked are listed in the first column of Table 1. The questionnaire took between two and five minutes to administer at each freeze point during role-play.

The KCHC helped the assessment team identify five questions that would suggest situation awareness quality. The questions included: 1) How many patients are in a significant state of discomfort currently? 2) Where are these patients located? 3) How many patients are currently in transit between the evacuating and receiving hospital? 4) How much more time will it require to fully evacuate the existing hospital given ideal circumstances? 5) How much more time will it require to fully deliver all evacuating patients to their receiving hospital given ideal circumstances?

The assessment team calculated a metric of situation awareness quality by comparing the

Table 1. Session 1 v. Session 2, Mean (Std.Dev), Diff, t-test

Question	Session 1 n=30	Session 2 n=30	Difference n=30	t-test data
1	2.933 (3.423)	0.333 (0.606)	2.600 (2.817)	t_{29}=4.347 p=0.0002
2	8.133 (8.253)	4.267 (4.362)	3.867 (3.891)	t_{29}=2.610 p=0.0142
3	2.467 (3.350)	0.533 (1.042)	1.933 (2.308)	t_{29}=3.537 p=0.0014
4	2.533 (1.655)	1.133 (1.042)	1.400 (0.614)	t_{29}=5.558 p<0.0001
5	19.667 (12.888)	15.9 (6.583)	3.767 (6.305)	t_{29}=2.055 p=0.049

hospital evacuation team role-players' answers to the questions compared to the state of the scenario as contained in the simulation. The metrics were calculated for each of the two role playing sessions, comparing situation awareness with use of the RSR tool interface to situation awareness attained previously without use of the RSR tool interface. Three players answered the five questions ten times each for each session, providing a sample-size of 30 questions to compare between sessions.

Session 1 did not use the RSR tool interface but provided role-players the opportunity to verbally inquire about the state of the simulation and make their own visualizations using pen and paper. The second column in Table 1 shows the mean and standard deviation of how far off the subject's answer to a question was from the state of that variable in the simulation for session 1. Subjects then spent the next two days getting familiar with the RSR tool interface before a second simulation session took place.

Session 2 required the role-players to use the RSR tool interface to track the state of the simulation without the help of any verbal inquiries to the assessment team. The third column in Table 1 shows the mean and standard deviation of how far off the subject's answer to a question was from the state of that variable in the simulation for session 2. The simulation engine remained the same in both sessions and managed the state of key simulation variables in response to role-player and simulated player actions. The roles of the players were kept identical, as was the beginning state of the simulation they encountered at the start of the role-playing session.

When comparing the level of distributed situation awareness accuracy between the game play session with and without the RSR tool interface, a paired-t statistic comparing situation awareness accuracy showed significant improvement when using the RSR tool interface compared to the session without. For question one, the mean number of improvement (M=2.6, SD =3.28, N= 30) was significantly greater than zero, t(29)=4.347, two-tail p = .0002, providing evidence that the situation awareness of discomforted patients was higher with the RSR tool interface.

The t(29) statistic for question two showed a mean number (M=3.867, SD =8.114, N= 30) that was significantly greater than zero, t(29)=2.610, two-tail p = .0142, providing evidence that the situation awareness of discomforted patient location was higher with the RSR tool interface than without.

The t(29) statistic for question three showed a mean number (M=1.933, SD =32.99, N= 30) that was significantly greater than zero, t(29)=3.537, two-tail p = .0014, providing evidence that the situation awareness of in-transit patients was higher with the RSR tool interface than without.

The t(29) statistic for question four showed a mean number (M=1.400, SD =1.380, N= 30) that was significantly greater than zero, t(29)=5.558, two-tail p < .0001, providing evidence that the situation awareness of the time to complete the scenario was higher with the RSR tool interface than without.

The t(29) statistic for question five showed a mean number (M=3.767, SD =10.040, N= 30) that was significantly greater than zero, t(29)=2.055, two-tail p = .0049, providing evidence that the

situation awareness of discomfort patient location higher with the RSR tool interface.

To summarize Table 1, all five questions showed significant improvement with 95% confidence when using the RSR tool interface compared to when not as developed for the June 2010 use sessions.

Insight Generation

In the June trials, evaluators asked the role-playing team in training to verbalize the insights they had gained in regards to meeting the objectives of the hospital evacuation scenario. Evaluators recorded all verbal statements and game interface actions and transcribed them into a format appropriate for visualizing the data in the RSV tool. Players could then use the RSV tool to review their game play outside of the time pressure the simulated game-playing environment suggested. With the RSV tool, the role-play team identified all insights they had experienced throughout the game play session.

The definition of insight continues to be challenged more by researchers than the definition of situation awareness. Insight has varying definitions that appear to be honing in on two different definitions pursued by two distinct research groups — computer scientists and cognitive scientists (Chang & Ziamkiewicz, 2009). Computer scientists investigate insight as a contribution to knowledge building whereby each insight contributes to a relationally semantic knowledge base that enables problem solving and reasoning heuristics. In this regard, each insight is a describable incremental piece that adds value to the whole knowledge base — insight as a noun. Cognitive scientists investigate insight as a neurological function of the brain's left hemisphere where a new perspective on a problem is gained through a burst of brain activity — insight as a verb. Evaluators asked role-playing participants for a listing of all insights irrespective of whether they met the requirements of either or both definitions.

Along with help from the evaluators' notes taken during game play, the role-players listed all the insights they had that helped them make progress toward hospital evacuation objectives. The game role-players provided a time stamp and score on a scale from 1 to 100 as to the significance of each insight to overall scenario objectives attainment. Players discussed the scores at length if necessary to reach a consensus on the value to their overall shared performance. Evaluators watched the process and asked questions they believed could help the process become more accurate without suggesting new insights not identified by the game role-players themselves.

Evaluators noted how players interacted with the RSV in order to remember what had been said and the time the remarks had been spoken. Players asked a facilitator to scroll the chronological messages widget on the RSV Communications page to remind them of exact words spoken and cross-referenced the chronological glyph representation to review their thoughts during long periods of quiet on all communication channels. Patients also reviewed the timings of patient evacuations to gain a sense of how well they were grouped for transport to other hospitals. Although evaluators watched both the Communications and Decisions graphs change dynamically as other widgets were scrolled, the subject team did not discuss them at all to help them confirm their insights.

Role players challenged themselves to remember what they were thinking about when there were long absences between spoken words as if they were justifying the long gaps in interpersonal communication. The observers found that process to be one of the most important in team-building and in generating insights that could lead to a better distributed situation awareness in future game sessions. Participants suggested that the gaps in the presentations of scrollable timelines jumped out at them first and foremost before they considered the detailed distribution of glyphs overlaid upon the views associated with the timelines.

One of the more interesting results to the RSR development team came about when comparing insight generation scores between a paper-based interface that did not use the RSR tool interface (but did use the back-end simulation to run the scenario) and the RSR tool game interface. Evaluators found that the total insight generation score increased from 1,889 to 2,487.

Evaluators found it interesting that the RSR tool interface use trial generated 49 insights that were scored at less than ten points by the players versus only 17 insights scored less than ten during the paper-based trial. As a result, it appeared to the RSR development team that perhaps the computer-based interface lets players consider the evacuation at a higher level of resolution than their usual paper-based interface. Evaluators also found that five higher scored insights were generated significantly earlier in the RSR tool interface trial and three key scored insights were made by different role-players in the RSR tool interface trial.

Because both game play periods were exactly two hours in duration, a normalized insights metric on a per minutes played basis can be calculated by dividing the insight scored values by 120 minutes. The insights generation metric appeared to be highly dependent on the range of activities that took place before the game was played by role-players. The paper-only trial identified the physical cognitive aids (all consisting of words and symbols written on pieces of paper) players would have chosen to use in the absence of an available computer-based interface. Further details about the objectives, cognitive aids, and outcomes of role-playing trials are available in Campbell (2010).

DISCUSSION AND CONCLUSION

The RSV tool is available for game players and response coordinators to use to evaluate performance of first responders during a simulated emergency response game play session. Evaluation requires an evaluator to develop the metrics by which an emergency response effort is considered successful. Metrics vary greatly by the different constituents who judge first responder performance in a community. Some organizations in a community have significant investment in physical assets. Some organizations, like a museum, may have fewer assets but the assets may be of priceless value due to age or significance to human culture. Most constituents agree on the priority of saving human life, but don't agree on the relative priority of saving pets or livestock. And, we observed, there is a wide difference in opinion regarding where the cost of gasoline used to transport responders and resources fall within a list of response priorities.

The RSR tool allows a scenario developer to determine a scoring algorithm and show a team score at all times based on the algorithm at runtime. An analyst can refine the scoring algorithm by analyzing its impact on performance in order to determine its effectiveness in generating desired behavior from game players. Alternatively, an analyst can start by reviewing a role-play scenario session with the RSV tool and find an example of team behavior that appears to be most successful and then use that example to build a scoring algorithm based on recording scenario-appropriate behavior during RSR tool use. Ideally, the RSR and RSV tools can be used in unison to iterate upon a better scoring algorithm with which a player can play with software-based agents and get a sense of how well he or she is doing.

A development team built the RSV tool to support a metric of insight generation – the more insights generated from interacting with the RSV, the better. Although insight may appear to be just one valuable metric, raising its priority is consistent with recent goals of the visual analytics research community in general. Overall, the RSV

tool should allow anyone to get a deeper sense of how an emergency response effort performed just by interacting with simple widgets that accumulate additional value when use is coordinated in groups.

The RSV developers observed that RSV users who were familiar with the RSR tool could not agree on a single scoring algorithm as being sufficient for building an optimal perspective on an emergency response effort to many scenarios. Instead, a visual tool like the RSV tool lets an analyst discuss a response team's performance with changing metrics associated with changes in the nature of the unfolding scenario being analyzed. A discussion of metric relevancy can be facilitated through prolonged RSV consideration as well.

Participants in the RSR and RSV tool development teams have remarked how many insights they have gained by participating in both the software development process and scenario development process. No one on either development team had much exposure to the roles involved with hospital emergency response logistics teams so both teams developed many insights about the nature of emergency response just by working with experienced individuals who have the responsibility of the community to be prepared for responding to crisis scenarios. The RSR tool team developers gained insights into potential RSR interface components that were successfully deployed in the RSV tool. The RSV tool development team gained insights into potential RSV interface components that were already implemented in the RSR tool. As a result, the two teams have been discussing a closer-knit development coordination process moving forward. Both teams agree they could continue to fine-tune the RSR and RSV tools through simulating other mid-size hospital evacuation scenarios being run with hospitals and regional fire departments elsewhere in the world. Other scenarios could help verify the usefulness of the approach of maintaining a core software base from which new scenarios could be encoded.

FUTURE WORK

In order to scale up to larger and more complex scenarios, the visualization data model will need to be updated with more field data attributes as attributes are identified as critical to track in a simulation. The RSR visualization components will need to keep pace with new enhancements and the interactive RSV visualization tool will need to be enhanced to include coordinated views for new attributes added to support better scenario realism.

As the information visualization community works to transform the art of information presentation views into a better-founded science, highly relevant research publications suggest new widgets, views, and interaction techniques to be considered in both real-time simulation game interfaces and sense-making tool interfaces. Many potential countywide emergency response scenarios can benefit from detailed geospatial and temporal analysis. As the skill level rises in visual literacy of complex interfaces, the RSR and RSV teams can work to incorporate and assess emergent suggestions from the literature. Already, the RSV team is considering a more detailed presentation of geospatially moving assets in the mapping display views of the RSV.

As the PARVAC office is located in Seattle and researchers have a history of collaborating with other institutions up and down the west coast of North America, the PARVAC researchers are interested in the Cascadian Subduction zone earthquake threat to communities in the Pacific Northwest that looms as a potential disaster similar in scope to the Katrina hurricane event along the Gulf Coast of North America. Already, various workshops have been held to help communities at risk share plans and coordinate mutual support agreements in preparation for a potential Richter 9 earthquake affecting communities along the length of the tectonic plate boundary from Northern California north to Vancouver Island (Pacific North West Economic Region, 2006). Earthquake

visualization techniques improve in lockstep with ground motion and structural response data simulators (Meyer & Wischgoll, 2007). As a result, the RSR development team foresees a challenging opportunity to include natural phenomena visualizations in both the RSR and RSV tools to help with game play and evaluation.

The software development teams see some benefit to cross-pollinating the visual components between the RSR and RSV tools in order to provide a richer experience during game play. Perhaps there is potential to analyze past actions during role-play in a manner similar to post-simulation analyses performed today. In that case, RSV widget controls and views would need to be integrated into the RSR tool to enable such evaluations to take place by the first response team members or a specialized team member who trained on that specific skill within an emergency operations center. Real-time simulation controls could be added to the RSV tool in order to allow the emergency response team to replay a part of the simulation differently as part of a what-if analysis. Research processes of the nature presented in this paper will continue to be refined as a part of a large global community working on first responder emergency response scenario planning and training.

ACKNOWLEDGMENT

The authors wish to thank the Human Interface Technology Laboratory for providing resources to accomplish this work.

REFERENCES

Adams, M. J., Tenney, Y. J., & Pew, R. W. (1995). Situation awareness and cognitive management of complex systems. *Human Factors, 37*(1), 85–104. doi:10.1518/001872095779049462

Campbell, B. (2010). *Adapting simulation environments for emergency response planning and training.* Unpublished doctoral dissertation, University of Washington, Seattle, WA.

Campbell, B., Mete, O., Furness, T., Weghorst, S., & Zabinsky, Z. (2008, May 12-13). Emergency response planning and training through interactive simulation and visualization with decision support. In *Proceedings of the IEEE International Conference on Technologies for Homeland Security*, Waltham, MA (pp. 176-180).

Campbell, B., & Schroder, K. (2009). Training for emergency response with RimSim:Response! In *Proceedings of the Defense, Security, and Sensing Conference*, Orlando, FL.

Campbell, B., Schroder, K., & Weaver, C. (2010). RimSim visualization: An interactive tool for post-event sense making of a first response effort. In *Proceedings of the 7th International Conference on Information Systems for Crisis Response and Management*, Seattle, WA.

Chang, R., Ziamkiewicz, C., Green, T. M., & Ribarsky, W. (2009). Defining insight for visual analytics. *IEEE Computer Graphics and Applications, 29*(2), 14–17. doi:10.1109/MCG.2009.22

Endsley, M. R. (2004). Situation awareness: Progress and directions. In Banbury, S., & Tremblay, S. (Eds.), *A cognitive approach to situation awareness: Theory, measurement and application* (pp. 317–341). Aldershot, UK: Ashgate Publishing.

Hutchins, E. (1996). *Cognition in the wild.* Cambridge, MA: MIT Press.

Klein, G. A. (1998). *Sources of power: How people make decisions.* Cambridge, MA: MIT Press.

MacEachren, A. M., Gahegan, M., Pike, W., Brewer, I., Cai, G., Lengerich, E., & Hardisty, F. (2004). Geovisualization for knowledge construction and decision support. *IEEE Computer Graphics and Applications, 24*(1), 13–17. doi:10.1109/MCG.2004.1255801

Pacific North West Economic Region. (2006). *BLUE CASCADES III: Critical infrastructure interdependencies exercise: Managing extreme disasters.* Bellevue, WA: Pacific North West Economic Region.

Pea, R. D. (1993). Practices of distributed intelligence and designs for education. In Salomon, G. (Ed.), *Distributed cognition: Psychological and educational considerations.* Cambridge, UK: Cambridge University Press.

Thomas, J. J., & Cook, K. A. (Eds.). (2005). *Illuminating the path: The research and development agenda for visual analytics.* Richland, WA: National Visualization and Analytics Center.

Weaver, C. (2004, October). Building highly-coordinated visualizations in improvise. In *Proceedings of the IEEE Symposium on Information Visualization*, Austin, TX (pp. 159-166).

Weaver, C., Fyfe, D., Robinson, A., Holdsworth, D., Peuquet, D. J., & MacEachren, A. M. (2007). Visual exploration and analysis of historic hotel visits. *Information Visualization, 6*(1), 89–103.

Wischgoll, T., & Meyer, J. (2007). *Earthquake visualization using large-scale ground motion and structural response simulations.* Retrieved from http://imaging.eng.uci.edu/~jmeyer/PAPERS/c-25.pdf

This work was previously published in the International Journal of Information Systems for Crisis Response and Management, Volume 3, Issue 3, edited by Murray E. Jennex and Bartel A. Van de Walle, pp. 1-15, copyright 2011 by IGI Publishing (an imprint of IGI Global).

Chapter 10
Knowledge–Based Issues for Aid Agencies in Crisis Scenarios:
Evolving from Impediments to Trust

Rajeev K. Bali
Coventry University, UK

Russell Mann
Coventry University, UK

Vikram Baskaran
Ryerson University, Canada

Aapo Immonen
*Coventry University, UK, & Emergency Services
College, Finland*

Raouf N. G. Naguib
Coventry University, UK

Alan C. Richards
Coventry University, UK

John Puentes
Télécom Bretagne - Campus de Brest, France

Brian Lehaney
University of Wollongong in Dubai, UAE

Ian M. Marshall
Coventry University, UK

Nilmini Wickramasinghe
RMIT University, Australia

ABSTRACT

As part of its expanding role, particularly as an agent of peace building, the United Nations (UN) actively participates in the implementation of measures to prevent and manage crisis/disaster situations. The purpose of such an approach is to empower the victims, protect the environment, rebuild communities, and create employment. However, real world crisis management situations are complex given the multiple interrelated interests, actors, relations, and objectives. Recent studies in healthcare contexts, which also have dynamic and complex operations, have shown the merit and benefits of employing various tools and techniques from the domain of knowledge management (KM). Hence, this paper investigates three distinct natural crisis situations (the 2010 Haiti Earthquake, the 2004 Boxing Day Asian Tsunami, and the 2001 Gujarat Earthquake) with which the United Nations and international aid agencies have been and are currently involved, to identify recurring issues which continue to provide knowledge-based

DOI: 10.4018/978-1-4666-2788-8.ch010

impediments. Major findings from each case study are analyzed according to the estimated impact of identified impediments. The severity of the enumerated knowledge-based issues is quantified and compared by means of an assigned qualitative to identify the most significant attribute.

INTRODUCTION

The United Nations (UN) is an international organization that was founded in 1945 to maintain global peace and security. Since its inception, the UN has expanded its role. These additional roles relate to developing friendly relations, promoting social progress, elevation to higher standards of living and better living standards, and upholding human rights (United Nations, 2010). In 1994 the UN introduced the concept of human security as part of the United Nations Development Programme (UNDP) Human Development Report (UNDP, 1994). The report was published in an effort to commend the virtues of sustainable human development which not only generates economic growth, but also promotes equal distribution of benefits to those worst affected by the crisis (UNDP, 1994). The purpose of such an approach is to empower the victims, protect the environment, rebuild communities, and create employment (UNDP, 1994). This concept was born out of a need to react to the changing elements of crisis situations and includes new security threats, international failures, civilian impact, a lack of preventative measures, globalization, and international advocacy (Comprehensive Crisis Management, 2008).

In recent years, and in accordance with human security, the UN has directed its focus away from peacekeeping and more towards peace building (Benner et al., 2007). With respect to this, the UN has looked to implement preventative measures in order to avoid crises. Crisis management can be multi-faceted and involves the prevention of crises, the development of preparedness measures, the course of relief and recovery, and the identification of a redevelopment process in the crisis aftermath.

This paper will discuss the efficacy of contemporary KM for crisis scenarios, the need for increased (associated) information sharing, greater teamwork and reduced duplication of effort. Finally, the paper lists a number of impediments that need to be broken down before a knowledge-based solution can be developed.

THE CYCLICAL NATURE OF CRISES

The crisis management cycle is best captured by a schematic which shows the severity of the each incidence against a given timeline, plotting all contributing factors in a pre and post-crisis situation (see Figure 1). The curve also identifies when weak signals are felt and any escalation into early warning signs thereafter; culminating in full-blown crisis which needs management to contain and/or minimize resultant impacts on society. Other work (Jennex & Raman, 2009) has described crises as a series of four phases: situational analysis (SA), initial response (IR), crisis response (ER), and recovery response (RR); and five decision/hand off points: the initiating event (IE), the control event (CE), the restoration event (RE), the normalizing event (NE), and a terminating event (TE).

CHALLENGE

When society discovers *a priori* signals that can lead to crisis, then the governing authority reacts through preventive diplomacy to return order to the system if the crisis is manmade. If the crisis is natural, then society instigates preventive strategies, i.e., building earthquake proof buildings, tsunami resistant coastlines, etc. These preventive steps may minimize loss of life and damage to

Figure 1. Crisis and disaster management curve (adapted from Immonen et al., 2009)

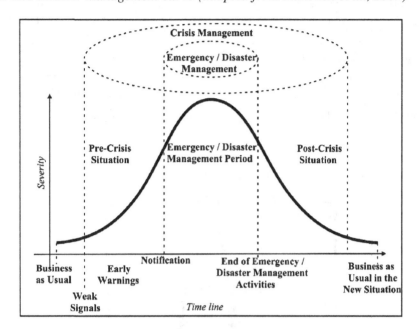

property whilst ensuring that day-to-day activities continue as unaffected by change as possible. This ideal situation is seldom achieved; rather than being proactive, society is usually reactive, resulting in ineffective and inefficient responses to crisis situations in most instances. Many major crises have been experienced over time; such major crises attract a multitude of aid agencies hoping to provide assistance and alleviate suffering for those worst affected by such situations. This mass influx of agencies can often add further complexity to an already difficult situation. Lack of ownership, trust, coordination, communication, and knowledge transfer between international aid agencies, International Inter-governmental Organizations (IGOs) and Non-Governmental Organizations (NGOs) can often be complicated or nonexistent: as a result the entire mission can be compromised.

Aid agencies and NGOs play vital roles in managing such crisis situations. Inadvertently they may conceal information which they regard as insignificant which is in fact vital to another stakeholder within the crisis domain. Without

necessary systems, ownership and trust being in place, opportunity to provide appropriate aid may be severely hampered. Furthermore the clarity of an agency's role within a crisis situation can often be ambiguous and undefined. With no clear definition as to what they are attempting to achieve, and without the necessary guidance, problems can become apparent. The identification of clear leadership, coordination, and trust between organizations is of paramount importance: as such the ability to measure performance of vital importance.

This paper intends to investigate real-world crisis management situations with which the United Nations have been, and are currently involved, in an attempt to identify recurring issues with these knowledge-based impediments. In particular, this paper presents three distinct natural crisis situations – the 2010 Haiti Earthquake, the 2004 Boxing Day Asian Tsunami, and the 2001 Gujarat Earthquake. These crisis situations have been presented as cases and thus selected due to their time periods and geographical dispersion. The cases underline ongoing problems apparent in each situation and which show little sign of mitigation. Particular focus is placed on the UN's role as a coordinating body during the crises. Figure 2 illustrates generic issues and their linkages for crises; it identifies major barriers that impede organizations like the UN, IGOs, NGOs and the state from delivering peace and human security.

There are four layers enumerated in Figure 2: the upper layer (layer one) identifies the situation as currently identified; elements of the layer contain risk, crisis, and post-crisis elements but are so entangled as to provide seamless transition between elements. This layer also highlights that there is currently little value placed upon imple-

Figure 2. Crisis management - current and desired situation

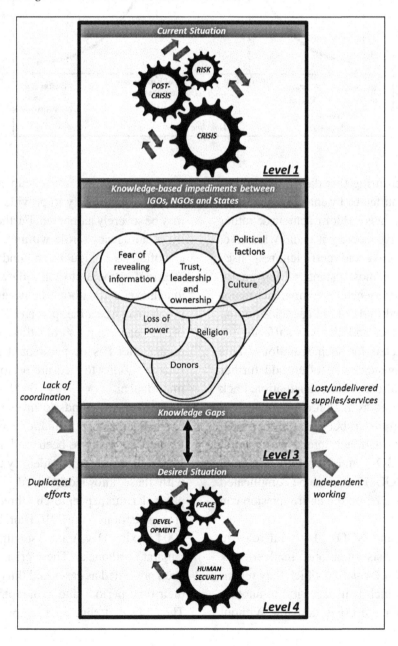

menting measures of preparedness. Layer two identifies the impediments to knowledge sharing in a crisis situation:

- **Trust, Leadership, and Ownership:** A lack of, or poor implementation of, any of these elements can be detrimental to crisis management situations.
- **Donors:** Stipulations and guidelines laid out by donors as to where finance is distributed can prevent organizations from sharing data.
- **Religion:** Long-standing feuds based upon religious beliefs can make it difficult for organizations to openly share information.
- **Political Factions:** Interstate feuding and organizational disagreements can promote a reluctance to share information. This may be particularly apparent where a state is required to reveal information which may be highly secretive, or cause embarrassment to the state in question.
- **Fears of revealing information which may be valuable to the organization:** As has been witnessed in the Human Genome Project, many organizations have uncovered valuable information which they have then been unwilling to reveal. This may be as a result of future opportunities to use the acquired knowledge to generate revenue streams.
- **Loss of Power:** High ranking IGOs and NGOs may feel their power base is weakened due to sharing knowledge. By revealing information they possess, other organizations can find themselves on an equal footing.
- **Organizational Culture:** If the IGO or NGO lacks a culture of knowledge sharing it is unlikely to advocate this change.

Layer three highlights the gaps which are caused as a direct result of these impediments to knowledge sharing. Such knowledge gaps are widened by a lack of coordination, duplicated efforts, lost or undelivered supplies and services, and the urge of organizations to work independently. The fourth layer describes the desirable situation. When knowledge gaps are bridged and impediments removed, then basic components for development, peace, and human security will be acquired cyclically. In addition, when the final layer is fulfilled, given organizations will be able to identify risks and weak signals at the beginning of the *Crisis and Disaster Management Curve* (Immonen et al., 2009) presented earlier and hence put in place measures of preparedness (Raman & Jennex, 2010). However, the aforementioned desirable final layer is hardly ever reached. The reasons behind this ineffective approach can be attributed to a number of factors. The next section outlines three cases to enumerate these knowledge-based issues and the impediments that have negatively impacted crisis and post-crisis situations.

CASE 1: HAITI EARTHQUAKE, 2010

On 12th January 2010 an earthquake of magnitude 7.0 struck the Haitian capital of Port au Prince. The quake left up to 230,000 people dead, and another 200,000 injured. Additionally, up to a million people were left homeless (Bilham, 2010). Six months on from the disaster those affected continue to be in dire need of support, and the aid process has been heavily criticized in many quarters. This case study aims to understand the issues which have made the delivery of aid so difficult. The arrival of aid to Haiti was prompt and abundant. However, this created its own problems. Due to a lack of preparedness the arrival of aid was described as incredibly disorganized (BBC News, 2010). As increasing numbers of aid agencies arrived bringing with them drastically-needed supplies, vital equipment was left sitting next to the runway for a number of days. In the immediate aftermath of the disaster when healthcare equipment was most needed, a lack of organization disrupted its distribution.

Médecins Sans Frontières complained of delays to supplies arriving at the expense of foreign troops. GOAL blamed the UN and USA for the delay, claiming it was their failure to work together which caused such disruption. The UN responded by blaming what it termed as underestimated logistical problems (BBC News, 2010). One of the reasons behind some of the chaotic beginnings to the aid situation can be attributed to a loss of personnel and systems belonging to the UN during the earthquake. The Inter Agency Standards Committee (IASC), however, who produced a report on the disaster in July 2010 criticized the early lack of co-ordination whilst praising subsequent efforts to improve this. The report also went on to describe the following issues (Inter-Agency Standing Committee, 2010):

- The need to work more proactively with the various actors involved and with local governments,
- The need for aid agencies to identify the necessary expertise, tools, knowledge and partnerships to operate effectively,
- A need to improve communications with disaster-affected populations,
- A need for strong, decisive, and empowered leadership,
- The need for a shared strategic vision,
- The loss of information due to the earthquake was unrecoverable,
- The lack of coordination of over 400 humanitarian groups,
- A lack of information on national capacities,
- A lack of preparedness on the part of the Haitian government and the UN,
- No exit strategy,
- The delay in the collection and sharing of information between agencies,
- A lack of willingness of agencies to priorities the need for information sharing, and

- The lack of an accepted system by which to record and share information.

The IASC reserves its biggest criticism in its report for the lack of leadership and organization displayed, highlighting the lack of coordination between aid agencies in the early stages. This is blamed upon the high number of independent agencies involved and the shortage of available information. However, the report goes on to suggest that the cluster approach taken towards the development enabled communication to significantly improve. The report also highlights the failure of some aid agencies to engage with the authorities, hence restricting their strategic effectiveness. Furthermore, other organizations were hampered by their inability to access and thus benefit from the knowledge-based possessed by these agencies (Inter-Agency Standing Committee, 2010).

IASC identifies two barriers to inter-organizational knowledge sharing; one being the language barrier, the other a restricted access to coordination centers' headquarters (Inter-Agency Standing Committee, 2010). In addition, incompatible systems may have prevented agencies from working together effectively. The report further highlights the failure of inexperienced, well intentioned agencies to share and coordinate relief operations. A report carried out by the Disaster Accountability Project on aid agencies working in Haiti was also highly critical of the lack of transparency displayed. Of 200 organizations interviewed only 6 regularly provided factual updated reports (Disaster Accountability Project, 2010). This report calls for increased transparency in all aid agencies in order to ensure accountability and monitoring, and proposed that that policies should be put in place as a matter of course to ensure the enforcement of such guidelines. Disaster Accountability Project (2010) blames many of the problems on ineffective coordination strategies. Such problems may have been caused by the sheer scale of such a relief effort.

CASE 2: BOXING DAY ASIAN TSUNAMI, 2004

On December 26th 2004 an earthquake of magnitude 9.0 struck 250 km off the coast of the Indonesian city of Banda Aceh (Pickrell, 2005). The subsequent tsunami caused substantial damage to towns and cities in thirteen countries as far away as Somalia. The death toll was estimated to be around 225,000 with a further 500,000 injured. An additional 150,000 people were killed by the resultant spread of infectious disease (Pickrell, 2005). The subsequent after effects of the disaster left five million homeless and a further one million unable to make a living. The estimated total cost of the disaster was placed at approximately $7.5 billion.

Such previously unprecedented geographical dispersion of the disaster made the distribution of effective aid a near impossible prospect and proved a huge drain on resources, particularly to large organizations like the UN. Takeda and Helms (2006) discussed how the United Nations Disaster Relief Agency managed the situation and argued that the highly bureaucratic system employed for disaster preparedness actually added to the complexity of the disaster relief efforts due to its failure to include any scope for adaptability. The same authors argue that a system was in place to coordinate the work of NGOs and aid agencies under the UN umbrella (Takeda & Helms, 2006) and argue that the rational approach employed by these systems is unworkable in an irrational situation. They additionally call for the use of a more fluid system capable of changing according to the needs of the situation (a holistic management system). They state their belief that, whilst management information systems are vital for sharing information, in an ever-changing world, it is equally important that these systems are fluid and adapt to change. In addition to the problems with the systems employed by the UN, a number of other impediments to the delivery of aid have been identified. Table 1 highlights some of these issues (grouped according to source).

Table 1. Impediments to the delivery of aid

Issue	Source
• Slowness in arranging meetings • The under preparedness of the states to deal with disaster • An insufficient supply of external information • A refusal of some organizations to accept outside help	Takeda and Helms (2006)
• Some NGOs operated independently making coordination more complex • Arriving UN staff had little understanding of countries or cultures • A lack of information flow resulted in a duplication of efforts • Information was available but inaccessible to those who needed it • The language barrier • Inexperienced NGOs were unaware of the need to coordinate with other actors	UN (2005)
• A failure on the parts of governments and aid agencies to work in close proximity • Refusal of Indian government to allow access to the Andaman Islands due to military sensitivity	BBC News (2006)
• Aid being delivered to the same spot twice whilst others were left desperately waiting unattended • Conflicting directions being given by differing sources	Gonsalves (2005)
• A lack of infrastructure with an uncoordinated response to the arrival of aid agencies	CNN (2005)
• No adequate distribution system resulted in goods and supplies waiting at airport	Paddock and Magnier (2004)
• Government protocols prevented access to certain countries or areas	Takeda and Helms (2005)
• Numbers of refugees in camps were inflated by government officials in order to gain greater levels of aid	Solomon and Higgins (2005)

Despite the reported problems highlighted, the general consensus as to the relief efforts experienced following the tsunami is largely positive, which has resulted in a belief that much has been learnt from the situations encountered (Shaw et al., 2010). However, some reports suggest that similar errors to those made during the tsunami relief effort are still being made today. This indicates that mistakes are not being appropriately engineered out of the relative organizations' relief strategies.

In the immediate aftermath of the tsunami the UN elected to set up Humanitarian Information Centers (HICs) which contained information on ownership, maps, surveys, meetings and funding. The HIC was largely considered a success and currently remains operable in some affected countries. One point of note which has arisen from the UN's work in Indonesia is the reported change in relationship between the UN and NGOs. Whereas previously NGOs acted as a 'sub-contractor' to the UN, increasingly NGO funding has changed the dynamic. The UNDP reports one NGO asking the UN to carry out work on their behalf (UNDP, 2005). This blurring of established relationships sets a precedent which could compromise the UN's ability to influence government policy making. The UNDP also reports success in collaborations between Islamic and Western NGOs (UNDP, 2005) which had previously been an area of much sensitivity. The mass geographic dispersion of aid agencies during the tsunami relief effort created unprecedented problems for the UN in coordinating an efficient response.

CASE 3: GUJARAT EARTHQUAKE, 2001

On January 26th 2001 a major earthquake measuring 6.9 on the Richter scale struck the Western Indian state of Gujarat. Reports suggested the earthquake killed in excess of 20,000 people and left a further 165,000 injured. Over 16 million people were affected by the disaster which left around 7,000 homes destroyed (BBC News, 2001). In the immediate aftermath of the disaster, relief work was largely undertaken by bilateral agencies and NGOs. UNDAC (United Nations Disaster and Co-ordination) reported nearly 250 agencies were active in the region by February. The UN response to the disaster was relatively minor, with the organization endeavoring to develop a coordination centre. The centre became operational a week after the initial disaster, which, by the UN's own concession proved largely ineffective; the UN also confesses that the team sent was largely ill-equipped and insignificant (Harland & Wahlstrom, 2001).

The UN also highlights a lack of clarity in the role which each division within the organization should be undertaking, resulting in poor relations between divisions. This in turn led to external organizations and NGOs expressing their disappointment at the UN's input and their failure to coordinate a response to the crisis (Harland & Wahlstrom, 2001). The DEC (Disaster Emergency Committee) reported well-established NGOs had long-standing uncomfortable relations with the UN. The resultant effect of this situation was that most NGOs were relying on the UN to coordinate operations, but believed that they failed to in this task. As a result the UN and other organizations operating at the scene were accused of being *'weak, slow, and lacking in direction'*; the UN itself was accused of absorbing lots of information but delivering little in return and of acting in an 'elitist' manner (Harland & Wahlstrom, 2001).

More worryingly for the UN was the view taken by many that the organization was a group of disparate agencies, each responsible for doing its own thing. This lack of collaboration and failure to work together significantly hindered the UN response. The lack of information flow in place during the disaster resulted in further problems. The UN reports that its divisions were reliant upon UNICEF's experience in the region to provide contacts and information regarding how to operate in Gujarat. Logistical and administrative support, along with a supply of information, was left desperately lacking (Harland & Wahlstrom, 2001).

A distinct lack of management capacity and good relations with the Indian government further hampered attempts to develop successful strategies. This was attributed to India's longstanding directive of managing disasters internally. The result was a slow response by the UN to the disaster and a subsequent lack of information sharing between the state and the UN. A slow response to the disaster and the UN's late entry into the situation resulted in UN departments being forced to spend much of the early days catching up. Harland and Wahlstrom (2001) reiterated the need for leaders to be appointed and how fast they should respond to a given crisis situation: "Leadership can only be provided if the aspiring leader is the first on-site".

Failure to produce field reports was put down to a problem in centralized decision making. Various UN agencies in the field failed to agree upon whose views the report should include, thus causing further in-fighting. Geneva (the UN Headquarters) failed to rectify this situation which, as a result, was allowed to perpetuate, further hampering the relief effort (Harland & Wahlstrom, 2001). The UNDAC systems used by the UN during the disaster to disseminate information were generally

regarded functioning to an acceptable level whilst being made available to all relevant actors within the crisis domain. Nevertheless it is iterated that a system is only as good as its users and the input they provide. As such, the system was criticized for being user unfriendly (Harland & Wahlstrom, 2001). As is apparent, criticisms of the UN were abundant. However, NGOs, the state and other actors were also responsible for a number of problems which are presented in Table 2.

ANALYSIS AND DISCUSSION

This section provides an analysis of the case studies discussed previously, commencing with major findings from each, supplemented by discussion regarding the cause and effect of various impediments. Each impediment will be assigned a level of severity as to the problems it caused during each situation; this section will additionally align impediments with the effects created to highlight those which generated the most problems.

Leadership, Trust, and Ownership

A significant lack of leadership, trust and ownership is apparent in all of the case studies analyzed.

Table 2. Problems of NGOs

Issue	Source
• A lack of experience, partnerships, and relationships with other actors • A lack of a strategic plan, or exit strategy • A lack of clarity as to the actors' roles within the situation • NGOs created disparate communities who shared information only between themselves due to the UN's failure to make central decisions • Mistakes made during previous disasters were repeated • Failures on the part of the UN to heed lessons learnt in previous scenarios and apply them to new situations	Harland and Wahlstrom (2001)
• A failure by the Indian government to provide the necessary information • A lack of preparedness by all sides to respond in an area prone to natural disasters • Weak assessments of the damage inflicted • A failure to engage with local communities during rebuilding	Shah (2002)
• Failure to develop effective local partnerships • Duplication of efforts	Disasters Emergency Committee (2001)

In Haiti, particularly during the early phases of the relief effort, there were a number of reports to suggest that leadership was an issue: many were highly critical and blamed a lack of leadership for the problems which arose during the first few weeks. The concerns with leadership identified during the Haiti situation, and indeed the other case studies analyzed, were exasperated by the huge influx of aid agencies who attended the disaster. This provided significant problems with effective coordination and engagement. There were also elements concerning trust during the Haiti situation, evidenced by the refusal of the UN to allow access to its command headquarters.

Leadership proved problematic during the tsunami, in large part due to the vast geographical dispersion of the disaster compounded by a need for the UN to operate in a different manner in the various affected countries. The Indonesian government, for example, requested that the UN take responsibility for the aid, and close ties with the Sri Lankan government were also established. In other countries the UN has to operate under state government legislation. The main criticisms leveled at the UN concerning leadership were focused upon the lack of, and slowness in, arranging regular meetings in order to gauge process and coordination of effort. The Gujarat earthquake provides the most compelling evidence of how poor leadership can influence the outcome of crisis situations. The relief program experienced massive problems which threatened to undermine the entire effort and seriously damage aid agency relations. The study highlights failures of the UN in establishing good governance, effective leadership and ownership of the crisis situation. Not only was leadership of other organizations exposed; the study also demonstrated significant issues in ownership and leadership within the UN's own departments.

Alongside ownership and leadership, the study showed that centralized decision-making can be considerably damaging to a project, and seriously delay decisions and impede information flows.

The cases also demonstrate the issues which can become apparent if relations between the state and the UN become strained where outright ownership of the situation is not established. The element of trust between the UN and other organizations can play a pivotal role in delivering aid in a timely manner. Aid agencies accused the UN of absorbing much information and delivering little, failing to achieve what was expected of them. This had the effect of forcing aid agencies to operate independently and create disparate communities. The recent Chilean earthquake has highlighted the benefits of good governance during crisis situations where government then swiftly took ownership of the problem, largely refusing help from various aid agencies.

The effects of a lack of leadership, trust, and ownership had underlying implications for the relief efforts and significantly hampered the delivery of aid in all of the case studies analyzed. Aid agencies were often forced to repeat work already accomplished, with reports suggesting aid was delivered more than once to the same location. The lack of leadership and ownership again highlights how poor coordination forced aid agencies to act under their own recognizance. This was caused by a failure to provide regular meetings and, therefore, a knowledge base for sharing information across agencies. A failure to provide leadership also badly affected the opportunity to develop any form of strategic plan. Therefore agencies had little to no clarity as to their specific role and no opportunity to monitor or evaluate performances. The need for effective leadership was further exaggerated by the vast number of aid agencies drawn to the crisis.

Under Preparedness

One of the most important findings from the Haiti case study was the under-preparedness of the aid organizations in dealing with a crisis in this area. With Haiti being highly prone to a series of natural disasters, including hurricanes and earthquakes, one would expect aid agencies to have some level of infrastructure permanently set up in this area. In addition, the preparedness of the Haitian state has been criticized for its failings to instigate the necessary measures for such predictable crises. Although the UN tragically reported losing personnel and systems during the earthquake, the failure to back-up data and to share knowledge among aid workers considerably hampered the effort. This illustrates the UN's failure to become a learning organization and disseminate knowledge gained in the field.

Evidence of under preparedness is also apparent during the tsunami case study. In addition to criticisms being leveled predominantly at aid agencies, the states involved have been largely condemned for their lack of preparedness, insufficient infrastructure and slowness to respond. However, the UN has not been exempt from criticism with attending staff admitting to having little knowledge of the culture and values of the countries to which they were seconded. The unprecedented geographical dispersion of the tsunami provides some mitigation as to the unpreparedness of the UN response. Nevertheless, it is feasible to expect that the UN would have greater preparatory measures in certain affected regions. Furthermore, the lack of cultural awareness of UN staff can be largely attributed to insufficient training and sharing of regional knowledge. State awareness remains a difficult area to quantify. The majority of the thirteen nations affected were criticized in some quarters for their slow and ineffective response to the disasters. Whilst national preparedness is largely the remit of the state, it appears there is scope for the UN to become involved with some form of training or preparation scheme to help states deal with disasters.

The Gujarat earthquake provides further evidence as to the unpreparedness of both the nation state and the UN in responding to the disaster. The UN's failure to respond adequately to the event until a week after the earthquake is attributable to India's previous predilection of responding to di-

sasters internally. The subsequent failure of the UN to provide a sufficiently equipped and adequately staffed team cannot, however, be attributed to this short-coming. This provides further support for the view that the UN was underprepared. This under preparedness shown by the UN, states, and other aid agencies led to significant issues during the early phases of the crisis, perhaps the most detrimental being the lack of coordination. The early stages of a relief effort are widely regarded as some of the most important; effective distribution of medical supplies, clean water, shelter and food at this time can be the difference between life and death for many. Proactive preparation for such situations should be a driver for change within the UN.

Systems and Communications

The failure of aid agencies to adopt widely accepted, compatible systems has contributed to major problems with information flow across the three case studies. In Haiti, the failure to back-up systems which had been in place prior to the disaster meant that essential information was lost which could have been vital to early stages of relief. This issue was further exaggerated by the UN's failure to employ an adequate number of staff and resources in the development of information management. As a result, delays in collecting and sharing information became apparent. Criticisms were also leveled at other aid agencies for their failure to prioritise the need for information exchange. This may have been as a result of incompatible systems, or a lack of experience among agencies.

Systems used by the UN during the tsunami also encountered a great deal of criticism, mainly due to their bureaucratic and user-unfriendly nature. Evidence has been provided which suggests the systems employed were chronically unsuitable for purpose and that there is a need for a complete overhaul in the systems used. However, there is also a requirement for better methods of data collection and analysis to be implemented in order to make any new system successful. During the Gujarat earthquake there were complaints that, although information flow was successful during the early phases of the relief effort, future flow was disrupted by the UN's failure to empower users. Also, complaints were once again raised with regard to the user unfriendliness of the systems.

The effects of poor and disparate systems and the failure of aid agencies to prioritise any need to share information had far reaching effects across the three case studies. Inaccessible and reduced flow of information during the relief efforts caused problems in coordinating an effective response and keeping aid agencies informed of the latest developments: the knock-on effect of which was duplication of effort and difficulties in tracking the distribution of aid and supplies. Furthermore, this led to immense difficulty in monitoring and evaluating performance.

Lack of Collaboration and Transparency

Collaboration from agency to agency, and government to agency, provided significant impediments to effective relief efforts. In addition, the lack of transparency displayed by some agencies further hampered the opportunity to share information. Some of the major criticisms leveled at the UN have surrounded their failure to actively collaborate with governments, other agencies, and even among their own departments. The result of this has been a failure on the UN's part to gather vital information from governments and other agencies, prompting accusations that some aid agencies have been allowed to act on their own recognizance. Furthermore, some aid agencies have as a result failed to engage with the UN in any way.

A lack of collaboration with the disaster affected populations, particularly during the redevelopment phase, has resulted in accusations of arrogance and thus populations being unhappy with the results. The level of transparency shown

by some agencies has also been called into question (Murphy, 2010). This has led to some aid agencies refusing to accept external help and security issues during the delivery of aid. It is important to note that aid agencies are not completely responsible for the lack of collaboration; state governments have also been guilty of failing to engage and demonstrate sufficient transparency. This has been evidenced by the deliberate overestimation of Internally Displaced Persons (IDPs) in refugee camps in order to gain access to greater levels of aid. In addition, as witnessed during the Gujarat earthquake, the Indian government's reluctance to disseminate information hampered relief efforts with the resultant effects of failing to collaborate once again manifested in a lack of coordination between agencies. This in turn led to some agencies acting independently, outside of the jurisdiction of either the UN or the state government. Some agencies felt forced into forming disparate communities, resulting in conflicting directions being given and supplies failing to be delivered.

Proliferation and Inexperience of Aid Agencies

One trend witnessed during all three of the case studies analyzed has been the mass proliferation of aid agencies, and in particular NGOs, who attended the relief effort. This has been mirrored by the number of NGOs now in operation around the globe. Despite the undoubted good intentions of NGOs, both new and old, the inexperience of some has caused problems during the aid effort. This has been witnessed by a failure to collaborate, an unawareness of protocols, and failure in reporting to governing bodies. In one extreme case there was the example of the American Baptist Church being accused of kidnapping Haitian orphans. However, apportioning blame solely towards NGOs is not necessarily justified; lack of leadership and good governance must also be addressed.

The effects of NGO inexperience and mass proliferation again bred difficulties in establishing a sufficient level of coordination. The attendance of over four hundred NGOs to the Haitian earthquake would make coordination difficult in any circumstance, and particularly when a host of inexperienced NGOs are involved. This situation can be further exaggerated when good governance and leadership are absent. As a result significant issues were recorded in terms of information flow, reporting, monitoring and evaluation.

Organizational and State Culture

The culture of an organization or state is typically something which is deeply ingrained, particularly in well established organizations within strongly independent countries. The results of cultural differences in the three case studies are widely visible. It is noticeable that a number of aid agencies have refused to work with other organizations and as such have been reluctant to share any information. This also leads to the issue of strategic visions becoming disparate and some organizations having different goals which may contradict or disrupt the entire process. There are accusations that the UN themselves have been guilty of developing an 'elitist' culture which has resulted in some NGOs viewing them with suspicion and as being unreliable in their ability to oversee the aid effort. Questions have also been asked about integrity, embraced as a core competency, within the UN (Deloitte Consulting LLP, 2004).

Furthermore, there are reports that the various UN departments have been viewed as being a group of disparate agencies rather than one enterprise focused upon on a shared vision (Olara, 2009). This has resulted in failure to adequately share information between departments. Additionally, centralized decision-making and failures to empower staff in the field have provoked further issues. As such we can decipher that the UN is some way from becoming a learning organization. The problems of state culture have also been apparent in the case studies analyzed. Reports of the Indian government refusing to allow access to the Andaman Islands following the tsunami due to military sensitivity highlight this problem. There have also been reports of some states refusing to accept aid from certain foreign countries. One positive note relating to the culture of certain organizations has been the example, during the tsunami, of Islamic and Western agencies working in unison. Due to well publicized reasons cooperation has not always been evident.

The major effect associated with disparate cultures is the difficulty in establishing working relationships. This can lead to agencies being unwilling to share information and work in partnership with other organizations. As such, the entire aid mission can be compromised as goals and strategies begin to contrast. This can lead to problems in establishing a mission strategy by which all aid agencies abide, significantly hampering the aid process. Some organizational cultures have also dictated that there is a regular turnover of staff both organizationally and by crisis situation. As such, knowledge of processes, procedures and expertise is lost from the business. The work of agencies can become unregulated due to a culture of operating secretly which may lead to duplication of effort. Once again the problem of coordinating a response is compromised by an ingrained culture.

Language Barrier

The language barrier between agencies and affected populations can provide a considerable hurdle to successful crisis management. In Haiti where the majority of the population is French speaking there is reliance upon aid agencies to dispatch staff fluent in the language. However, this is not always possible due to the resources available to many organizations. In the other case studies analyzed it is easy to imagine this situation being further exacerbated due to the increased number of languages spoken, predominantly in the tsunami example. The effects that a language barrier can create include difficulties in communication between agencies and populations. Therefore the problem of developing a system with a common language runs the risk of alienating some NGOs.

Donor Stipulations and Aid Distribution

The Haiti case study has provided evidence that distribution of aid is not being entirely used to encourage development within the affected area. Reports suggest that a percentage of the aid being donated has been redistributed to NGOs' command locations in order to finance the employment of highly-paid consultants; essential aid has been lost by those who need it most. Donor stipulations can have a major effect on the way in which aid agencies operate, in particular NGOs. The Gujarat case study has shown that some aid agencies have been attracted to high profile disasters whilst disasters with less press coverage have been largely ignored. This has resulted in accusations of aid agencies using high-profile disasters in order to boost their own public image.

There have also been examples of high-budget NGOs being disingenuous to coordination of efforts in conjunction with other agencies. This may be due to the fact that the high level of finance available to them enables them to undertake and employ sufficient staff to manage an entire project single-handedly. In terms of the distribution of aid many of the problems have been linked to the difficulties in tracking its supply and in some cases aid was delivered to the same area twice. In order to combat this problem the UN has developed two new systems which have largely been proven to be successful during the tsunami aid distribution. Accusations have also become apparent that aid has only targeted certain demographics and largely ignored others. This goes against the mandates of many aid agencies that aid should be delivered regardless of age, race, gender or social standing.

The effect of aid agencies failing to engage with other organizations leads to monumental problems. Once again the issue of duplicated ef-

forts comes to the fore as independent organizations fail to notify others of their achievements. In terms of aid distribution, without suitable systems in place and sufficient information sharing, there is considerable scope for aid deliveries being repeated and certain areas in need being overlooked. In addition, tracking the finance and available supplies becomes very difficult.

Agency Relations

Relations between agencies are fluid and regularly need to adapt and change in accordance with the situation. Serious dysfunction was witnessed between the UN and other aid agencies during the Gujarat earthquake, and indeed between UN departments. The tsunami provided an example instance of an NGO with a great deal of financial backing attempting to subcontract work to the UN. The continued growth of NGOs and the example mentioned set a dangerous precedent for the UN whilst threatening to compromise their position as a leading body. As such there may be scope for the UN to look to re-evaluate their position in the aid agency hierarchy in order to reaffirm their status as a leading body.

The possible effect of poor relations threatens to further complicate the coordination of aid agencies. The Gujarat earthquake provided the UN with an exemplar as to the threats poor relations can pose; some NGOs broke away to form disparate communities. The importance of establishing long-standing and trustworthy partnerships is vital, particularly in an age when the number of NGOs continues to increase. The blurring of the UN's standing threatens to create further problems. The UN is currently regarded by many as a leading body in crisis management; allowing agencies to dictate terms to the UN could undermine their position of leadership and ownership in crisis situations. To summarize the severity of enumerated knowledge-based issues, a severity score is associated with each of these factors in the cases discussed earlier as shown in Figure 3. A total of these scores provides a cumulative severity index for these impediments. For the said cases, leadership, systems and communications, collaboration and transparency, under preparedness, and lack of trust provide the largest impediments, closely followed by ownership, inexperience, and culture. Due to the fact that lack of leadership, trust and ownership play a vital role in any crisis situation, and are mainly associated with a cascading effect on other factors, Figure 3 shows them as separate entities for more emphasis. The remaining attributes are coupled together to provide a consolidated score. The scale used to categorise the severity is as follows:

- **5:** Cause is highly apparent
- **3:** Cause is moderately apparent
- **1:** Cause is apparent

Assigning the severity rating for each of the factors has been accomplished through a qualitative analysis of the case.

APPLICATION OF KNOWLEDGE MANAGEMENT CONSTRUCTS

The incorporation and effective use of Knowledge Management (KM) could help aid agencies in overcoming many of the issues which they currently face. However, before KM can become a fundamental aspect of crisis management within aid agencies, there is a pressing need to overcome the current causes of issues and impediments to KM which currently exist (Immonen et al., 2009). Only by making changes at each stage of the current processes can KM become a truly worthwhile venture. There is an urgent requirement for making provisions to encourage greater information sharing, collaboration, leadership, and transparency among aid agencies. An industry-wide acceptance of KM could help achieve this. KM has bore witness to some outstanding achievements in recent years and has helped to make significant progress in the world of healthcare. The result has been that KM is now valued as a key corporate resource by many global organizations (Bali et al., 2009).

Figure 3. Severity of impediments for each of the cases

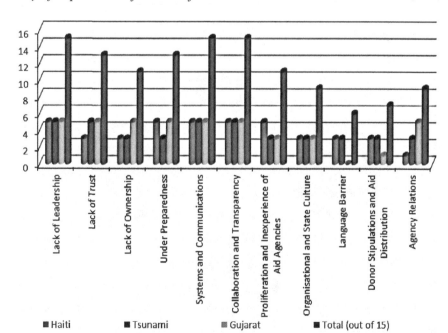

The effective use of KM in crisis management can help to retain expertise, train new employees, and analyze and disseminate vital statistics and information to operatives working in the field (Wickramasinghe & Bali, 2008). Figure 4 illustrates the current status of the Crisis Management cycle and the impediments which exist.

Figure 4 depicts the cycle of impediments from one crisis management situation to another. If we take the upper circle (under preparedness) to be the start of a new crisis management situation the diagram reads as follows:

- A lack of preparedness for the new situation.
- Resultant delay in establishing effective leadership or ownership.
- Collaboration, relations, and transparency issues come to the fore which results in inexperienced agencies being left to their own devices.
- Lack of meaningful data being collected and analyzed - meaning that systems re-

main poor, disparate, individual to the organization, and with no scope for knowledge sharing.
- Ultimate failure to record information resulting in agencies being unable to reuse information and apply it to a new scenario or for training purposes (the cycle begins again with the same disparities apparent in the new crisis).

Well executed KM initiatives to improve this cycle offer many benefits which could be reaped by aid agencies during the management of crisis situations. Such benefits include:

- Establishment of better working relationships.
- Reduction in organizational competitiveness.
- Increased transparency.
- Identification of the UN as a strong leader, therefore negating fears over agency concerns.

Figure 4. The "Circle of Impediments"

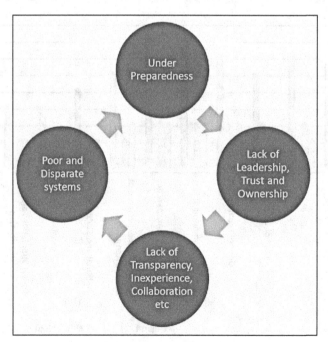

- Improved performance measurement.
- Establish a shared vision.
- Strengthen partnerships.
- Increase coordination and reduce duplicated efforts.
- Increase information flow.

Successful KM can only be achieved if all actors embrace the process. It is imperative that the UN plays a key role in driving this change. However, as the case studies have evidenced, the UN incorporates its own issues when it comes to knowledge-sharing within departments. As a result, one of the fundamentals of aid agencies embracing KM must begin with the UN becoming a knowledge-based "Learning Organization" (Bali et al., 2009). Schematically, such changes improve the "Circle of impediments" to a KM-enabled "Circle of Trust" (see Figure 5).

CONCLUSION

Figure 2 captures the current management state of the crisis and provides a perspective on how impediments can occur in crisis situation, additionally highlighted by analysis of the cases. Such impediments result in widening the knowledge gaps between stakeholders (IGOs, NGOs and states) and, in turn, lead to undesirable and chaotic crisis management. Many of the problems which aid agencies have experienced in their attempts to facilitate successful crisis management have been as a result of similar factors, regardless of the situation. Therefore, the failure to learn from these mistakes and to implement measures to prevent their reoccurrence is unacceptable, and necessitates an urgent response to the problem. Lessons learned from each disaster and exercise need to be captured and disseminated to those responsible for creating formal or ad hoc emer-

Figure 5. The "Circle of Trust"

gency response systems (Murphy & Jennex, 2006). The recent success of knowledge management in the healthcare industry and corporate world has prompted questions as to the benefits which it could bring to crisis management. These benefits could include improved information sharing, greater teamwork, better preparedness, reduced duplication of efforts and increased coordination. All of these benefits would to some extent reduce the recurring issues which have been witnessed across recent crisis management situations. However, a number of impediments need to be broken down and studied to identify their root cause before a knowledge-based solution can be developed for crisis management.

The UN has largely accepted that knowledge-based solutions should become the core of their operations. Yet their initial efforts to achieve this have been met with a less than encouraging response. It is imperative that the UN focuses upon efforts to become a learning organization and implement knowledge-based solutions effectively. When such KM-based tools and techniques are integrated into the UN's management of crisis situations, they should facilitate crisis management working at level 4 (see Figure 2). As a result the benefits and subsequent effects on other agencies could change and improve the way crisis management is delivered globally.

REFERENCES

Bali, R. K., Wickramasinghe, N., & Lehaney, B. (2009). *Knowledge management primer*. New York, NY: Routledge.

Benner, T., Binder, A., & Rotmann, P. (2007). *Learning to build peace? United Nations peace-building and organizational learning: Developing a research framework*. Berlin, Germany: Global Public Policy Institute.

Bilham, R. (2010). Lessons from the Haiti earthquake. *Nature, 463*, 878–879. doi:10.1038/463878a

CNN. (2005). *Rivers still strewn with bodies: Special Report – After the tsunami*. Retrieved from http://edition.cnn.com/2005/WORLD/asiapcf/01/03/otsc.aceh.chinoy/index.html

Comprehensive Crisis Management. (2008). *Human security in peacebuilding*. Kuopio, Finland: Crealab Oy.

Deloitte Consulting, L. L. P. (2004). *United Nations organizational integrity survey*. Retrieved from http://whistleblower.org/storage/documents/UN_Integrity_Survey.pdf

Disaster Accountability Project. (2010). *Report on the transparency of relief organizations responding to the 2010 Haiti earthquake*. Hartford, CT: Disaster Accountability Project.

Disasters Emergency Committee. (2001). *Independent evaluation: The DEC response to the earthquake in Gujarat (Vol. 1)*. London, UK: Disasters Emergency Committee.

Gonsalves, C. (2005, January 28). The deadly bureaucracy in the Andamans. *The Indian Express*, p. 8.

Harland, D., & Wahlstrom, M. (2001). *The role of OCHA in emergency United Nations operations following the earthquake in Gujarat, India - 26 January 2001*. Geneva, Switzerland: United Nations.

Immonen, A., Bali, R. K., Naguib, R. N. G., & Ilvonen, K. (2009). Towards a knowledge-based conceptual model for post-crisis public health scenarios. In *Proceedings of the IEEE International Conference on Humanoid, Nanotechnology, Information Technology, Communication and Control, Environment, and Management*, Manila, Philippines (pp. 185-189).

Inter-Agency Standing Committee. (2010). *Response to the humanitarian crisis in Haiti: Achievements, challenges, and lessons to be learned*. Geneva, Switzerland: Inter-Agency Standing Committee.

Jennex, M. E., & Raman, M. (2009). Knowledge management is support of crisis response. *International Journal of Information Systems for Crisis Response and Management*, *1*(3), 69–82. doi:10.4018/jiscrm.2009070104

Murphy, T. (2010). *Transparencygate!* Retrieved from http://www.huffingtonpost.com/tom-murphy/transparencygate_b_695382.html

Murphy, T., & Jennex, M. E. (2006). Knowledge management, emergency response, and Hurricane Katrina. *International Journal of Intelligent Control and Systems*, *11*(4), 199–208.

News, B. B. C. (2001). *UK offers help to quake victims.* Retrieved from http://news.bbc.co.uk/1/hi/uk/1139081.stm

News, B. B. C. (2006). *Tsunami disaster.* Retrieved from http://news.bbc.co.uk/1/hi/in_depth/world/2004/asia_quake_disaster/default.stm

News, B. B. C. (2010). *Haiti earthquake.* Retrieved from http://news.bbc.co.uk/1/hi/in_depth/americas/2010/haiti_earthquake/default.stm

Olara, S. (2009). *Critique: Who will police the United Nations?* Retrieved from http://www.blackstarnews.com/?c=135&a=5536

Paddock, R. C., & Magnier, R. C. (2004, December 30). Tsunami relief efforts mired in chaos. *Los Angeles Times.*

Pickrell, J. (2005). *Facts and figures: Asian tsunami disaster.* Retrieved from http://environment.newscientist.com/channel/earth/tsunami/dn9931-facts-and-figures-asian-tsunami-disaster-html

Raman, M., & Jennex, M. E. (2010). Knowledge management systems for emergency preparedness: The way forward. *Journal of Information Technology Case and Application Research*, *12*(3), 1–11.

Shah, A. (2002). *Relief, rehabilitation and development: The case of Gujarat.* Retrieved from http://www.jha.ac/articles/a097.pdf

Shaw, J., Mulligan, M., Nadarajah, Y., Mercer, D., & Ahmed, I. (2010). *Lessons from tsunami recovery in Sri Lanka and India.* Melbourne, Australia: Monash University.

Solomon, J., & Higgins, A. (2005, January 24). Indonesia reviews claims of Graft in tsunami relief. *Wall Street Journal*, p. A16.

Takeda, M., & Helms, M. (2006). Bureaucracy, meet catastrophe: Analysis of the tsunami disaster relief efforts and their implications for global emergency governance. *International Journal of Public Sector Management*, *19*(2), 631–656. doi:10.1108/09513550610650446

United Nations. (UN). (2005). *National post tsunami lessons learned and best practices workshop.* Retrieved from http://www.un.or.th/pdf/6months-govidn-idn-01jun.pdf

United Nations. (UN). (2010). *The United Nations at a glance.* Retrieved from http://www.un.org/en/aboutun/index.shtml

United Nations Development Program (UNDP). (1994). *Human development report.* New York, NY: Oxford University Press.

United Nations Development Program (UNDP). (2005). *The millennium development goals report 2005.* New York, NY: United Nations Development Program.

Wickramasinghe, N., & Bali, R. (2008). Controlling chaos through the application of smart technologies and intelligent techniques. *International Journal of Risk Assessment and Management*, *10*(1-2), 172–182. doi:10.1504/IJRAM.2008.021061

This work was previously published in the International Journal of Information Systems for Crisis Response and Management, Volume 3, Issue 3, edited by Murray E. Jennex and Bartel A. Van de Walle, pp. 16-35, copyright 2011 by IGI Publishing (an imprint of IGI Global).

Chapter 11
Supporting the Allocation of Traumatized Patients with a Decision Support System

Tim A. Majchrzak
University of Münster, Germany

Oliver Noack
University of Münster, Germany

Philipp Neuhaus
University Hospital Münster, Germany

Frank Ückert
University Hospital Münster, Germany

ABSTRACT

In this paper, the authors present a business rules-based decision support system for the allocation of traumatized patients. The assignment of patients to vehicles and hospitals is a task that requires detailed up-to-date information. At the same time, it has to be carried out quickly. The authors propose supporting medical staff with an IT system. The proposed system could be used in cases of mass incidents, as it is problematic, but essential, to provide all injured with adequate healthcare as fast as possible. The contribution is a system based on business rules, which is a novel approach in this context. Its feasibility is proven by prototypic implementation. In this paper, the authors describe the development project's background as well as the system's requirements and implementation details. The authors present an exemplary scenario to show the strengths of the proposed approach.

INTRODUCTION

The assignment of patients to hospitals is an every-day job for emergency call center staff. However, it is no task that is free of hassle. Emergency calls mark incidents that were not expected. In most cases, accidents or medical incidents that require fast action are reported. Offering medical aid to the affected patients is not only needed for their comfort. In case of e.g., a heart attack or especially after serious accidents, fast help can save lives.

Decisions are delayed by many determinants that have to be taken into account. In general,

DOI: 10.4018/978-1-4666-2788-8.ch011

several transportation possibilities are available if an incident is reported. Vehicles such as ambulances or helicopters are positioned at various places and have no uniform equipment; in fact, some vehicles are equipped for *special forms of emergencies* such as sick babies. The same applies to hospitals; not all hospitals are ready for all forms of injuries. Most of them also specialize on some forms of treatment such as infant treatment or neuro surgery. Therefore, transport decisions are highly non-trivial. A patient with a severe injury might require immediate transportation to a nearby emergency room instead of a specialized hospital in order to safe his live before detailed care is possible. The allocation of traumatized patients is a task that requires medically trained personnel to work quickly on complex decisions.

The allocation of patients to vehicles and hospitals is not decided in one step. It rather has to be decided continuously *over and over* again. This requires taking into account that vehicles could be occupied and that hospitals have limited capacities both with regard to short-term availability of staff and to unoccupied beds. Circumstances could even require rescheduling of vehicles. A special form of patient allocation is required in cases of *mass incidents*. A mass incident is an event in which due to disastrous circumstances a high number of people is wounded and thus require medical care. In general, such incidents are *local* and *do not span a long time*. Consequently, many patients require medical care at the same place almost immediately. An example from Europe is the *Ramstein airshow disaster* (Martin, 1990), in which inadequate handling was demonstrated that led to 70 dead and over 1000 wounded. Apparently, the transportation of patients was extremely uncoordinated and many of them received treatment later than it would have been required. More recent examples are the *train disaster of Eschede* (Oestern, Huels, Quirini, & Pohlemann, 2000) and the *Enschede fireworks disaster* (Woltering & Schneider, 2002; van Kamp et al., 2006).

We propose to support the allocation of traumatized patients using geographically enabled decision support systems. In collaboration with a regional network for the care of traumatized patients (Spitzer, Verst, Juhra, & Ückert, 2009; Traumanetzwerk-Nordwest, 2011) we developed a system for the allocation of patients. It incorporates a business rule management system, which allows it to suggest patient allocation while being versatile and expandable.

Our paper makes the following contributions. Firstly, it describes requirements for a patient dispatching system. Secondly, it explains how these requirements can be fulfilled with a business rules-based approach. Thirdly, it highlights how further capabilities such as using geographical data can enhance patient dispatching. And fourthly, it demonstrates the effectiveness of using business rules in the healthcare context by discussing a scenario.

This article is structured as follows. In the next section we give an overview of the project and related work. We sketch the methodology used. Then the decision support system is introduced. We illustrate exemplary usage and early evaluation results. Eventually, we draw a conclusion and discuss future work.

PROJECT BACKGROUND AND RELATED WORK

Emergency calls are answered locally in Germany (Pohl-Meuthen, Kochanda, & Kuschinsky, 1999). Larger cities and districts set up one or more *emergency call centers*. Incoming calls are answered personally. Vehicles are assigned and sent to the patient's location; also, the hospital is informed and preparations can be made. IT support is limited to information about available vehicles and similar data. Decisions are made as arrangements after phone, radio, and sometimes fax communication. Occasionally, an ambulance takes an emergency physician with it or joins him at the destination; not all ambulance stations are near to hospitals and ambulances often depart without taking physician with them. In this case, any further decisions are usually delayed until the patient's condition has been checked.

In cases of mass incidents patient allocation is decided on the disaster site. Injured are brought to a nearby place where *triage* is done (Kennedy, Aghababian, Gans, & Lewis, 1996). The basic medical condition of the patient is checked to determine most viable needs and prioritize transportation. Usually, a larger number of vehicles are sent to the incident's location. There is no immediate determination of actually needed vehicles. The order in which patients are transported is based on the triage groups they were assigned to. As the number of patients provided with medical care increases and the situation get clearer, organization is successively adapted to it. For example, more vehicles are ordered or drivers are informed that they are not needed at the incident site any more.

Both for the daily allocation routine and for mass incidents improvements by using decision support systems were considered. The Trauma-netzwerk Nordwest (TNNW) supports projects to improve the quality of service offered to trau-matized patients (Juhra, Ückert, Weber, Hentsch, Hartensuer, Vordemvenne, & Raschke, 2010). It consists of 41 hospitals, which allow develop-ing new solutions under realistic conditions and applying them in a complex environment. The prototypic implementation of a system for patient allocation was set up as a joint interdisciplinary project between the Department of Medical Infor-matics and Biomathematics and the Department of Information Systems of the University of Münster as well as the University Hospital Münster.

There is a plethora of work on systems for decision support in general. A general overview is e.g., drawn in the standard work of Turban, Aronson, Liang, and Sharda (2006). Decision sup-port for the medical sector is no new emergence, though. Newer titles focus on this topic directly (Berner, 2006; Greener, 2006). Most techniques

used for building decision support systems are well-understood; the same applies to Web 2.0 information systems (Vossen & Hagemann, 2007) and to the usage of geospatial mashups. Mashups combine external sources to create a new service. Despite the hype around mashups in the context of web applications, they are just a structured way of applying the *facade* design pattern designed by Gamma, Helm, Johnson, and Vlissides (1995) in their groundbreaking work on reusable software design.

Our work combines various techniques in the context of crisis management and healthcare. Therefore, we only discuss systems related to healthcare. A detailed discussion of similar tech-nical approaches is not only out of scope of this article but we explicitly do not want to exaggerate technical aspects besides motivating the usage of a rule-based system. Technology is a requirement to reach our goals rather than the purpose of our contribution.

Andrienko and Andrienko (2005) describe the OASIS project which is a system for "knowledge-based decision support" and "advanced methods for handling decision complexity". In contrast to our system it is not especially focused on patient dispatching. The research on this project is ongo-ing; Andrienko and Andrienko (2007) for example describe techniques for information visualization in the context of crisis management. There are numerous articles on vehicle location and route planning in general. For the medical sector there e.g., is a study of models and ways to locate ambu-lances (Brotcorne, Laporte, & Semet, 2003). Work on solving locating problems can be found that has been published over a decade ago (Gendreau, Laporte, & Semet, 1998). Newer work discusses patient transportation strategies (Kiechle, Doerner, Gendreau, & Hartl, 2009) but does not do so in the light of supporting transportation decisions by IT.

An active community is researching on geo-graphically supported tools for almost any pur-pose. Liu and Palen (2009) present a survey of 13 crisis related mashups using Web 2.0 techniques and geographical support. Cai, MacEachren, Sharma, Brewer Fuhrmann, and McNeese (2005) developed a system for the usage of geographical visualization for the collaborated handling of crisis situations. It has a much more general focus than our work. Cai (2005) discusses how to extend geo-information support systems to crisis management. Even though there are a lot other approaches to using geo-information support, there are none that can be directly related to our system.

METHODOLOGY

The project that led to the development of the prototype has a special character. Firstly, it is conducted between the contexts of emergency response, patient treatment in hospitals, research on medical informatics, and information systems development. A high level of interdisciplinarity offers the possibility to acquire new areas of research and to get very qualified feedback. However, it also asks for coordination and a justified waging of interests. Secondly, the project did not mean to have a technical focus. Technology in the context of healthcare is a means to an end. Besides, the basic technology is well understood. Nevertheless, the technology's application and especially its combination is novel and it has not yet been applied to healthcare information systems in the given form.

The research approach chosen is *design science* (Simon, 1996). We focus on practitioners' problems (March & Smith, 1995) and try to solve them "in unique or innovative ways" respectively "in more effective or efficient ways" (Hevner, March, Park, & Ram, 2004). Design-oriented approaches are especially useful if new information systems are to be developed using existing technology. Technology development often races ahead of its actual application. A challenge lies in adopting adequate techniques to solve problems.

Another benefit of design science is to actually focus on a system's design. For future users only its design and therefore its functionality and performance matter.

Our main course of work can be sketched as follows. The Department of Medical Informatics cooperates with various medical service providers and tries to identify research opportunities based on practitioners' needs. Based on the work of the network for the care of traumatized patients the idea was born to support decision making in emergency call centers. Existing systems and possibilities were identified and the idea was refined. The next step was to specify the system. Development was carried out jointly with Information System (IS) researchers who combine a healthcare-focused view with competence in IS development. The system was implemented while keeping in mind the particularities of the project. It was first evaluated by the researchers and then initially shown to medical emergency personnel. The yet uncompleted step is a close examination by healthcare professionals. Ideally, feedback would lead to the development of a system that will be used in emergency call centers as well as in new research projects regarding routine patient transportation.

THE DECISION SUPPORT SYSTEM

In the following subsections development of our system is described. Following common practices from Software Engineering (SE), the process has been partitioned into requirements analysis, design, and implementation. Even though this presentation resembles the *waterfall model*, this sequential presentation has been chosen for reasons of clarity. Development was conducted in an *agile* manner. While iteratively refining and evaluating our prototype, we adjusted requirements accordingly.

Requirement Analysis

Our approach is different to existing systems. The prototype is not only supposed to sophistically support the allocation of patients but also to include as much information into a decision as possible. Thus, analyzing requirements for this novel tool is critical.

The basic idea is to have a system that can be used to assign patients to transportation possibilities and hospitals. Therefore, it has to be possible to manage vehicles such as ambulances or helicopters and hospitals. For each object managed by the system, typical characteristics that affect the allocation process have to be stored. For vehicles this could be the *home position*, *type of movement* (such as road usage or flying), and details about *equipment*. There also can be characteristics specific to a *group of possibilities*. Ambulances might be suited for off-road driving or for special forms of injuries. Hospitals are typically characterized by their *location*, *capacities* for various forms of treatment, and *specialization* on distinct forms of healthcare. Whereas this data should be manageable from within the system, it has a *static character*. While it might be changed in the course of time, it is static when a transportation decision has to be made.

To accommodate the dynamic nature of emergencies, the system should offer a real-time connection to both transportation and hospital information systems. This includes automatic checking whether vehicles are available and what their current position is. It is needless to say that unavailable vehicles do not need to be considered as transportation possibilities for a current incident. However, availability has to be checked carefully. Only vehicles that are broken or that are far out of reach are ultimately unavailable; occupied vehicles might be rerouted in cases of severe emergencies.

Status and position information are very helpful. Consider an occupied helicopter that almost reached the destination hospital. Once it finished the current assignment and departs from the hospital, it could immediately be sent to the next patient. If the patient is at a remote place and has a serious medical condition, this could be preferable over ambulance transportation. Interfaces to hospital information systems could ensure that capacities are checked and information about patients is exchanged. Even if this is not available, ambulance drivers would inform the call center about the patient's status. In serious cases preparation at the hospital can be made very early. Consequently, arriving patients can be given the best possible care.

The information base of the system should be used to support the decisions of emergency call center staff. If an incident is reported, it should be entered into the system including incident details such as the patient's location, general information about her such as age or weight as well as her medical condition. Of course, not all of this might be available imminently. The system should suggest a transportation possibility including a vehicle and a destination hospital based on the currently available data. It should also highlight alternatives. If adequate data is available, differences to the suggested possibility – e.g., longer transportation time but a better suited destination hospital – should be shown. Suggestions should be similar to those medical expert would make.

Ideally, communication with vehicles should take place wirelessly. Thereby location, assignment, and probably routing information can be submitted to ambulance drivers immediately. Tools for that purpose are in operation already (Liao & Hu, 2002). Therefore, the system should provide support for using them.

A major requirement is geographical support. From considerations it is clear that patient allocation is based on geographical particularities. Choosing a vehicle depends on the patient's location, the vehicle's location, and the location of hospitals suited for the patient. Routing based on the vehicles abilities is required. Moreover, geographical support should also be available to the dispatching personnel. Showing the locations on a map helps humans to judge about transportation possibilities as well as the overall situation especially in occurrence of a mass incident.

Besides these general requirements a number of further characteristics are demanded:

- The system needs an intuitive user interface (UI). Personnel have to be enabled to work with it very quickly.
- Adaptability and flexibility are major concerns. The system must not be static but allow upcoming changes. It also has to be flexible so it can be adapted to a variety of different situations. This includes the possibility to change the reasoning behind the decision support. Decisions also have to be revisable at any time.
- The system has to respond timely to make it suitable for decisions in real-time. Allocation suggestions should be calculated in not more than a few seconds. An increased survival rate has been observed if transportation of injured is fast (Biewener, Aschenbrenner, Rammelt, Grass, & Zwipp, 2004); fast decisions contribute to an accelerated transportation process.
- It has to be robust. If single components fail, the system as a whole must not fail. Adjustable components that can be switched immediately are preferable. An example is the availability of several route planners with a differing level of quality.
- The system has to be able to make suggestions even in problematic cases. If a severely injured person is reported but no hospital that provides the desired level of care can be reached in time, the patient should be assigned to a nearby hospital and staff warned about the non-fulfillment of a requirement. Even though optimal healthcare cannot be guaranteed in any case, the system should be prepared for a large number of limitations that it has to find acceptable temporary solutions for.
- Future version of the system should be able to dynamically process additional data and make suggestions about transportation assignments that could be adjusted. This is no requirement for the first prototype, though.
- It should not be possible to use the system without authorization such as authentication of the user.

The system should be able to both allocate single patients and to support the management of mass incidents. There is no change to the basic requirements for either case. However, the system's user interface has to be suitable for displaying more complex allocation suggestions. It would otherwise be impossible to keep an overview. It has to be especially noted that the system has to respond quickly even if calculating plans for a higher two-digit number of patients. Suggestions have to be as precise as they are for single allocation. At the same time, they should reflect reasonable trade-offs. In a case in which two injured require fast transportation but one patient is in danger of life, the first suggestion could be to assign a helicopter to him. However, if an ambulance can reach him much quicker, he might be transported by it while the helicopter is assigned to the less injured person. Particularly if suggestions seem counterintuitive, visualization of the suggestions should help personnel in judging them.

A final requirement and a maxim behind our idea of IS for healthcare is to *support* personnel rather than to *relieve* them from decision making. No human can compete with a computer in mathematical terms or process redundant tasks without becoming exhausted. It is a good idea to relieve personnel from time-consuming tasks that can be automated. However, critical decisions that could be harmful to humans should *at least* be checked by human experts. Therefore, an emergency plan has to be drawn up for situations where the system fails. Even with high robustness, redundancy, and self-repairing functionality dispatching of patients' needs to be possible in any case. Failures of system parts would cause trouble for emergency call center staff but must never endanger patients' lives.

Specification and Design Decisions

There are no particular functional needs of the user interface. Taking future developments into consideration, extending the system to mobile devices is an option. Therefore, we decided to implement it Web-based. It theoretically is accessible by any device that has Internet access but still can be used locally. The robustness of the authentication mechanism and its security has to be verified before deploying it outside of closed intranets. The benefit of enabling physicians to view and change transportation options at the patient's location would justify additional security effort. Security in unsecure environments is out of scope of the current work, though.

We decided to use a system based on a *rule engine*. Business rules systems and the underlying Business Rule Management are used in corporate contexts for a long time and have proven to be feasible (Ross, 2003; Turban et al., 2006). The basic idea behind business rules is to model workflows i.e., sequences of actions using simple rules. Most of these rules obey either a if condition than action or a if condition than action else alternative action scheme. Rules are easy to understand and easy to manage. Systems of rules, so called *rule sets*, can be used to automate complex decisions. They are executed in a rule engine that automates *inference* and helps to achieve good performance. Additionally, rule engines offer support in coding rules, consistence checks and so on.

Medical personnel demanded a very flexible and adaptable system. Hard-coding the allocation process would have made a programmer intervention mandatory for even the slightest changes. By using a rule engine instead, any decision processes can be described as rules. It even is possible to

support the creation of rules by e.g., offering drop-down fields of possible actions and conditions. Rules should ideally be maintainable by domain-experts with limited mathematical and rule-writing knowledge (Grzenda & Niemczak, 2004).

With the rule system as much information can be used as possible. For example only helicopter or off-road ambulances should be sent to places of incidents that cannot be reached via roads of sufficient quality. At the same time, non-mandatory rules that are out of interest due to a lack of information will simply be neglected by the rule engine.

A rule system also is flexible enough to handle particularities and special needs. To give an example: special care is required for infants (Leslie & Stephenson, 1998). However, the number of patients that is, say, under 2 years old is relatively small. Only a small number of ambulances are hence equipped for baby rescue. This can be reflected by the rule system. If the reported patient is less than 2 years old and a nearby ambulance with baby equipment is available, it would be assigned. In cases were other ambulances without baby equipment could reach the location significantly faster, the rule system could select such an alternative while also showing the slower but better equipped ambulance. If the age of the reported patient is greater than 2, the system can be set up to recommend ambulances without baby equipment as long as availability or location do not require such an ambulance to be sent. Taking into account detailed information thereby not only helps to better aid current patients but also supports to be suited for emergencies reported in the near future.

The rule system can be used for further refinements. One possibility is to align it to duty rosters. Consider that a physician specialized in infant healthcare is on duty. Assigning him to an ambulance would enable it to be used for baby rescue. The rules have to be specified in order to reflect the complex and dynamic nature of healthcare. Rule sets have to enable suggestions that take as much information into account as possible to offer adequate patient care.

We decided against a mathematical allocation approach. An implementation only relying on linear optimization did not convince us; besides being limited in functionality, it seemed to be inhuman and thus less appropriate. Depending on the modeling, the system would e.g., prefer accelerating the transport of 100 slightly injured to saving one live – this could happen since it hardly is possible to embed *ethics* into an optimization approach. Furthermore, changing the system's behavior is more laborious. It also is harder to understand than simple rules.

The decision for a Web-based system leads to a number of further technical decisions with regard to what programming language to use, which third-party software to embed and so on. However, these decisions are rather unspectacular from a technical standpoint and will not be further discussed in this paper. We encourage having a look at rule-based systems for additional areas of application as they allow technical details to be hidden from users.

One notable design decision is the specification of a role system. The prototype is used by various people such as call center personnel or emergency physicians. A role system enables setting up distinct roles and aligning them with the operational permissions. Call center personnel could e.g., get the right to make transportation decisions. Entering new hospitals could be limited to a supervisor and altering the rule set could be allowed for experts only.

Implementation

The system is implemented as a 4-tier web application following the *Model View Controller* design pattern. It is backed by a database system, has distinct tiers for its logic (i.e., the computation) and the generation of output, and is generating pages that can be displayed on Web browsers. The separation into tiers offers several amenities. For example, other means of representation such as a *rich client* can be added easily. The business logic can be accessed via a *Web service* which surpasses the presentation tier. Therefore, other systems can directly communicate with ours.

As we wanted to keep the system flexible, we put special attention on its modularization. Instead of having hard-coded interfaces to third-party systems, we use *wrappers*. The cartographic service is not accessed directly by the distinct components of the system, but a module combines access to it. It has a single interface it accesses and it receives well-formatted data from it. The wrapper module can be amended to access a variety of services. It could also be extended to access services based on priorities (e.g., if the local service does not provide adequate data, a Web-based service is polled) or to send parallel requests to several services. Modularization helps to keep the system flexible. Another example is the retrieving of location information of vehicles. Instead of having fixed interfaces, the system allows modules to be installed for each type of vehicle. These modules wrap the location requests and for example forward them to a secured Web service. This is e.g., used to get helicopters' statuses. Any service can be replaced easily. Since the rules can be replaced as well, the system is adaptable to any changed circumstances or endeavors.

Providing our system's logic tier on an *application server* offers great flexibility (see Figure 1). It is possible to create a mobile version of our system that can be installed and used by a mobile command office near the disaster area. Actual information about helicopters, hospitals and rescue vehicles are used if an uplink is available. Regardless of the uplink status the system will be accessible by the rescue operation dispatchers at the accident site. Furthermore, rescue workers are able to get a compact view on the disposition status by using smartphones or netbooks connected to a local network.

Instead of giving technological details of the rule system, we describe its exemplary usage in

Figure 1. Architecture of the system

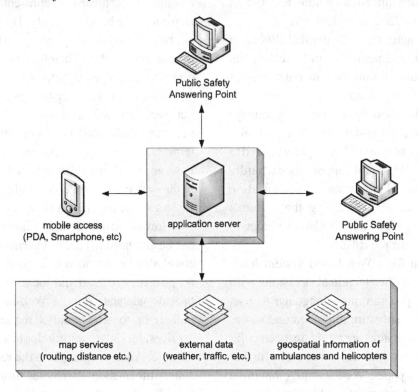

the next section. In general, the system is initialized and gets an input condition. It then determines rules meeting the input conditions. Those rules are fired i.e., executed. The results are saved and inference is continued until any rules that could fire have been fired. The sequence of simple rules yields complex results. It is not uncommon to have a four-digit number of rules executed before a result is gained.

Input conditions are information available for an allocation decision such as the position of patients, their condition, and their age. *Meeting conditions* means that a condition in the if-part of a rule is present. *Firing* a rule is the consequence of meeting a condition; the corresponding *action* is taken. Actions may change conditions leading to additional rules being fired. Eventually, there are no rules that could fire since all conditions have been met. Results can be derived from the actions taken from system initialization until inference completion.

Compiling the initial rule set was done in close consultation with domain experts. In general, we followed the requirements and made assumptions about how decisions have to be reflected by rules. We then checked back with experts to verify these assumptions. After testing a rule set in a fictional scenario, it was refined until results were satisfying. The verification goal was to create rules in such a way that dispatching suggestions resembled decisions that human experts would make after carefully studying the available data – while providing these suggestions dramatically faster than humans could.

The chosen process was feasible since our prototype should show the producibility of the approach. Due to the fact that rules can be modified and replaced easily, finding a rule set for using the system in the actual dispatching of patients is by no means a technical challenge. Nevertheless, it will require a structured, quality-assessed process and careful evaluation in consultation with experts from the medical domain.

The actual rule execution is not done naively. The execution of rules *can* be done in a very simple manner, which is sufficient for demonstration. However, optimizations ought to be applied to speed up execution. Algorithms based on the *Rete Algorithm* (Forgy, 1982) are used in our prototype. These algorithms increase inference speed dramatically. They are embedded in *JBoss Drools*, the rule engine employed by our prototype. Thereby, results are equal to those naive execution would yield but they are gained much faster.

Additionally, *Drools Guvnor* is used. It offers to manage rules and to access them via a Web-based console, which helps to easily alter them. Guvnor not only checks syntactic correctness of rules but can be used to run exemplary scenarios.

With regard to mass incidents the rule system allows setting up special rule sets. This does not only include rule sets to handle mass incidents in general. Larger events are planned in cooperation with police, medics, and crisis managers. In addition to conventional emergency plans special rule set can be defined that are tailored to the events (e.g., with regard to geographical particularities) and actively support crisis management in cases of actual incidents.

We not only implemented the rule-based system but combined it with a *value benefit analysis*.

Rules alone do not lead to fixed suggestions in any case. For example, a patient with minor injuries would still benefit from being cured in a hospital that is highly specialized even though a general care would be sufficient. If several transportation options are available – e.g., one to a better suited hospital and one to a standard one – and transportation times differ, the system will not be indifferent between options. The actual calculation is done by assigning *weights* (i.e., value benefits) to the components a decision is made of. The weighting can be adjusted. Initially, we set a high impact on transportation time. Routinely using our system might lead to readjusting weights.

The Web-based interface allows using so called *mashups*. Instead of implementing commonly used functionality, it can simply be integrated. Figure 2 shows the position of a rescue helicopter. There is no sense in implementing a geographical system from scratch; high quality general purpose solutions already exist. Thus, a system for geographical visualization is included and the required data is passed to it. The underlying components of the Mashups could both be accessed via the Internet or from a local server.

Figure 2. Map view of a rescue helicopter at its base

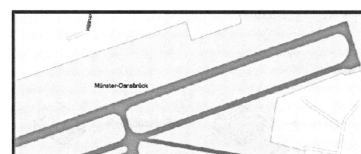
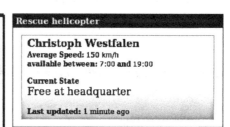

EXEMPLARY USAGE AND EVALUTION

The standard workflow the prototype is based on is illustrated in Figure 3. The incident is registered by the executive emergency physician with a mobile device or over a 2-way-radio system and is recognized by the corresponding emergency call center. After registration of the necessary data, the predefined rules in the system are evaluated with regard to the current data situation by the integrated inference engine. Every transportation possibility will be selected and a benefit value is assigned to every possibility. The best one is allocated to the considered patient. If there is more than one patient at a time, an optimization mechanism is executed to improve the result.

Of course, the above description is simplified. The granularity and the quality of the decision depend on the predefined rules in the system. Rules are arranged in a so called *ruleflow*, which is a similar construct to a workflow. Rules can be part of *rule groups* and the order in which the different groups are evaluated can be arranged. The evaluation of the different possibilities is also mostly done with rule sets. For example, the presence of specific equipment in the vehicle or in the hospital can be put into relationship to specific situations with the help of rules. Consider again a baby to be involved in an incident: a specific rule can be defined so that vehicles that provide corresponding equipment to medicate babies are identified as advantageous for this situation. The benefit system includes that unnecessary equipment strikes negative in the benefit. This is done since another vehicle that does not have the equipment and that is reachable in a similar radius should be preferred. Consequently, no equipment is sent out unused and special equipment will be available for a future emergency (in which it might be needed). A very simple rule which describes this situation can be seen in Figure 4. In this screenshot the editor Guvnor is used to edit the rule set. It is not obligatory to use this editor but it already has a high degree of maturity and a broad list of features (Browne, 2009). The system itself can also be accessed via a Web-based interface. The proposed decision can be retrieved in the interface as well.

The rule inference is based on actual data. Current positions of vehicles can be determined with the help of a *Global Navigation Satellite System* (GNSS) device. An example project for communicating positions of emergency helicopters is Rescue Track (DRF Lutrettung, 2009). With this infrastructure the system is able to calculate realistic durations of the transportation and take status information like a helicopter's occupancy into account. By employing systems like Rescue Track, current surrounding conditions like traffic and weather situations can be included in the decision to get a more detailed calculation of driving and transportation times of the different alternatives.

Figure 3. Illustration of the standard workflow

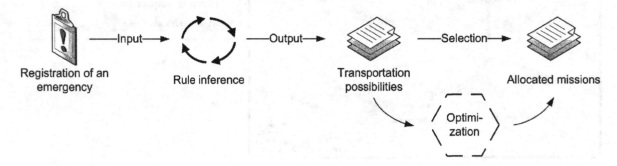

Figure 4. Usage of the rule editor Guvnor to edit the rule sets

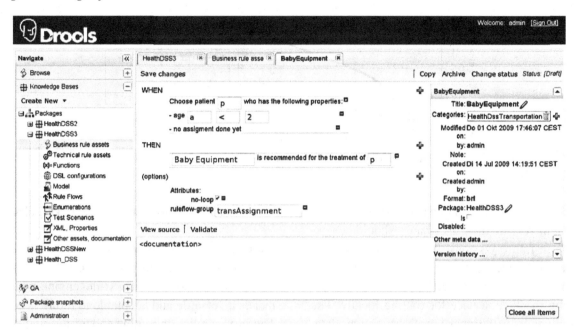

After assigning a patient to a destination and a vehicle, results can be analyzed. The rules that were fired can be retraced. Thereby, consecutive examinations are possible and the rule set can be optimized to better suit future incidents. This enables a more sophisticated consideration of the selection of ambulances and hospitals. With an elaborated rule set the system can help to use resources more efficiently. It is possible to define different rule sets for different types of emergency incidents. It is even possible to create temporary rule sets for special occurrences e.g., to be prepared for situations that are constructed before. In many cases the emergency medical services take a planning period to prepare for emergency situations at events with many participants (such as music festivals or funfairs). The described system has the advantage of being very flexible; it is possible to create new rule sets to enable it to react to specific situations within a short period of time without changing the program itself.

As exaggerated earlier, a system for patient allocation has to respond quickly and offer an acceptable performance. We thus evaluated two alternative implementations on a computer with a 1.6 GHz Intel Core2Duo processor and 1 GB of DDR2 SDRAM. Even though the system would be deployed to more powerful servers for its actual application, it should perform well on relatively weak hardware. Results are shown in Figure 5. Execution time of an implementation using a rule set only and an implementation with some rules replaced by hard-coded calculations are compared. Example calculations have been run for up to 7 000 transportation possibilities. In general, the number of possible options should be far lower in actual scenarios. It can be observed that the hard-coded implementation performs better and execution time only increases slightly with the number of transportation possibilities. However, even with the implementation that is based on rules only, execution time merely increases linearly with the number of transportation possibilities. A calculation time of far less than 1.5 seconds for 7 000 possibilities is acceptable. There-fore, we prefer the rule-bases approach since it offers greater flexibility with a performance trade-off that can be neglected.

Figure 5. Performance evaluation results

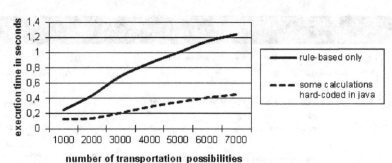

Since the current implementation is a prototype, increasing the performance should be possible in future versions. In rule systems not all rules are depended; hence, parallel execution could be realized. A major speedup is possible by offering the Web service for routing locally. The same applies to the inclusion of cartographic data. The system makes extensive usage of geographical information and displays maps to the users. Generating them locally hence offers a speedup. Several further actions such as intelligent caching are possibly but out of scope of this paper. The important conclusion is that the systems performance is satisfying

Even though our implementation has the character of a proof of concept, early results are promising and we deem its real-world application feasible. We therefore evaluated it in cooperation with medical personnel. While changes to parts of the system are inevitable, the general idea was received positively. It was argued that for the actual implementation the rule set has to be refined. Moreover, connection to third-party systems must be set up redundantly; it has to be ensured that the system will not fail even in conditions were data is only available partly or incorporated systems such as *positioning* fail. While a lot of future work has to be done (cf. with the next section) it is likely to see decision support systems that are capable of handling both routine tasks and particularities such as mass incidents enter the emergency call centers.

CONCLUSION AND FUTURE WORK

We presented a system for the support of the allocation of traumatized patients to vehicles and hospitals. It utilizes a business rules engine, which makes it versatile and adaptable to future needs. The system relies on routing and positioning services and provides personnel with cartographic views. It can both be used to support routine decisions on the dispatching of single patients and to enable quick action in case of mass incidents.

We have shown the background of the project it was developed in and discussed differences to related systems. There are many other systems that use mashups for applications in healthcare but we did not identify projects with the same goals that we had. After explaining our design science research method, we discussed requirements, the specification, and implementation details of the system. It particularly had to be flexible, stable, and adaptable. Out prototype is versatile and its underlying logic can be altered without programming since it is based on business rules. We eventually showed an exemplary scenario and gave an idea of the early experiences medical personnel gained. The general perception was positive.

The decision support system we presented is a prototype and meant to be a proof-of-concept. However, early results are promising and the system offers actual benefits to call center personnel – and, eventually, to traumatized patients. Therefore, we encourage the usage of similar systems in general. Emergency call centers should be equipped with systems that use real-time data on vehicle availability and position as well as hospital capacity information. This information should be amended with detailed data on vehicle equipment and hospital specializations. We also suggest offering decision support both for normal dispatching and especially for mass incident handling. Implementing a system based on a rule engine has proven to be a feasible approach. It offers the flexibility to adapt to changed conditions. Approaches based on business rules could also be implemented for other contexts within the field of healthcare.

While the prototypic implementation is finished, we will continue our research. For example, the rules can be refined. More real-world scenarios have to be tested. It also has to be investigated on how to effectively and efficiently support medical staff. The decision support system should seamlessly integrate into the emergency process. It could also help to improve communication between medical field staff, call center personnel and hospitals. The system should furthermore automate any medical decisions that are simple enough to relieve humans of them; it should provide any information needed by medical personnel to make the remaining ones. And it should make sure decision processes stay transparent and can be checked – and revised – by humans at any time.

ACKNOWLEDGMENT

This article greatly extends the short papers by Majchrzak et al. (2010) and Neuhaus et al. (2010).

REFERENCES

Andrienko, N., & Andrienko, G. (2005). A concept of an intelligent decision support for crisis management in the OASIS project. In van Oosterom, P., Zlatanova, S., & Fendel, E. M. (Eds.), *Geo-information for disaster management* (pp. 669–682). Berlin, Germany: Springer-Verlag. doi:10.1007/3-540-27468-5_48

Andrienko, N., & Andrienko, G. (2007). Intelligent visualisation and information presentation for civil crisis management. *Transactions in GIS, 11*(6), 889–909. doi:10.1111/j.1467-9671.2007.01078.x

Berner, E. S. (2006). *Clinical decision support systems: Theory and practice*. Berlin, Germany: Springer-Verlag.

Biewener, A., Aschenbrenner, U., Rammelt, S., Grass, R., & Zwipp, H. (2004). Impact of helicopter transport and hospital level on mortality of polytrauma patients. *The Journal of Trauma Injury Infection and Critical Care, 56*(1), 94–98. doi:10.1097/01.TA.0000061883.92194.50

Brotcorne, L., Laporte, G., & Semet, F. (2003). Ambulance location and relocation models. *European Journal of Operational Research, 147*(3), 451–463. doi:10.1016/S0377-2217(02)00364-8

Browne, P. (2009). *Jboss drools business rules*. Birmingham, AL: Packt Publishing.

Cai, G. (2005). Extending distributed GIS to support geo-collaborative crisis management. *Geographic Information Sciences, 11*(1).

Cai, G., MacEachren, A. M., Sharma, R., Brewer, I., Fuhrmann, S., & McNeese, M. (2005). Enabling geocollaborative crisis management through advanced geoinformation technologies. In *Proceedings of the National Conference on Digital Government Research* (pp. 227-228).

Forgy, C. (1982). Rete: A fast algorithm for the many pattern/many object pattern match problem. *Artificial Intelligence, 19*, 17–37. doi:10.1016/0004-3702(82)90020-0

Gamma, E., Helm, R., Johnson, R., & Vlissides, J. (1995). *Design patterns: Elements of reusable object-oriented software*. Reading, MA: Addison-Wesley.

Gendreau, M., Laporte, G., & Semet, F. (1997). Solving an ambulance location model by tabu search. *Location Science, 5*(2), 75–88. doi:10.1016/S0966-8349(97)00015-6

Greenes, R. A. (2006). *Clinical decision support: The road ahead*. Orlando, FL: Academic Press.

Grzenda, M., & Niemczak, M. (2004). Requirements and solutions for web-based expert system. In L. Rutkowski, J. H. Siekmann, R. Tadeusiewicz, & L. A. Zadeh (Eds.), *Proceedings of the 7th International Conference on Artificial Intelligence and Soft Computing* (LNCS 3070, pp. 866-871).

Hevner, A. R., March, S. T., Park, J., & Ram, S. (2004). Design science in information systems research. *Management Information Systems Quarterly, 28*(1).

Juhra, C., Ückert, F., Weber, T., Hentsch, S., Hartensuer, R., Vordemvenne, T., & Raschke, M. J. (2010). Improving communication in acute trauma. *ElectronicHealthcare, 8*(3), 3–8.

Kennedy, K., Aghababian, R., Gans, L., & Lewis, C. (1996). Triage: Techniques and applications in decisionmaking. *Annals of Emergency Medicine, 28*(2), 136–144. doi:10.1016/S0196-0644(96)70053-7

Kiechle, G., Doerner, K. F., Gendreau, M., & Hartl, R. F. (2009). Waiting strategies for regular and emergency patient transportation. *Operations Research Proceedings, 2008*, 271-276.

Leslie, A. J., & Stephenson, T. J. (1998). Transporting sick newborn babies. *Current Paediatrics, 8*(2), 98–102. doi:10.1016/S0957-5839(98)80127-9

Liao, T., & Hu, T. (2002). A CORBA-based GIS-T for ambulance assignment. In *Proceedings of the IEEE International Conference on Application-Specific Systems, Architectures, and Processors* (pp. 371-380).

Liu, S. B., & Palen, L. (2009). Spatiotemporal mashups: A survey of current tools to inform next generation crisis support. In *Proceedings of the 6th International Information Systems for Crisis Response and Management Conference*, Gothenburg, Sweden.

Lutrettung, D. R. F. (2011). *Besetzung der hubschrauber und ambulanzflugzeuge*. Retrieved from http://www.drf-luftrettung.de/rescuetrack.html

Majchrzak, T. A., Noack, O., Kuchen, H., Neuhaus, P., & Ückert, F. (2010). Towards a decision support system for the allocation of traumatized patients. In *Proceedings of the 7th International Conference on Information Systems for Crisis Response and Management*.

March, S. T., & Smith, G. F. (1995). Design and natural science research on information technology. *Decision Support Systems, 15*(4), 251–266. doi:10.1016/0167-9236(94)00041-2

Martin, T. E. (1990). The Ramstein airshow disaster. *Journal of the Royal Army Medical Corps, 136*(1), 19–26.

Neuhaus, P., Noack, O., Majchrzak, T., & Ückert, F. (2010). Using a business rule management system to improve disposition of traumatized patients. *Studies in Health Technology and Informatics, 160*(1), 759–763.

Oestern, H. J., Huels, B., Quirini, W., & Pohlemann, T. (2000). Facts about the disaster at Eschede. *Journal of Orthopaedic Trauma, 14*(4), 287–290. doi:10.1097/00005131-200005000-00011

Pohl-Meuthen, U., Koch, B., & Kuschinsky, B. (1999). Rettungsdienst in der Europäischen Union. *Notfall & Rettungsmedizin, 2*(7), 442–450. doi:10.1007/s100490050175

Ross, R. G. (2003). *Principles of the business rule approach*. Reading, MA: Addison-Wesley.

Simon, H. A. (1996). *The sciences of the artificial* (3rd ed.). Cambridge, MA: MIT Press.

Spitzer, M., Verst, H., Juhra, C., & Ückert, F. (2009). Trauma Network North-West - Improving holistic care for trauma patients by means of internet and mobile technologies. In *Proceedings of the European Congress for Medical Informatics* (pp. 371-375).

Traumanetzwerk-Nordwest. (2011). *Startseite*. Retrieved from http://www.traumanetzwerk-nordwest.de/

Turban, E., Aronson, J. E., Liang, T., & Sharda, R. (2006). *Decision support and business intelligence systems* (8th ed.). Upper Saddle River, NJ: Prentice Hall.

van Kamp, I., van der Velden, P. G., Stellato, R. K., Roorda, J., van Loon, J., & Kleber, R. J. (2006). Physical and mental health shortly after a disaster: First results from the Enschede firework disaster study. *European Journal of Public Health, 16*(3), 252–258. doi:10.1093/eurpub/cki188

Vossen, G., & Hagemann, S. (2007). *Unleashing Web 2.0: From concepts to creativity*. San Francisco, CA: Morgan Kaufmann.

Woltering, H. P., & Schneider, B. M. (2002). Das Unglück von Enschede am 13.05.2000. *Der Unfallchirurg, 105*(11), 961–967. doi:10.1007/s00113-002-0526-0

This work was previously published in the International Journal of Information Systems for Crisis Response and Management, Volume 3, Issue 3, edited by Murray E. Jennex and Bartel A. Van de Walle, pp. 36-51, copyright 2011 by IGI Publishing (an imprint of IGI Global).

Chapter 12
Visualizing Composite Knowledge in Emergency Responses Using Spatial Hypertext

José H. Canós
Universitat Politècnica de València, Spain

M. Carmen Penadés
Universitat Politècnica de València, Spain

Carlos Solís
University of Limerick, Ireland

Marcos R. S. Borges
Federal University of Rio de Janeiro, Brazil

Adriana S. Vivacqua
Federal University of Rio de Janeiro, Brazil

Manuel Llavador
Universitat Politècnica de València, Spain

ABSTRACT

Having the right information at the right time is crucial to make decisions during emergency response. To fulfill this requirement, emergency management systems must provide emergency managers with knowledge management and visualization tools. The goal is twofold: on one hand, to organize knowledge coming from different sources, mainly the emergency response plans (the formal knowledge) and the information extracted from the emergency development (the contextual knowledge), and on the other hand, to enable effective access to information. Formal and contextual knowledge sets are mostly disjoint; however, there are cases in which a formal knowledge piece may be updated with some contextual information, constituting composite knowledge. In this paper, the authors extend a knowledge framework with the notion of composite knowledge, and use spatial hypertext to visualize this type of knowledge. The authors illustrate the proposal with a case study on accessing to information during an emergency response in an underground transportation system.

DOI: 10.4018/978-1-4666-2788-8.ch012

INTRODUCTION

Emergency response is among the most critical activities performed by humans: a process where decisions affecting lives and properties must be made in short time. These decisions must be made from information coming from different sources, which must be accessed and combined adequately to avoid both information lacks and overloads. Moreover, different actors may require different information elements or, at least, different views of them. For instance, decision makers at a control room may have more sophisticated means to access to the information than the responders working at the emergency location, carrying mobile devices with reduced graphical capacity. Thus, information management is becoming a key aspect of modern emergency management systems, as illustrated by recent cases described in Jennex and Raman (2009) and Murphy and Jennex (2006).

The emergency response plan is a document that includes the procedures to be activated in response to any type of incident, plus all the information required to make decisions (such as maps, pictures, videos, etc.). The advance of information technologies has enabled the development of rich-content emergency response plans, going beyond the classical printed documents to become sophisticated hypermedia structures integrating text and multimedia content to provide decision makers and responders with the most accurate information. Typical cases of rich information are the use of Geographical Information Systems to calculate the optimal road to an emergency location, or video recordings of the different sections of a subway tunnel. Nevertheless, a valuable part of the information required to solve emergencies cannot be available in advance, as it must be gathered from the emergency location; this information is known as the context of the emergency, and may be very relevant for decision making. For instance, routes calculated by route planners may

be unusable in case of avalanches or earthquakes or even traffic jams, so they should not be used by response teams.

In general, contextual information is complementary to the context-independent one; that is, there are parts of emergency response plans considered contextual, and the plans include the necessary actions to gather such information and make it available to decision makers. In some cases, context overrides non-contextual information previously available. For instance, if a road is closed, some request should be sent to the route planner to recalculate and find a clear way to the place. In other cases, overriding is not recommendable, as previously recorded information can still be valuable. If a tunnel has collapsed, the video of the tunnel should not be shown as an optional escape way, but could be still available to look for valuable information such as possible obstacles, or just to analyze properties of tunnels similar to the damaged one for which there is no video recording available.

Managing contextual information poses several challenges, from its capture to its visualization. Specially challenging is how to have access to both contextual and non-contextual information related to the same object (e.g., the road and the tunnel in the aforementioned cases). Having the appropriate mechanisms is important, as wrong information can lead to wrong decisions. In this paper, we tackle the problem of combining contextual and non-contextual information to make it available to decision makers in emergency responses.

The main contribution of the paper is the definition of a framework for knowledge representation and visualization. The framework builds on previous work on knowledge management and hypermedia engineering to provide a full-lifecycle solution for knowledge organization and visualization. Starting from a context-enabled knowledge model, that includes the so-called composite knowledge, we create emergency response plans whose components can be labeled with the different types of knowledge (contextual, non-contextual or formal, and composite). This is important since different types of components require different visualization mechanisms. To build actual plans, a transition from knowledge models to interface models must be performed. To achieve the transition, we define transformation rules to generate hypermedia navigational structures that, being technology-independent, provide structure and navigation to the information space of an emergency response plans. Finally, specific plan realizations can be made using different hypermedia languages. We have selected ShyWiki (Solis & Ali, 2008), a spatial-hypertext Wiki system which enables the superposition of different types of knowledge in terms of spatial properties such as position, color, etc. In this way, all types of information can be defined and used collaboratively following the Wiki paradigm.

Our proposal provides a methodological guide to obtain emergency plans from abstract knowledge models, which is missing in previous works on knowledge management for emergencies. We illustrate the use of our approach with the case of a subway transportation system; our experience in the development of a hypermedia emergency response plan revealed the need to cope with composite knowledge. We show how ShyWiki allows the superposition of different types of information to enrich the expressiveness of previous plans.

The paper is structured as follows. In the next section, we define a knowledge management model which extends the one presented in Diniz et al. (2008) with the notion of composite knowledge. Then we show how composite knowledge can be included in emergency response plans via the use of stereotypes in the definition of an emergency response plan model. Later, we show how to turn emergency response plans into navigational structures, to conclude with the generation of spatial hypertext plans using ShyWiki. A discussion of our approach and a mention to further work concludes the paper.

KNOWLEDGE MANAGEMENT IN EMERGENCY RESPONSE

The emergency response phase starts when a dangerous situation needing immediate action occurs. Response teams, composed of well-trained members who may belong to more than one organization - for instance, firefighters and policemen – perform diverse activities oriented to mitigate the effect of the emergency on people and property. The diversity of actors makes decisions mostly collaborative, as the different organizations involved must communicate with each other, creating a large body of shared knowledge and using it to make most decisions during the response process. From the knowledge managed during the emergency response, decisions are made that result in actions to mitigate the effect of incidents. Managing and accessing to such knowledge is not an easy task due to several reasons:

- The sources of knowledge are in general heterogeneous and distributed, which require the implementation of some interoperability mechanisms as part of the emergency management systems.
- The sources may be static or dynamic, depending on whether the information they provide is stable or may change during the development of the response.
- The sources may be explicit or implicit; by explicit we mean that the information source is identified (e.g., the emergency response plan) and access mechanisms provided; conversely, an implicit source is such that it is not known in advance, and hence must be dealt with immediately after its discovery. This is often the case of natural disasters.

Knowledge can be available in different forms, and be of different nature, as pointed out by Diniz et al. (2008) and illustrated in Figure 1, which summarizes the decision making process in emer-

Figure 1. Conceptual map of knowledge support during an emergency response phase (Diniz et al., 2008)

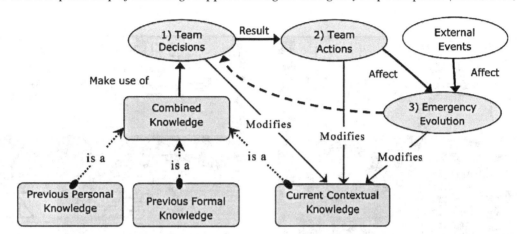

gency response. The experience and background of emergency responders constitute the so-called Previous Personal Knowledge (PPK). In general, a good PPK reduces the time needed to make decisions as autonomy of responders is enforced. However, it is difficult to handle as it is tacit, highly personal and hard to formalize (Nonaka & Takeuchi, 1995). As a complement to PPK, explicit knowledge is originated in some information sources, and may belong to two categories: on one hand, the Previous Formal Knowledge (PFK) is generated in advance, as a result of the prevention activities, and does not change during the development of the emergency; in general, PFK is contained in the emergency response plan. On the other hand, Current Contextual Knowledge (CCK) is composed of all the information which cannot be compiled in advance because it is mostly generated during the development of the emergency, and may even change during the emergency evolution. Sometimes the CCK pieces are known to be needed (e.g., the location of train running through a tunnel), and hence some type of placeholder could be inserted in an emergency response plan (e.g., "…request the location of the train…"); other times, however, CCK elements are not known in advance as they are generated during the emergency (e.g., a responder may communicate to the command and control the

presence of a toxic leak near a hospital, which requires immediate evacuation of the building).

In general, information sources can be classified in one of the categories included in the framework. However, there are cases in which the nature of the information source introduces some overlaps. Figure 2a shows a screenshot of a hypermedia emergency response plan on a subway service (Canós et al., 2004). There, information of different types is used to represent part of the actions to be performed by responders when a train catches fire inside a tunnel: instructions for the train driver expressed in natural language, a surface street map, and a video of the tunnel to help emergency coordinator to find obstacles for the evacuation of passengers. The video is shown in a region that includes playing controls. All the information pieces shown can be classified as PFK, since they are explicit and known in advance. However, there may be situations where PFK should not be used: if the fire in the train has been originated by a larger event such as an earthquake, the tunnel may have collapsed, and hence the video showing a clean tunnel should not be used at all because passengers can be driven to a dangerous place. In this case, some alternative information should replace the video (e.g., a warning sign picture preventing the evacuation through

Figure 2. Screenshots of hypermedia emergency response plan on a subway service (adapted from Canós et al., 2004)

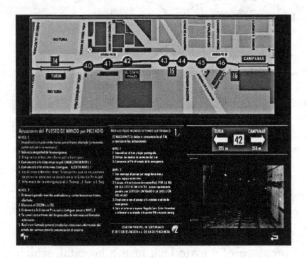

(a)

(b)

the tunnel) and then, the screenshot of the emergency response plan changes, as Figure 2b shows.

This example illustrates what we call in this paper *composite knowledge*: information that is formalized, but may be replaced (or, at least, updated) with contextual information. Unlike Situated Knowledge (Gahegan & Pike, 2006), which refers to the influence of context during the PFK creation processes, we extend the PFK with information generated at the knowledge use time. In general, replacing a formal source by a contextual one does not mean overriding. Even in the case of an unusable tunnel, the emergency coordinator may still be interested in having access to the tunnel to study some characteristics which may be helpful to make a decision. In this case, the ability to access to both the formal and contextual information must be provided by the information visualization tools of Emergency Management Systems. In this paper we will focus on the representation and access to composite knowledge during emergencies.

INCLUDING COMPOSITE KNOWLEDGE IN EMERGENCY RESPONSE PLANS

An emergency response plan is a complex document that includes the coordination mechanisms among responders, the procedures to be executed by the different actors involved, and the information to be exchanged between them. As mentioned earlier, emergency management systems use these plans as the main source of knowledge to drive the emergency response and the information contained is mostly composed of instances of PFK, as they have been defined in advance. CCK sources can also be included in emergency response plans, whenever they are identified in advance. Obviously, PPK is out of the scope of the plans as we are focusing on explicit knowledge. Specifically, we are interested in including and supporting visualization of composite knowledge pieces. With this aim, we define a knowledge model that will be combined with domain models to create rich emergency response plan models.

Figure 3 shows the UML class diagram representing our Knowledge Model. Knowledge is stored in *InfoElements*, which can be formal (*FormalIE* class), contextual (*ContextualIE* class) or composite (*CompositeIE* class). A composite element has a formal component associated, and may have associated one or more contextual elements. In general, formal and contextual elements are represented as digital objects which can be visualized using one or more disseminators, e.g. image viewers, video players, etc. (Kahn & Wilensky, 1995).

InfoElements can be composed. The pages shown in Figure 2 can be described in terms of the Knowledge Model as an *InfoElement* composed of the following elements:

- A surface map combined with the tunnel map (top of the screen). It is a formal element, represented as an image (a digital object of type image) and visualized using some image viewer;
- Three text boxes, one at the bottom left and other two composing the compound element shown in the bottom center of the figure. All boxes are formal elements, too. The user may switch from one to the other one clicking on the numeric labels associated to each text (which is part of the behavior of its associated disseminator);

- A distance indicator (the red box in the center, right part of the screen), indicating the distance of the train to the closest stations. It is a contextual element since the distances are calculated during the development of the emergency, and is represented by an image generated on demand by a specialized disseminator; and
- A composite element representing the tunnel; on one hand, a piece of formal knowledge, a video, which corresponds to the usual tunnel's appearance and can be disseminated by means of the appropriate video player; on the other hand, a contextual element, the warning picture, which should be shown in case of a tunnel collapse. In Figure 2b, the picture is hiding the video; however, it may be useful to access to the video if some valuable information can be extracted from it.

There are other elements in the page which cannot be described in terms of the knowledge model, as they have been included to give structure to the plan, allowing the navigation throughout its structure; they are described in further models.

Figure 3. Knowledge model including composite knowledge

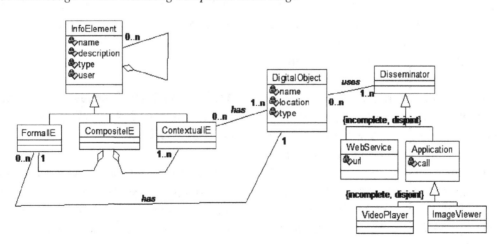

Developing Rich-Content Emergency Response Plans

Emergency response plan designers specify the mechanisms to respond to the different incidents that may affect the organization, as discussed above. However, the structure of the emergency response plan, as well as the specific information it contains may vary depending on issues like legal regulations or the type of organization considered. Whatever the particular issues are, the process to develop knowledge-aware emergency response plans can be summarized as follows: first, an emergency response plan model is built which defines the structure and content of them according to the particular settings. Second, this model is combined with the Knowledge Model defined in the previous section classifying the different emergency response plan components as formal, contextual or composite information pieces. To illustrate our approach, we have specified an emergency response plan model based on the legal regulation in Spain, called "Norma Básica de Autoprotección" (NBA) (http://www.boe.es/boe/dia/2007/03/24/pdfs/A12841-12850.pdf), that specifies the structure and minimal content of an emergency response plan. According with this regulation, an emergency response plan belongs to an organization, has a responsible, identifies a set of risks and contains a collection of emergency procedures. A complete description about the business activities is required, including the organizational structure. The emergency response plan must also contain a systematic warning system enumeration and a complete set of maps about the organization location and infrastructure. The emergency procedures, called action plans, not only contain the activities to perform by the response teams, but also the resources used and their responsible, the alarm mechanism and evacuation plans.

Figure 4 shows a simplified view of the emergency response plan model as UML class diagram. The Knowledge Model has been included in the form of *stereotypes*. Each element of the emergency response plan model is considered a piece of knowledge -an *InfoElement*- and can be labeled as formal, contextual or composite knowledge (<<*formal*>>, <<*contextual*>> and <<*composite*>> *stereotypes*, respectively). Notice that in the example, the knowledge about the organization, their business activities and their employees is considered formal. Many other classes are labeled as composite, assuming the possibility of having associated contextual information. This may vary from one emergency response plan model to another one, and therefore, assigning a stereotype to each emergency response plan component is a modeling decision.

Table 1 lists the different components appearing at the screenshot of the hypermedia emergency response plan on a subway service shown in Figure 2. For example, the *ActionPlan* instance is shown as three different views of formal knowledge, one per each different actor involved in the response. Fragment a) represents the procedure to be executed by the emergency coordinator, and fragments b) and c) represent the activities which are responsibility for the train driver and the station chief, respectively. In the case of the *Map* instance called *TunnelView*, the emergency response plan designer decided to assign the *composite stereotype* and hence both the digital objects that represent the formal knowledge- in this case, a recorded video of the tunnel- and the contextual knowledge, a warning sign picture that must be shown if a tunnel is not a safe way- must be specified.

TURNING EMERGENCY RESPONSE PLANS INTO NAVIGATIONAL STRUCTURES

The emergency response plan model described is technology-independent. To build actual plans, some path must be followed from modeling languages to implementation languages. Assuming

Figure 4. Emergency response plan model based on the NBA

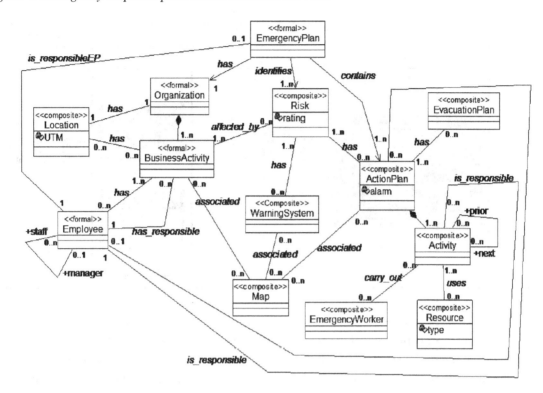

that our goal is to generate rich-content, hypermedia emergency response plans, we need to turn the class-based description into a navigational structure. Due to the variety of hypermedia formats available, we decided to create an abstract representation of such structure, which can be easily transformed into actual hypermedia models, as we will show in the following section. Figure 5 shows a partial view of the so-called Knowledge Graph Model; we represent the emergency response plan as a graph composed of nodes. The notion of node corresponds to that of page in hypermedia documents, that is, a collection of information pieces that are grouped and shown in a coherent way. These pieces (the InfoElements) are the components of the emergency response plans, some of which can include hyperlinks to other nodes in the graph.

Classical Hypermedia Engineering techniques (Ceri et al., 2000; Schwabe & Rossi, 1995; Solís et al., 2006) can support the graph generation

process. Specifically, navigational design techniques include heuristics to generate a navigable structure from class diagrams, transforming classes into navigable elements and class relationships such as associations into links between them. Additionally, some abstract interface techniques can help grouping the different elements into nodes to generate consistent and effective user interfaces. In the example about "Fire Inside a Tunnel" in a subway service, we have considered all *InfoElements* compound the same graph node.

VISUALIZING EMERGENCY RESPONSE PLANS USING SPATIAL HYPERTEXT

Abstract navigational emergency response plan graphs must be realized into some existing technology to make them accessible to emergency responders. As mentioned earlier, hypermedia (or

Table 1. Emergency response plan components in "Fire Inside a Tunnel" in a subway service

Class	ActionPlan	Map		EvacuationPlan
Instance	Fire Inside a Tunnel	Surface Street Map	Tunnel	Distance Indicator
Stereotype	<<formal>>	<<formal>>	<<composite>>	<<contextual>>
Digital Object			<<formal>> / <<contextual>>	

hypertext) appears as the natural target model as their models are similar to the navigational model introduced in the previous section. Since its conception, hypertext has been thought as a way for helping people to access to knowledge, in a kind of augmentation of the human intellect (Bush, 1945; Engelbart, 1963). However, knowledge creation tasks were not supported by hypertext tools until very recently. Wikis (Leuf & Cunningham, 2001) have emerged as an effective hypertext technology for enabling organizations and individuals to both capture and share knowledge. The advantage of using Wikis for knowledge management is that the management tasks can be performed in an open, collaborative and distributed way. Wiki philosophy fits very well with the common collaborative way of defining and utilizing emergency

response plans, as pointed out by Raman et al. (2010). However, wikis have not overcome all the problems of original hypertext. Specifically, the documented-centered nature of hyperdocuments made users get lost using hyperlinks as the only means to navigate in large networks. This is the case of emergency response plans, which may be large and complex documents.

Spatial hypertext (Marshall & Shipman 1995) is based on using visual and spatial characteristics of hypertext elements for defining the relationships among them. The elements of a spatial hypertext document are seen as cards that are classified through visual clues, or spatial positions. When these characteristics are used to relate elements, the hyperlinks become implicit, and related elements can be represented by sharing the same

Figure 5. Knowledge graph model (partial)

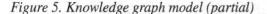
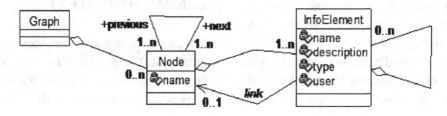

visual and spatial characteristics: a color, borders, font types, adorns layouts, position, proximity, geometric relations, etc. Using spatial hypertext features, superimposed information can be visualized easily, bringing the right way to visualize the composite knowledge. Merging these features with a wiki-like model for information capture and visualization, we will have the key to building rich-content emergency response plans.

ShyWiki (Solís & Ali, 2008) is a wiki which uses spatial hypertext for representing its content. It was designed to support users in creating, storing, editing, and browsing knowledge structures, understood as interconnected networks of information. ShyWiki manages a network of wiki pages (the nodes of the network). Each page is a hypermedia document that is identified by its name and is made up of an unordered set of named attributes called notes. Wiki page notes can contain text, images, and hyperlinks, or a mix of them. The main function of the notes is to define the attributes that characterize the concept represented by a wiki page.

The content of the wiki pages is spatially organized: notes may be placed in different regions of the page, moved around, and may be of different sizes and colors. The notes can contain text, hyperlinks, and images. Composite notes may be created from simpler ones, helping to organize knowledge hierarchically.

The components of the ShyWiki Hypertext Model are summarized in Figure 6. The root of a ShyWiki document is the ShyWikiWeb, which is composed of information and knowledge stored in WikiPages, connected by hyperlinks. The WikiPages are composed by notes. The AbstractNote class includes the properties which are common to the two types of notes: on one hand, ContentNotes hold content of different types (text, images, video and hyperlinks) and can be composed of other ContentNotes; and on the other hand, the so-called TranscludedNotes. Transclusion is the inclusion of notes already defined in other wiki pages by reference, i.e., without duplicating them in the including note. Essentially, a TranscludedNote is a note whose content is defined by another note. ShyWiki supports this model providing the basic operations to create or modify wiki pages. In the edition mode, a user can perform a number of knowledge management actions, namely creating wiki pages, and creating, editing, moving grouping and transcluding notes. These actions permit to represent easily the composite knowledge.

ShyWiki is a service oriented wiki. ShyWiki web client interacts with the server using Asynchronous JavaScript and XML (AJAX) web services. These services can be used by other

Figure 6. ShyWiki hypertext model

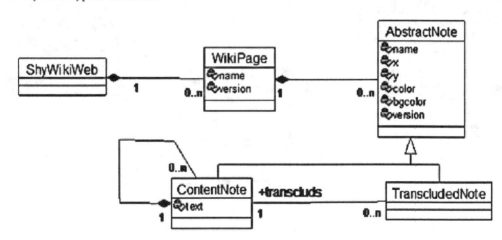

agents different than a web browser to interact with ShyWiki. ShyWiki's web services allow external agents to write or read the content of the wiki pages. So, emergency management systems can connect to ShyWiki servers to generate personalized information for the different responders.

Generating ShyWikis from Knowledge Graph Models

Elements in the Knowledge Graph Model are transformed in ShyWiki spatial hypertext elements. Figure 7 shows the wiki page generated from the emergency response plan components shown in Table1. The following paragraphs illustrate the process followed to obtain it.

A knowledge graph is represented as whole Wiki. Nodes in the graph are transformed into wiki pages taking their names from the nodes' names. Each *InfoElement* in a node is transformed

into a *ContentNote* in the corresponding page. The name of the *InfoElement* is given to the *ContentNote*. The type of the *InfoElement* is used to build the wiki text that represents the *InfoElement* content. Compound *InfoElements*, that is, those made up from the aggregation of one or more *InfoElements*, are transformed into *ContentNotes* that act as containers of other *ContentNotes* generated from the component *InfoElements*. Other mappings are defined, but we omit them due to space limitations.

The page in Figure 7 contains the different *ContentNote*s generated from each *InfoElement* defined. The "FireInsideTunnel" *InfoElement* is showed on the left of the page (emergency coordinator view) and on the right (train driver and station chief views). On the top-center, we have the *ContentNote* generated to "SurfaceStreetMap" *InfoElement,* and below it, the one corresponding to the "DistanceIndicator" *InfoElement*. Finally,

Figure 7. "Fire Inside a Tunnel" ShyWiki page

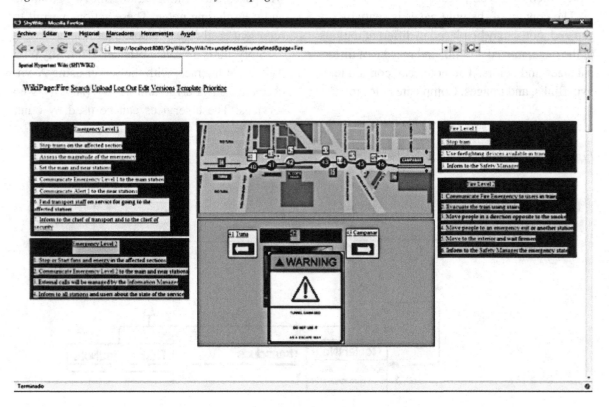

the "Tunnel" *InfoElement* appears at the bottom-center of the page; this component represents composite knowledge, and it was transformed into a compound *ContentNote* that contains the digital objects associated to the formal (tunnel video) and the contextual knowledge (warning image) elements. Both knowledge elements can be visualized in the same page using the spatial and visual features of ShyWiki.

CONCLUSION AND FURTHER WORK

Managing composite knowledge is a key issue in emergency responses. Giving responders the option to access to formal knowledge, even when it has been overwritten by contextual information, may improve the situational awareness of decision makers, which may result in more agile and confident instructions sent to responders. We have shown how a knowledge model including composite knowledge has allowed us to stereotype the classes in an emergency response plan model to classify its components in formal, contextual or composite. Then, strategies to derive implementation-independent navigational structures make the adoption of a specific platform easy.

Composite knowledge can be seen as a case of what other authors called Superimposed Information (Murthy et al., 2004); in Superimposed Information architecture, a base layer can be combined with the superimposed one to create complex knowledge structures. To visualize superimposed information, however, some middleware is required. Another related approach was the Multivalent Browser (Phelps & Wilensky, 2000), which allowed to insert annotations and other extra information to documents. Again, specific software was needed to insert and access to such extra information.

ShyWiki, however, allows a simpler approach as its documents are visualized in regular web browsers. Moreover, the spatial features of its elements make it easy to access to both the base and contextual elements of a composite knowledge piece. The different types of multimedia information can be integrated in the hypermedia emergency response plans using a simple, Wiki-like language.

Future work includes the implementation of a tool supporting the framework, so that emergency response plan designers can define conceptual plan models which can be labeled with the different stereotypes and used to generate the navigational structures in a way as automated as possible. Then, different generators can obtain final emergency response plan versions using automated transformations. We also plan to evaluate the approach in other domains where composite knowledge is present.

ACKNOWLEDGMENT

A preliminary version of this paper was presented at the 7th International Conference on Information Systems for Crisis Response and Management (ISCRAM 2010), held from May 2 to 5, 2010 in Seattle, WA (USA). The work of J. H. Canós, M. C. Penadés and M. Llavador is partially funded by the Spanish Ministerio de Ciencia y Tecnología (MICINN) under projects DEEPEN (TIN2009-08084) and TIPEx (TIN2010-19859-C03-03). The work of C. Solís is funded by Science Foundation Ireland, grant 03/CE2/I303_1, to LERO. The work of M. R. S. Borges is partially supported by grants No. 304252/2008-5 and 480461/2009-0, respectively, from CNPq (Brazil). The work of A. S.Vivacqua is partially supported by FAPERJ (Brazil).M. Llavador was the holder of the MEC-FPU grant no. AP2005-3356. The cooperation between the Brazilian and the Spanish research groups was partially sponsored by the CAPES/MECD Cooperation Program, Project #169/PHB2007-0064-PC.

REFERENCES

Bush, V. (1945). As we may think. *Atlantic Monthly, 176*(1), 101–108.

Canós, J. H., Alonso, G., & Jaen, J. (2004). A multimedia approach to the efficient implementation and use of emergency plans. *IEEE MultiMedia, 11*(3), 106–110. doi:10.1109/MMUL.2004.2

Canós, J. H., Penadés, M. C., Solís, C., Borges, M. R. S., & Llavador, M. (2010). Using spatial hypertext to visualize composite knowledge in emergency responses. In *Proceedings of the 7th International ISCRAM Conference*, Seattle, WA.

Ceri, S., Fraternali, P., & Bongio, A. (2000). Web modeling language (WebML): A modeling language for designing web sites. *Computer Networks, 33*(1-6), 137-157.

Diniz, V. B., Borges, M. R. S., Gomes, J. O., & Canós, J. H. (2008). Decision making support in emergency response. In Adam, F., & Humphreys, P. (Eds.), *Encyclopedia of decision making and decision support technologies* (1st ed., pp. 184–191). London, UK: Information Science Reference. doi:10.4018/978-1-59904-843-7.ch021

Engelbart, D. C. (1963). A conceptual framework for the augmentation of man's intellect. *Vistas in Information Handling, 1*, 1–29.

Gahegan, M., & Pike, W. (2006). A situated knowledge representation of geographical information. *Transactions in GIS, 10*, 727–749. doi:10.1111/j.1467-9671.2006.01025.x

Jennex, M. E., & Raman, M. (2009). Knowledge management is support of crisis response. *International Journal of Information Systems for Crisis Response and Management, 1*(3), 69–82. doi:10.4018/jiscrm.2009070104

Kahn, R., & Wilensky, R. (1995). *A framework for distributed digital object services*. Retrieved from http://www.cnri.reston.va.us/k-w.html

Leuf, B., & Cunningham, W. (2001). *The Wiki way: Quick collaboration on the Web*. Reading, MA: Addison-Wesley.

Marshall, C. C., & Shipman, F. M. (1995). Spatial hypertext: Designing for change. *Communications of the ACM, 38*(8), 88–97. doi:10.1145/208344.208350

Murphy, T., & Jennex, M. E. (2006). Knowledge management, emergency response, and Hurricane Katrina. *International Journal of Intelligent Control and Systems, 11*(4), 199–208.

Murthy, S., Maier, D., Delcambre, L., & Bowers, S. (2004). Putting integrated information into context: Superimposing conceptual models with SPARCE. In *Proceedings of the First Asia-Pacific Conference of Conceptual Modeling*, Denedin, New Zealand (pp. 71-80).

Nonaka, I., & Takeuchi, H. (1995). *The knowledge-creating company: How Japanese companies create the dynamics of innovation*. Oxford, UK: Oxford University Press.

Phelps, T., & Wilensky, R. (2000). Multivalent documents. *Communications of the ACM, 43*(6), 83–90. doi:10.1145/336460.336480

Raman, M., Ryan, T., Jennex, M. E., & Olfman, L. (2010). Wiki technology and emergency response: An action research study. *International Journal of Information Systems for Crisis Response and Management, 2*(1), 49–69. doi:10.4018/jiscrm.2010120405

Schwabe, D., & Rossi, G. (1995). The object-oriented hypermedia design model. *Communications of the ACM, 38*(8), 45–46. doi:10.1145/208344.208354

Solis, C., & Ali, N. (2008). ShyWiki-A spatial hypertext Wiki. In *Proceedings of the International Symposium on Wikis,* Porto, Portugal (p. 10).

Solis, C., Canós, J. H., Penadés, M. C., & Llavador, M. (2006). A model-driven hypermedia development method. In *Proceedings of the World Wide Web Internet Conference* (pp. 321-328).

Chapter 13
Emergency Management, Twitter, and Social Media Evangelism

Mark Latonero
University of Southern California, USA

Irina Shklovski
IT University of Copenhagen, Denmark

ABSTRACT

This paper considers how emergency response organizations utilize available social media technologies to communicate with the public in emergencies and to potentially collect valuable information using the public as sources of information on the ground. The authors discuss the use of public social media tools from the emergency management professional's viewpoint with a particular focus on the use of Twitter. Limited research has investigated Twitter usage in crisis situations from an organizational perspective. This paper contributes to the understanding of organizational innovation, risk communication, and technology adoption by emergency management. An in-depth longitudinal case study of Public Information Officers (PIO) of the Los Angeles Fire Department highlights the importance of the information evangelist within emergency management organizations and details the challenges those organizations face engaging with social media and Twitter. This article provides insights into practices and challenges of new media implementation for crisis and risk management organizations.

INTRODUCTION

And the same way I worry most about new media, people will not embrace it. People see it in a monocular focus, they think it's about distribution, it's about talking, it's about yelling. It really is about listening. People see this as one dimensional, and they don't see their need to be part of that community.—Brian Humphrey, PIO, Los Angeles Fire Department

DOI: 10.4018/978-1-4666-2788-8.ch013

The use of computer-mediated communication in times of emergency is gaining momentum and is the focus of much existing research. Social media allow users to generate content and to exchange information with groups of individuals and their social networks. First gaining attention in the aftermath of large-scale disasters such as the Banda Aceh Tsunami, networked, online conversations among the affected publics and onlookers offering help have been especially in focus during extreme events (Palen, Vieweg, Liu, & Hughes, 2009; Scaffidi, Myers, & Shaw, 2007; Majchrzak, Jarvenpaa, & Holingshead, 2007; Liu, Iacucci, & Meier, 2010; White, 2011). Twitter, the popular micro-blogging site, has gained particular attention due to its increasingly widespread adoption. A recent study by the Pew Internet & American Life Project found that 19% of all Internet users share updates about themselves on Twitter or another similar service (Fox, Zickuhr, & Smith, 2009). While there is much media hype and excitement over the use of Twitter during times of emergency, researchers are just beginning to examine the value and logic behind its usage (Starbird, Palen, Hughes, & Vieweg, 2010).

There are two primary streams of research investigating the use of social media in emergency response. One stream is concerned with how emergency management organizations use such technologies to coordinate in response to disaster as they conduct rescue activities (White, Plotnick, Kushman, Hiltz, & Turoff, 2009; Bharosa, Appelman, & de Bruin, 2007; van de Ven, van Rijk, Essens, & Frinking, 2008). The other stream is concerned with how those affected by disaster and those who volunteer to help utilize social media to locate information and to seek or provide support (Liu, Iacucci, & Meier, 2010; Hughes & Palen, 2009; Starbird & Palen, 2011; Sutton, Palen, & Shklovski, 2008). Few studies have considered how emergency response organizations utilize currently available technologies both to communicate with the public in emergencies and to potentially collect valuable information using the public as sources on the ground. In this paper we describe the use of social media from the emergency management professional's viewpoint with a particular focus on the use of Twitter.

As emergency management professionals add social media to the range of tools they use to communicate with the public in times of crisis, a critical investigation of how and why these tools are adopted is crucial: Adoption and implementation of technology requires allocation of precious time and resources. We argue that the public's usage of Twitter differs from its usage by emergency management professionals in significant ways. We discuss these differences and focus on how and why officials in emergency response organizations responsible for communication with the public implement social media at the organizational level. We rely on conversations with emergency management professionals in New York City and Los Angeles and elaborate on an in-depth case study of the PIOs at the Los Angeles Fire Department and their use of Twitter and other social media.

BACKGROUND

Micro-blogging is a form of lightweight, mediated communication where users can broadcast short messages to their networks and direct these messages to specific people within networks. Users of Twitter send short (up to 140 characters) messages or "tweet" to their networks of "followers" – people who chose to be updated when the person they "follow" adds a new message to the stream. Twitter users send "tweets" to their followers, and users can also "retweet" or pass along messages originating from others. Twitter includes search functions so users can search the site for prevalence of keywords, phrases, topics, trends, or individuals. Other features of Twitter include options to add website links and geo-location information to tweets.

When Twitter first launched, the tweets were often personal and seemingly inconsequential updates on the goings- on of the everyday life which gained Twitter a reputation in the media for being an inane, narcissistic, whimsical medium with little value outside of mere entertainment (Cohen, 2009). Such opinions overlooked that early adopters in the Twitter community were building worldwide social networks accustomed to sending and receiving short messages in real time. Twitter gave individuals the unprecedented ability to rapidly broadcast and exchange small amounts of information with large audiences regardless of distance. Although Twitter is Internet based, functioning over the World Wide Web, its primary focus is on integration with mobile/cellular devices, which creates the potential of an alternative communications system apart from traditional telephony, radio and television. In retrospect such affordances of Twitter seem obviously useful during times of emergency and crisis as information changes unexpectedly and needs to be disseminated to the public rapidly.

Today Twitter has clearly captured the public imagination, particularly in light of Twitter usage to mobilize and inform political opposition to in countries with regimes that have curtailed public communications infrastructure. During the Iranian student protests of 2009, the US Department of State reportedly asked Twitter management to delay a scheduled maintenance so users of Twitter in Iran could continue their mobilizing of protests via the site. According to Gilad Lotan (2011), "Twitter served as an incredibly engaging mechanism to disseminate information on the riots and protests that were taking place around the world. Its real-time qualities enables information to rapidly spread between users, while its personal style drives a sense of emotional involvement to the events." During the revolutionary protests in Egypt in spring of 2011, users reportedly employed Twitter (and Google) as an ad hoc distributed communication system, until the government shut down Internet access country wide (Idle & Nunns, 2011).

Despite an increasing number of high profile uses of Twitter, what is rarely discussed is the use of Twitter by official organizations during crisis events. Thus the question remains – to what extent can Twitter be repurposed successfully to meet the needs of crisis response organizations? We are also interested in the extent to which Twitter is used by organizations not only for broadcasting information, but also for information gathering during crisis situations?

There is a small but growing research literature focused on how the public uses Twitter in times of emergency. Palen et al. (2010) have been conducting extensive studies of Twitter use during mass convergence or emergency events such as the Southern California Wildfires (Sutton et al., 2008), the Democratic and the Republican National Conventions (Hughes & Palen, 2009), and the recent flooding of the Red River Valley (Starbird et al., 2010). These authors report a large volume of conversations and substantial information exchange on Twitter during crisis and mass convergence events. Information exchange relied on extensive self-organizing and information vetting as well as on the emergence of personalities that became information hubs to the rapidly growing legions of their followers. Moreover, Starbird et al. (2010) clearly show that people seek out and even privilege official information, augmenting, rather than discounting statements issued by emergency services and mass media outlets (Starbird et al., 2010). This consistently observed privileging of information ostensibly gleaned from official sources on Twitter suggests that traditional broadcast media are not only retaining their importance for disseminating emergency information, but also that this information now can be given extra weight and legitimization through the word-of-mouth nature of Twitter communication. Yet Twitter-based communication comes with affordances of interactivity and audience choice in ways that traditional risk communication channels never did.

Members of the public sending and receiving messages are only one part of Twitter's communicative dynamic. Based on the literature briefly reviewed above, we propose that Twitter communication during times of emergency and crisis falls into four broad categories:

1.   Twitter users posting self-generated messages about the crisis to their social networks.
2.   Twitter users retweeting messages received from members of their social networks, traditional media, unofficial, and official sources.
3.   Emergency management professionals using Twitter in either official or unofficial capacities to send messages to the public in affected communities or the public at large.
4.   Emergency management professionals monitoring Twitter feeds from the public to gather information during times of emergency.

The first two categories represent the bulk of the existing research, which focuses on the public's use of Twitter during emergencies. In their paper on Twitter use in the Red River Valley, Starbird et al. (2010) openly acknowledge that although it is clear that the public will use Twitter for communication and information exchange in extreme events, the question is what does this mean for emergency management? Starbird et al. (2010) state: "One of the challenges for emergency management today is to know "what to do" with social media applications. The new digital world provides both an opportunity but also a real and understandable dilemma for emergency management: How can they make sure that the information that is "out there" is accurate during an emergency event?" (p. 9). We address this question by examining the latter two categories of Twitter usage. We focus on what one fire department in a large metropolitan area *does* with social media applications for official purposes. As officials these professionals

represent an organizational perspective on the utility and functions of Twitter during crisis. We ask what the logic is behind emergency management professional's adoption of Twitter as a channel to communicate with the public. In addition, we ask to what extent does emergency management utilize Twitter to monitor and use self-generated information from the public.

Organizational Innovation

The majority of research on technology use within crisis response organizations considers technologies that such organizations might use for collaboration and information exchange with other organizations involved in crisis response (Boersma, Groenewegen, & Wagenaar, 2009; van de Ven et al., 2008). Research that focuses on crisis communication with the public tends to implicitly expect full organizational support for technology adoption as part of the organizational stance toward risk communication (Gomez & Turoff, 2007; White et al., 2009). Improvisation is an important aspect of successful organizational response to emergencies and members of organizations often improvise by using available technologies (Kendra & Watchtendorf, 2003; Mendonca, Jefferson, & Harrald, 2007). Ad hoc usage of social media and mixing of media for situational purposes can lead to innovation, adoption, and repurposing of communication technologies (Yun, Park, & Avvari, 2011). Yet major emergency response organizations such as state or city fire and police departments are encumbered when it comes to large-scale adoption of technologies as they are part of government structures and much of the decision-making is dependent on political will. For example, Boersma et al. (2009) provide a good overview of the role of political will in technology adoption in their ethnographic study of emergency management organizations in Amsterdam.

Researchers have considered organizational technology adoption issues in the emergency management area. For example, Bharosa et al. (2007) examined the role of the information manager who brought IT expertise and technological innovation into a crisis response context. Their results suggest that such information managers or brokers are necessary to serve as the human experts who mediate between the technological system, information, organization, and audience. These information integrators often double as early adopters and innovators within organizations that may not understand the technological capabilities of the systems being implemented. In a survey of non-emergency management organizations that are nonetheless involved in crisis response and mitigation of effects, Milis and van de Walle (2007) showed that the presence of crisis management personnel with IT backgrounds is imperative for organizations that use IT for crisis management.

Organizational innovation has been extensively investigated in the organizational literature. More recent ideas on organizational innovation conceptualize it as a continuous process (Brown & Esienhardt, 1997). These studies show that while adoption of large-scale organizational and management IT systems requires top-down decision making, the majority of smaller-scale technology-use innovations are lead by innovators or 'evangelists' from within the organization (Lawrence, Dyck, Maitlis, & Mauws, 2006). From our initial informal conversations with emergency management professionals, it became clear that such evangelists are key to IT adoption, innovation, and use in crisis response and management. Traditionally, risk and crisis communication has been conceptualized as a one-way stream of information from emergency management organizations to the public. New media technologies, however, offer opportunities to change that dynamic toward a greater level of interactivity between emergency management professionals and the public. We investigate how the interactivity affordances of new media play out in innovative uses of these technologies in emergency management organizations.

Risk and Crisis Communication

Emergency management work with disasters, emergencies, crises and mass convergence events has always included some form of communication with the public and with mass media outlets (Sorenson & Sorenson, 2006). Risk communication is an extensive research area. In an overview, Reynolds and Seeger (2005) present several definitions of risk communication and conclude that "in practice, risk communication most often involves the production of public messages regarding health risks and environmental hazards" (p. 45). In the event of a crisis or an emergency, according to Reynolds and Seeger (2005), "crisis communication seeks to explain the specific event, identify likely consequences and outcomes, and provide specific harm-reducing information to affected communities" (p. 46). The general goal of risk and crisis communication then is to inform the public of potential or current events and to persuade the public to adapt their behavior in ways that would improve health and safety.

Traditionally in crisis communication, the news media serves as the intermediary between emergency mangers and the public at large. Generally, radio, television, and print messages reach the largest audience over the greatest distance. Thus emergency management organizations need these intermediary news organizations to disseminate information, and conversely, news organizations need emergency management for official information to report. The news organizations, in turn, capitalize on emergency information as news reports of disasters generate high appeal and maintain rapt audience attention. Media personnel often see their reporting as providing valuable service to people affected by the disaster, and crisis events tend to receive a lot of attention from news media, capturing the airwaves and newspaper space. At the same time, much of the reporting can focus on sensationalizing the news and repetitive use of striking imagery at the expense of provision of more mundane risk communication (Sood, Stockdale, & Rogers, 1987).

Despite information dissemination via news media outlets, and no matter the disaster, the people affected experience severe information dearth and take steps to alleviate it (Mileti & Darlington, 1997). Current research on public response to disasters or emergencies often ends with a call to action directed toward emergency management organizations for organizing and deploying their crisis and risk communication better, more up-to-date and more interactively (Sutton et al., 2008; Palen et al., 2009; Jennex, 2010). Many emergency organizations have indeed made efforts in this direction. For example, in the aftermath of Hurricane Katrina, the American Red Cross developed a centralized system helping people find each other (Scaffidi et al., 2007; Murphy & Jennex, 2006). More recently, the Department of Homeland Security approached MySpace with a request to develop a notification and communication widget during Hurricane Gustav (White et al., 2009). Sutton (2009) detailed the issues emergency management PIOs encountered during the DNC in Denver in their attempts to integrate the myriad of online news sources and conversations. In the study, however, PIOs did not use interactive means of data collection through social media but instead relied on more traditional methods of one-way information dissemination employed on blogs and alternative news sources.

Yet as emergency management organizations adopt technologies they have to contend with the nuances and dynamics of new technologies that may not fit the traditional organizational conception of risk communication. We identify the nuances of one such dynamic: that of managing an official and less official, more interactive voice in communication with the public. We explore this in considering the nature of Twitter which often involves a single individual in an organization sending messages in an official and sometimes unofficial capacity. This blurring of intentions and communications between the individual and organization is a central problematic in social media where private and public spheres often collide.

Interactivity and Participation

Media and communication technologies play a large role in managing emergency and crisis situations and managing community perceptions of risk and preparedness. The major differences between traditional channels of communication and social media are the capabilities for one-way, two-way, or interactive exchange of information.

Today, the primary means for emergency management organizations to communicate with the public remains the traditional media: one-way broadcast radio, television, and newspapers. Social media provides the potential for interactive, participatory, synchronic, two-way communication. Dissatisfaction with traditional media is one of the more frequent reasons cited for why people affected by crisis situations turn to social media in search of information (Shklovski, Burke, Kiesler, & Kraut, 2010; Sutton et al., 2008). The ability to augment existing media channels and to engage in more interactive and real-time communication might explain why emergency management innovators decide to implement social media. The possibility of interactive communication between emergency management organizations and large audiences would essentially remove the "middleman" or reduce the reliance on the news media.

Media and communication technologies play a large role in managing emergency and crisis situations and managing community perceptions of risk and preparedness. The major differences between traditional channels of communication and social media are whether there exist capabilities for one-way, two-way, or interactive exchange of information. Traditional media has made concerted attempts to increase audience interactivity. For example, radio stations are now routinely asking listeners to call not only with responses to trivia questions but also with current actionable traffic information, while many news stations ask audiences to send in their cell phone videos and pictures of crisis events or other newsworthy moments. Many news organizations also seek relevant information produced by the public on various social media sites and use it in their reporting, such as, for example, BBC using photos of the London bombings that people posted on the Flickr photo-sharing service. In a similar fashion social media give Emergency Management the means not only to communicate with the public directly but also to collect real-time actionable information from a myriad of on-location eyewitnesses.

As our earlier categorization illustrates, Twitter can be used as a one-way asynchronous medium, but it can also be used to update one's status in real-time, follow other's tweets, respond to tweets, retweet original posts, act based on the tweets of others, and organize calls to action based on posts from the Twitter community. Thus official emergency management use of Twitter is likely to span this spectrum, from disseminating one-way messages to monitoring Twitter messages during a crisis and acting to allocate resources based on that information. The perception of Internet users as smart mobs with collective intelligence (Rheingold, 2002) may be what can drive effective use of Twitter and other social media by emergency management organizations. In this paper we explore how one fire department manages this process.

Fire Departments and the Role of the Public Information Officer (PIO)

Much of disaster and emergency research focuses on major crises and sudden emergencies that activate various organizations involved in first-response activities. Yet certain emergency management organizations such as police and fire departments are maintained and trained to contend with a multitude of disturbances and small emergencies in everyday life along with emergency response in major events. These organizations are in a constant state of alert and the kinds of organizational improvisation lauded in times of major disasters, would likely be an ongoing process of innovation to accommodate frequent potentially

dangerous situations. The work of Lawrence et al. (2006) suggests that we are likely to find individual innovators and evangelists in such organizations bringing in and advocating for communicative solutions that involve social media.

We focus on a specific kind of innovation, that of risk and crisis communication and interaction with the public. Risk and crisis communication is an important part of the function of state and city fire departments. Public information officers (PIOs) are usually charged with coordinating communication activities and performing as spokespersons (for an in-depth description of PIO duties see Sutton, 2009). Typically, PIOs provided information to the public through mass media. In fire departments, the role of the PIO is performed by firefighters as part of their tour of duty, often for two years at a time. PIO is a specialist position but they are the rank of Firefighter. If a PIO is to be promoted to Captain, they leave their job as a PIO and go on to other duties. From our informal conversations with emergency management professionals, however, it seems that some PIOs remain on the job for years, gaining both experience and the social connections necessary to successfully manage crisis and risk communication. These specialists are positioned to become technological evangelists promoting the use of new forms of media and technology. We focus this investigation, asking questions based on the review of the extant literature. First, what is the logic behind innovation, adoption, and implementation of interactive and social media? Second, if emergency management is to talk and to listen to the citizens using social media, how do they go about verification of information they receive when taxpayer dollars are at stake both in unnecessary action and in action not taken when necessary? Third, what is the role of the organization as a whole in supporting innovative risk and crisis communication to the public through social media? Fourth, is the traditional role of information officers in emergency management changing as social media is implemented?

METHODOLOGY

In January through May of 2009, the first author conducted a series of informal visits and meetings with emergency management professionals in New York City and in Los Angeles County, exploring the use of social media in their work. In the course of these conversations it became clear that while social media certainly was a point of concern, use of these technologies was intimately tied to individuals pushing the envelope, to the organizational structures within which these individuals are positioned and to political will to change modes of crisis response. In order to get an in-depth view of the way emergency management is already using new media for communication with the public, we decided on a case study of an emergency management organization known for their extensive use of Twitter, blogs, email lists, discussion groups and a range of other communication modalities available through the Internet. In the course of our investigation one particular person emerged as the main innovator in the organization. We do not anonymize his quotes as the person in question is a public persona due to his innovative uses of social media and has explicitly given permission to use his name.

Case Study: Brian Humphrey, LAFD

We used an exploratory case study method to investigate the processes of social media innovation, adoption, and implementation at the organizational level following the methodology recommended by Yin (1994).

We identified the key personnel at the LAFD, Brian Humphrey, who is a paramedic-trained firefighter, a 24-year veteran of the LAFD, serving as a public information officer (PIO) for 17 years. Humphrey has received a lot of attention as an innovator in social media. Wired Magazine writes that Humphrey is "single-handedly hauling the city's fire department into the Web 2.0 era" (Tabor, 2008). The article quotes Humphrey

saying "Short of motorized fire apparatuses, this technology is the best thing that's happened to our department in 122 years…It holds more potential to save lives than any other civic tool."

We conducted an in-depth interview with Humphrey together with one former PIO and one PIO in training in their office in a decommissioned bunker underneath the Los Angeles City Hall complex during one of the regular 24-hour tours of duty. Although other PIOs chimed in, the majority of the conversation was with Humphrey. The interview was transcribed and coded using an open coding scheme for emergent themes. We monitored LAFD associated Twitter feeds for a two year period (June 2009 – June 2010). We coded both collections of Twitter feeds by hand, using the themes that emerged from the interview, explicitly drawing parallels over time and monitoring the data for developments and changes. Below is the description of the Twitter feeds we followed.

The Twitter account @LAFD sends official short messages of emergency dispatches and updates of calls throughout the City of Los Angeles; @LAFDtalk is for discussion and queries about the LAFD managed by the three LAFD PIO officers; @BrianHumphrey is Humphrey's personal twitter feed; @LAFDFIRECHIEF is Chief Peaks Twitter account. In September 2009 we observe @LAFD is following 3, has 7,399 followers, and has 3,700 recorded Tweets; @LAFDtalk is following 4317, has 3926 Followers, and 2583 Tweets; @BrianHumphrey is following 9, with 997 Followers, and 1045 Tweets 2/28/10). From the number of followers one can infer the size of the audience receiving messages. The number of members the user is following references the amount of potential interactivity or ability to receive messages. From these numbers we can already see that LAFD is mainly used for one-way communication where LAFDtalk can potentially receive messages from more accounts than it sends to. Observing these Twitter feeds over the two year period, we find a pattern of increasing Twitter usage and a notable addition of the LAFD Fire Chief (activated in Dec 2009). Two snapshots of our findings over time are seen in Table 1.

FINDINGS

Innovation Evangelism

Early in our interview it became clear that Humphrey is the driving force behind the technological innovations at LAFD. He was active since the days of "Telnet, Archie, and Veronica," and with the advent of the WWW he and another firefighter had learned HTML to create one of the first webpages for LAFD of any major Fire Department in the U.S. This do-it-yourself ethos reflects the findings by Mendonca et al. (2007) as the PIOs at LAFD innovated in an ad hoc way to meet specific practical needs. As Humphrey explains:

I have no formal education in any of this ... So we've used a variety of things, starting with Yahoo e-mail lists, evolving into displaying the same content on Blogger ... And we ultimately became the first agency to have a blog listed in Google, for example, as a news agency ... at the time, it was unheard of, that not only a blog, but a Blogspot blog would be indexed.

In joining the online conversation early on, Humphrey exhibited a prescient understanding of the power of search terms and online audiences by signing all of his messages with "Respectfully Yours in Safety and Service." A Google search for this signature signoff results in tens of thousands of his messages posted (and reposted) on various blogs and sites. Such consistency allowed him to develop a reputation and to maintain source credibility, crucial for giving weight to his voice in a networked conversation. Source credibility is essential during crisis communications as evidenced by Starbird and Palen's (2010) research that demonstrates that Twitter users overwhelmingly retweet messages from official information sources such as emergency management or news media organizations. For Humphrey, the explicit purpose for conveying credible information online is to benefit the community the LAFD serves:

Our online presence has this very simple purpose: to help people lead safer, healthier, and more productive lives. That's it. And that's the only reason we're out there, and we try to be very transparent in that.

This sentiment is very much within the ethos of risk communication described. Following the LAFD Twitter activity makes it patently clear that the PIOs of the LAFD are the embodiment of this purpose for large and small "crises." For example:

@faboomama: Hot liquid burn like fire. Just burnt the crap out of both hands with 250 degree oil. PAIN, 2:08 PM Sep 22

Table 1. Observation of Twitter feeds over time

September 2009				June 2011		
Twitter	Following	Followers	Tweets	Following	Followers	Tweets
@LAFD	3	7,399	3,700	4	10,542	6,039
@LAFD talk	4317	3926	3185	6,731	6,447	8,657
@BrianHumphrey	9	997	1045	9	1,792	1,524
@LAFDFIRECHIEF	-	-	-	1	701	48

In the presentation of our findings, Twitter users are denoted with @username.

@LAFDtalk posts this public response: @faboomama Please continue to cool your burned hands in water (no ice!) for at least 15 minutes. No home remedies, seek MD or 911 if doubt 2:11 PM Sep 22nd from web in reply to faboomama.

A three-minute response-time represents a level of interactivity between a member of the public and emergency management far surpassing any traditional broadcast medium. This kind of interactivity, however, implies constant vigilance and a level of attention that is unheard of from large government organizations. This kind of focus is impossible to keep up given the simple human and logistical limitations: there is only one PIO per 24-hour shift regardless of how many incidents are in progress and how large these incidents are. This ability to respond and examples of such quick response becomes a double-edged sword. On one hand, this builds rapport and a feeling of community among the followers of @LAFD, yet this can also create expectations that such rapid response is always possible. Humphrey is well aware of this as he expresses an overwhelming sense of exasperation in this post:

@BrianHumphrey: 270 voice mails and 2000+ non-spam emails expecting a reply. Dunno how or when I'll get back to you all. 6:47 AM Sep 28th from TweetDeck

While Humphrey might respond to individual tweets he monitors, he directs those on Twitter to use the traditional 911 mechanism to report emergencies:

@LAFDtalk: @StaceyWong Yes, @LAFD is on Twitter, but we ask that emergencies be reported to 9-1-1 and official business matters to [phone number] 8:00 PM Aug 25th from web.

Clearly LAFD's PIOs cannot monitor social media all the time. Despite Humphrey's enthusiasm and determination to integrate social media

into firefighting and increasing communication with the public via Twitter, we surmise that LAFD's official stance is more traditional in scope.

OLD V. NEW MEDIA

Part of the impetus to interact more with citizens directly derives from an explicit dissatisfaction with traditional media. According to Humphrey, the position of the PIO was created in 1968 to "keep reporters from bothering the dispatchers." In addition to protecting the dispatchers, the PIO would interact with the reporters in order to communicate with the public. Humphreys explicitly states:

You don't talk to the media, you talk through them

LAFD were at the forefront of using IT when they began sending text messages to reporters' pagers telling them fire locations in hopes of getting a news team out to report on an incident. The media enjoyed this inside information, but Humphrey states that the public would be better served if it received the information directly rather than through the media:

We began to realize, this is not media information, this is public information. We were already sending it out. And we went from using a special terminal to send it, to where we had an e-mail gateway where we could send e-mail and it would show up on the text pager. I could show you these big clunky things, and eventually it continued to evolve, where we had SMS on cell phones, and we began to realize we wanted to get it to the public.

Bypassing mass media as traditional intermediaries emerged as an overarching motivator behind innovation and adoption of new media. Old media, or legacy and traditional media like television, is what Humphrey calls "appointment media" – one had to make an appointment at 6pm to watch the 6 o'clock news. Humphrey states, "the appointment media is dead and dying, depending on the market, and the media has gone to be really be a real time media. We are, and our department has tried to evolve with that." He states "that shift between the appointment media and the real time media have brought us some great opportunities."

Despite a clear dissatisfaction with the old media intermediaries, for now the strategy seems to be augmentation not replacement of other media for communication. The ability to get legitimate and verified information out to the public in as many ways as possible meant that LAFD PIOs did not simply leave behind traditional media channels of information dissemination but augmented them with social media. Twitter and other social media were deemed extraordinarily useful by the PIOs for disseminating information to the public rapidly, yet these same PIOs clearly understand that only some portion of their intended audience was on twitter.

Listening: Legitimizing and Validating User-Based Information

In addition to dissemination of information, we heard of instances where LAFD leveraged Twitter to monitor and collect information. Humphrey explained:

we're using the new media to monitor, not just send our stuff out via Twitter, but monitor what other people are sending via micro-messaging services, what other people are sending pictures of, what their queries, what their questions are in real time.

This monitoring raises the question: how does an emergency management official validate information from a citizen communicating via a social media platform? As described, Humphrey still refers citizens to traditional LAFD communication channels to report an emergency. Yet there is

evidence that LAFD PIOs not only monitor and evaluate Twitter for reports of fire but also reward and encourage reporting by citizens:

@LAFD: @Leafstalk: @ChristineNia @Deigh-vydQahztio Thank you for reporting the grass fire alongside the freeway.–Brian 8:45 PM Sep 22nd from web.

Humphrey explained one method he uses to validate information with search and sorting technologies to personally monitor key words related to crises:

I don't have any training, but I use Yahoo Pipes … I dump all my stuff in there, Feed Rinse, all those tools, grind them up and spit them out, and if enough people inside a 20 kilometer area are saying, OMG, or OMFG, that draws my attention. If then I have a traditional media RSS source that says the word, death, explosion, I have a whole algorithm. And then, if it gets good enough, it will make my phone beep. It has to be really--I had a lot of false alarms. My wife wasn't too happy … the phone would buzz all night long, because somebody said something. But people will do certain things, and it lends some degree of credence as to where you want to look closer.

Although the quote suggests a substantial level of technical expertise, Brian and his PIO colleagues stressed being self-taught and largely unsophisticated in their use of these kinds of technologies throughout the interview. Yet the types of solutions they described were not ad hoc. The "algorithms" these PIOs mentioned came less from formal technological expertise and more from intuition based on many years of experience that enabled them to conceptualize the right levels of analysis and sensitivity with the kinds of keywords and potential applications of findings. These PIOs assumed the roles of both information managers (Bharosa et al., 2007) and technological evangelists (Lawrence et al., 2006), wherein their promotion of social media for use within the organization depended on their ability to utilize social media effectively.

Validation of information available on Twitter and via other social media is a persistent and difficult question. Palen et al. (2009) have written extensively on how citizens do a lot of work to validate and correct information in times of crisis (Palen et al., 2009; Starbird et al., 2010; Sutton et al., 2008). The pressure to validate information from unknown sources received from the public is far greater on emergency managers who have to choose whether to commit resources for investigation and response. While Humphrey and his colleagues openly agree that the public is a necessary and important source of information, they have developed their own ways to help validate. Echoing the uncertainty of seeing something in just one medium, Humphrey states:

We try to validate multiple sources. We would not commit life safety, you know, from one point A, to point B, based upon just what we see on Twitter.

Yet a question remains: at what point is information validated such that resources are sent to address a fire or a crisis? Humphrey gave a number of examples, but we present one that is most evocative here, that of the Griffith Park fire in Los Angeles, May 8th, 2007. Humphrey was monitoring Twitter for any mention of the fire:

And they were posting about some smoke and wind conditions and embers going toward homes, and this structure that they said were threatened … I said, "These people have something, but I don't want to take it at face value." So I went to the page they referenced. They had an e-mail there. I sent them an e-mail … and the message said, "Call me." So we take in the old media and move into the new media, moving into the old media, the telephone. And they call up, and, "Hello…?" "Hi, I'm Brian." "You're the guy on the radio." "No,

I'm the guy on the telephone talking to you." "I hear you on the radio all the time." I said, "I appreciate your time. Tell me what you have there." And in that case, I felt I was able to add reasonable validation of what they were seeing, relayed that information to our responders in the field, and it turned out that there were people and property in danger in that area, things we couldn't see, that were over the horizon away from us. In effect, this was a moving wildfire. The military has a model that every soldier is a sensor. Every soldier--we like to say that every citizen is a contributor.

This is Humphrey's vision for the future of social media for emergency management – a system where each citizen is enabled with the technological means to transmit information about a fire, crisis, or disaster directly to emergency management professionals. In this perhaps utopian vision, interactivity is a means to a mutual communicative relationship between citizen and emergency management professional with systems of information verification in place to facilitate mutual trust.

Organizational Challenges to Innovation

With respect to social media implementation, we found some disconnect between the activities of Humphrey as innovator and the organization support structure within which he works. Organizationally he is expected to manage both old and new media in an overwhelming capacity. Humphrey states:

We are drowning in data down here, and we're thirsting for knowledge, just like the people.

Yet the "we" he refers to in his statement is ambiguous – at times Humphrey speaks in line with the LAFD as an organization but not always. Where technology advocates often consider technological innovation synonymous with greater efficiency, we observed that in incorporating social media into their work-practices, PIOs of the LAFD dramatically increased their workload. This happened because social media activity did not replace their other job duties of communicating with mass media and creating informational reports, but augmented these activities with a greater interaction with the public. Humphrey expressed the sheer workload often both in the interview and on his Twitter stream:

It's an inhuman--it's an inhuman workload sometimes, it absolutely inhuman. I get 300 phone calls a day, on a busy day. It just goes, goes, goes, goes, goes, goes, goes. We have not changed our staff.

Not only has the LAFD not increased staff for PIO positions to meet the demands of both old and new media, there is a sense that the value of Humphrey's work is insecure. Since Humphrey serves at the pleasure of the administration, a change in management could change his role and thus the entire social media strategy of the LAFD. On Twitter the PIOs explain:

@LAFD: Since some have asked... @LAFD social media efforts will continue or change at the discretion of our new Fire Chief. 11:11 AM Aug 28th.

There is a sense that the LAFD leadership might not fully grasp the value of social media for assisting the daily activities of firefighters in the city. As the new fire chief established in his position, he too started a Twitter account, clearly signaling LAFD's commitment to the use of social media as an invaluable tool. A deeper look at the Fire Chief's feed, however, showcased a more traditional approach to public communication – where all of that communication was limited to broadcast announcements of events and pointers to crisis information resources. Although LAFD is involved in a highly interactive conversation with the public, this conversation remains the purview of PIOs, one to a 24-shift. For Humphrey the ability to leverage the interactive capabilities of social media for risk and crisis communication is at a critical juncture. The move from the traditional broadcast model to interactivity creates real opportunities to manage emergencies, but in creating expectations that citizens will be heard via social media creates risks too. He explains:

Why the city leaders don't see -- I call it TLC information. When something comes to us, it's TLC, either it's time-life critical or it requires tender loving care. I mean, that's the fork in the road you're at. And the time-life critical have expectations, and I'm already starting to get some people who are angry. The recent fire near the Getty Center, I was out of town. We were short on staffing and people wanted more information and they became angry. They have an expectation. But we can't--I can't hire people, we have one human being on duty. And the ability to gather and analyze and then disseminate really, you can't do that [with] one person. You can't be listening while you're talking.

There is evidence that LAFD is moving toward broader support of social media in its daily activities, by establishing more wireless hotspots, increasing the number of people with Twitter accounts up the chain of command and even putting forth a technology initiative as the Fire Chief had announced in his Twitter feed:

@LAFDFIRECHIEF: Just issued "Technology Initiative 2010 – Phase One" letter. Deploying technology to position the LAFD for the future.

Yet these changes are slow and broader support for the activities of the PIOs is limited by budgeting woes and political will that wanes up the chain of command.

DISCUSSION AND CONCLUSION

The limitations of this case study lie in whether or not other emergency management organizations follow (or perhaps should follow) a similar process of innovation and social media implementation as Brian Humphrey and the LAFD. Clearly we cannot generalize based on one case study and our future research is geared toward addressing this question more comprehensively. At the same time, this case study identifies processes emerging as a variety of emergency management organizations move to adopt Twitter and other social media.

We find that the LAFD is utilizing Twitter as a tool for emergency management both for sending one-way messages to the public and for monitoring and responding to Twitter posts. Monitoring and evaluating posts from the Twitter community comprise what Humphrey describes as "listening" activities, which leverage the interactive two-way affordances of social media. Yet, not all emergency management organizations share Humphrey's vision of social media's implementation for real time interactivity and listening to the public. For example, FEMA states it has been using Twitter "as a means to offer information about the agency's mission, efforts and perspective." FEMA's purpose seems to lean more on the one-way dissemination model of media usage, not "listening." This is reflected in their Twitter stream @FEMA, which is Following 400 and has 50,360 Followers (6/17/11) compared to @LAFDtalk (Following 7,113 with 6,849 Followers, 6/17/11). For some emergency management organizations realizing the potential of interactivity and participation may not be practical or even ideal for their risk communication strategies. Further research in the area of computer security concerns with regard to interactive communication technologies may offer insights in this area.

This study supports the literature that locates an information evangelist at the heart of technological change within organizations. Although self-taught and often seemingly ad hoc in approach, Humphrey is indeed the main driver for information and communication innovations at the LAFD. Our preliminary conversations with other emergency management organizations also support the idea of a small group of visionaries initiating social media implementation. Speaking with those innovators, we also find that management is often resistant or wary to implement social media, which they might not fully understand. We have respected requests not to quote those innovators, often critical in tone, for fear of reprisals from higher management.

This paper highlights the changing nature of risk and crisis communication in light of the proliferation of Internet-based social media technologies that far outpace the constraints of traditional media. The affordances of Twitter and other interactive social media give emergency mangers abilities to communicate, interact with, and respond to the public on a hitherto unseen scale. As we observe PIOs engaging with these technologies, we argue that the PIO's function at the LAFD has exceeded its previous role as primarily sending official messages to the public via traditional media. Indeed, as social media continue to proliferate we might reformulate questions of how emergency management utilizes social media to include questions about how emergency management organizations themselves are changing due to the innovations offered by the emerging communication technologies and the push from broad-scale public adoption of these technologies.

Adding the role of "listener" creates a new orientation for the PIOs who now must manage, filter, and verify incoming information from a host of new social media sites. While we see that resources can at times be allocated based on messages from Twitter and social media, the collective intelligence of the public as a smart mob is not taken for granted. The ad hoc and intuitive manner by which social media messages are vetted indicates a dynamic and flexible evaluation process; however, we find an increasing potential for PIOs to be overwhelmed by the amount and types of information. Information overload for emergency management professionals is not a new or diminishing phenomenon (Hiltz & Turoff, 1985, 2009). One might say that Humphrey's kind of dedication pushes the utmost limits of human capacity and cannot be expected of every PIO. But our purpose here is in considering possibilities. If one PIO with a will, a determination, and a severe lack of resources and organizational backing is capable of conducting this kind of work, then, with sufficient technological support and organizational backing, this kind of service is possible on a broad level. We argue that while technological innovation is possible in emergency management organizations, it often relies on the limited capacities of individuals, the information evangelists, who might not be supported by the organization as a whole. Organizational support and political will to initiate and to support change is paramount if we are to see these kinds of services provided broadly, but it is also important for such organizations to recognize the function and value of information evangelists in their midst.

ACKNOWLEDGMENT

The authors are indebted to Brian Humphrey and the other PIOs and innovators who participated in this research. The views of the participants are their own and do not necessarily reflect the views of their employers. This research was supported by a grant from the John Randolph Haynes and Dora Haynes Foundation.

REFERENCES

Bharosa, N., Appelman, J., & de Bruin, P. (2007). Integrating technology in crisis response using an information manager: First lessons learned from field exercises in the Port of Rotterdam. In *Proceedings of the International Conference on Information Systems for Crisis Response and Management*, Brussels, Belgium.

Boersma, K., Groenewegen, P., & Wagenaar, P. (2009). Emergency response rooms in action: An ethnographic case study in Amsterdam. In *Proceedings of the International Conference on Information Systems for Crisis Response and Management*, Gothenburg, Sweden.

Brown, S. L., & Eisenhardt, K. M. (1997). The art of continuous change: Linking complexity theory and time-paced evolution in relentlessly shifting organizations. *Administrative Science Quarterly*, *42*(1), 1–34. doi:10.2307/2393807

Cohen, N. (2009, June 20). *Twitter on the barricades: Six lessons learned*. Retrieved from http://www.nytimes.com/2009/06/21/weekinreview/21cohenweb.html

FEMA. (2009). *Use of social media tools at FEMA*. Retrieved from http://tinyurl.com/yhsv9xn

Fox, S., Zickuhr, K., & Smith, A. (2009). *Twitter and status updating*. Retrieved from http://tinyurl.com/yfmgg25

Gomez, E. A., & Turoff, M. (2007). Community crisis response teams: Leveraging local resources through ICT e-readiness. In *Proceedings of the 40th Annual Hawaii International Conference on System Sciences* (p. 24).

Hiltz, S. R., & Turoff, M. (1985). Structuring computer-mediated communication systems to avoid information overload. *Communications of the ACM*, *28*(7), 680–689. doi:10.1145/3894.3895

Hughes, A., & Palen, L. (2009). Twitter adoption and use in mass convergence and emergency events. In *Proceedings of the International Conference on Information Systems for Crisis Response and Management*, Gothenburg, Sweden

Idle, N., & Nunns, A. (Eds.). (2011). *Tweets from Tahrir* [Kindle DX version]. Retrieved from http://www.amazon.com

Jennex, M. E. (2010). Implementing social media in crisis response using knowledge management. *International Journal of Information Systems for Crisis Response and Management, 2*(4), 20–32.

Kendra, J., & Wachtendorf, T. (2003). *Beyond September 11th: An account of post-disaster research* (pp. 121–146). Boulder, CO: University of Colorado, Natural Hazards Research and Applications Information Center.

Lawrence, T., Dyck, B., Maitlis, S., & Mauws, M. (2006). The underlying structure of continuous change. *MIT Sloan Management Review, 47*(4).

Liu, S. B., Iacucci, A. A., & Meier, P. (2010, August). Ushahidi in Haiti and Chile: Next generation crisis mapping. In *Proceedings of the American Congress on Surveying and Mapping.*

Lotan, G. (2009). *A visual exploration of Twitter conversation threads in the days following the Iranian Elections of June 2009.* Retrieved from http://giladlotan.org/viz/iranelection/about.html

Majchrzak, A., Jarvenpaa, S. L., & Hollingshead, A. B. (2007). Coordinating expertise among emergent groups responding to disasters. *Organization Science, 18*(1), 147–161. doi:10.1287/orsc.1060.0228

Mendonca, D., Jefferson, T., & Harrald, J. (2007). Collaborative adhocracies and mix-and-match technologies in emergency management. *Communications of the ACM, 50*(3), 44–49. doi:10.1145/1226736.1226764

Mileti, D. S., & Darlington, J. D. (1997). The role of searching in shaping reactions to earthquake risk information. *Social Problems, 44*(1), 89–103. doi:10.1525/sp.1997.44.1.03x0214f

Milis, K., & van de Walle, B. (2007). IT for corporate crisis management: Findings from a survey in 6 different industries on management attention, intention and actual use. In *Proceedings of the 40th Annual Hawaii International Conference on System Sciences* (p. 24).

Murphy, T., & Jennex, M. E. (2006). Knowledge management, emergency response, and hurricane Katrina. *International Journal of Intelligent Control and Systems, 11*(4), 199–208.

Palen, L., Vieweg, S., Liu, S., & Hughes, A. (2009). Crisis in a networked world: Features of computer-mediated communication in April 16, 2007 Virginia Tech event. *Social Science Computer Review, 27*(4), 467–480. doi:10.1177/0894439309332302

Reynolds, B., & Seeger, M. (2005). Crisis and emergency risk communication as an integrative model. *Journal of Health Communication, 10*(1), 43–55. doi:10.1080/10810730590904571

Rheingold, H. (2002). *Smart mobs: The next social revolution transforming cultures and communities in the age of instant access.* Cambridge, MA: Basic Books.

Scaffidi, C., Myers, B., & Shaw, M. (2007). Trial by water: Creating hurricane Katrina "person locator" web sites. In Weisband, S. (Ed.), *Leadership at a distance: Research in technologically-supported work* (pp. 209–222). Mahwah, NJ: Lawrence Erlbaum.

Shklovski, I., Burke, M., Kiesler, S., & Kraut, R. (2010). Technology adoption and use in the aftermath of hurricane Katrina in New Orleans. *The American Behavioral Scientist, 53*(8), 1228–1246. doi:10.1177/0002764209356252

Sood, R., Stockdale, G., & Rogers, E. M. (1987). How the news media operate in natural disasters. *The Journal of Communication, 37*(3), 27–41. doi:10.1111/j.1460-2466.1987.tb00992.x

Sorenson, J., & Sorenson, B. (2006). Community processes: Warning and evacuation. In Rodríguez, H., Quarantelli, E. L., & Dynes, R. (Eds.), *Handbook of disaster research* (pp. 183–199). New York, NY: Springer.

Starbird, K., & Palen, L. (2010). Pass it on? Retweeting in mass emergency. In *Proceedings of the International Conference on Information Systems for Crisis Response and Management*, Seattle, WA.

Starbird, K., & Palen, L. (2011). "Voluntweeters:" Self-organizing by digital volunteers in times of crisis. In *Proceedings of the ACM Conference on Computer Human Interaction*, Vancouver, BC, Canada.

Starbird, K., Palen, L., Hughes, A., & Vieweg, S. (2010). Chatter on *The Red*: What hazards threat reveals about the social life of microblogged information. In *Proceedings of the ACM Conference on Computer Supported Cooperative Work*, Savannah, GA (pp. 241-250).

Sutton, J. (2009). Social media monitoring and the democratic national convention: New tasks and emergent processes. *Journal of Homeland Security and Emergency Management, 6*(1). doi:10.2202/1547-7355.1601

Sutton, J., Palen, L., & Shklovski, I. (2008). Backchannels on the front lines: Emergent use of social media in the 2007 southern California fires. In *Proceedings of the International Conference on Information Systems for Crisis Response and Management*, Washington DC.

Tabor, D. (2008, October 20). *LAFD's one-man geek squad brings Web 2.0 to firefighting.* Retrieved from http://www.wired.com/entertainment/theweb/magazine/16-11/st_firefight

Turoff, M., & Hiltz, S. R. (2009). The future of professional communities of practice. In C. Weinhardt, S. Luckner, & J. Stößer (Eds.), *Proceedings of the 7th Workshop on Designing E-business Systems, Markets, Services, and Networks* (LNBI 22, pp. 144-158).

van de Ven, J., van Rijk, R., Essens, P., & Frinking, E. (2008). Network centric operations in crisis management. In *Proceedings of the International Conference on Information Systems for Crisis Response and Management*, Washington, DC.

White, C. (2011). *Social media, crisis communications and emergency management: Leveraging Web 2.0 technology.* Boca Raton, FL: CRC Press.

White, C., Plotnick, L., Kushma, J., Hiltz, S., & Turoff, M. (2009). An online social network for emergency management. In *Proceedings of the International Conference on Information Systems for Crisis Response and Management*, Gothenburg, Sweden.

Yin, R. (1994). *Case study research: Design and methods* (2nd ed.). Thousand Oaks, CA: Sage.

Yun, J.-H. J., Park, S., & Avvari, M. V. (2011). Development and social diffusion of technological innovation. *Science, Technology & Society, 16*(2), 215–234. doi:10.1177/097172181001600205

This work was previously published in the International Journal of Information Systems for Crisis Response and Management, Volume 3, Issue 4, edited by Murray E. Jennex and Bartel A. Van de Walle, pp. 1-16, copyright 2011 by IGI Publishing (an imprint of IGI Global).

Chapter 14
A Distributed Scenario-Based Decision Support System for Robust Decision-Making in Complex Situations

Tina Comes
Karlsruhe Institute of Technology (KIT), Germany

Michael Hiete
Karlsruhe Institute of Technology (KIT), Germany

Nick Wijngaards
Thales Research and Technology & D-CIS Lab, The Netherlands

Claudine Conrado
Thales Research and Technology & D-CIS Lab, The Netherlands

Frank Schultmann
Karlsruhe Institute of Technology (KIT), Germany

ABSTRACT

Decision-making in emergency management is a challenging task as the consequences of decisions are considerable, the threatened systems are complex and information is often uncertain. This paper presents a distributed system facilitating better-informed decision-making in strategic emergency management. The construction of scenarios provides a rationale for collecting, organising, and processing information. The set of scenarios captures the uncertainty of the situation and its developments. The relevance of scenarios is ensured by gearing the scenario construction to assessing alternatives, thus avoiding time-consuming processing of irrelevant information. The scenarios are constructed in a distributed setting allowing for a flexible adaptation of reasoning (principles and processes) to both the problem at hand and the information available. This approach ensures that each decision can be founded on a coherent set of scenarios. The theoretical framework is demonstrated in a distributed decision support system by orchestrating experts into workflows tailored to each specific decision.

DOI: 10.4018/978-1-4666-2788-8.ch014

INTRODUCTION

The growing complexity of contemporary industrial production systems in conjunction with an increased vulnerability clearly indicates the need for a well-structured risk and emergency management (Cruz, Steinberg, & Vetere-Arellano, 2006). In industrial production, emergencies are frequently caused by singular large-scale high impact events and confront society, economy and environment with substantial consequences (Cutter, 2003). Decision-making in emergency management presents all experts involved with demanding challenges, which stems to a great extent from the situation's complexity (Papamichail & French, 2005). This complexity arises from several sources. First, information from different domains and disciplines, such as natural and engineering sciences, medicine, law and economy, about the concerned systems or processes needs to be combined. Additionally, the information available is prone to be uncertain: it is frequently not (yet) confirmed, noisy, uncertain, or lacking (Cutter, 2003; Wright & Goodwin, 2009). Second, a consensus taking into account the objectives and value judgments of numerous actors must be found. Often, the objectives are conflicting and tradeoffs need to be made.

Two types of decision support systems have been developed to handle both types of complexity: the first due to the complexity of and the uncertainty about the system under scrutiny, the second due to the complex structure of the decision makers' preferences and objectives. *Scenario-based approaches* help the decision makers in structuring the information on the emergency and its possible developments. *Decision support systems* modelling all actors' perception of the decision problem have proven useful for reducing the complexity in the evaluation of decision alternatives (Papamichail & French, 2005).

This paper presents a method exploiting the integration of scenarios and Multi-Criteria Deci-

sion Analysis (MCDA) in a distributed decision support system addressing medium to long term emergency management. This approach has several phases. First, scenarios tailored for the decision at hand need to be constructed. Second, these scenarios must be evaluated and the results must be presented to the decision makers in an easily understandable manner. Both steps of the process are complementary in the sense that the development of scenarios is geared to assessing decision alternatives, while MCDA serves as a mechanism to evaluate and prioritise scenarios. Thus the integrated SBR & MCDA approach helps avoiding information overload, time-consuming analysis and processing of irrelevant information while ensuring that each single scenario as well as the set of scenarios is sufficiently rich to be a valid basis for the decision-making. Furthermore, unlike in standard scenario approaches, constraints with respect to the time, information and expertise available are respected.

The paper is structured as follows. The next section discusses the notion of robustness and discusses briefly some approaches used to support robust-making in emergency management. We give a general explanation on how a decision problem can be structured via directed acyclic graphs (DAGs). Particularly, it is highlighted that the use of DAGs facilitates the implementation of the scenario construction on basis of a service-oriented approach. Next, we describe the construction of scenarios. This procedure includes two steps. First, workflows that link locally available expertise and thereby structure the flow of information are created. Second, the resulting DAG is used as the basis for the construction of scenarios that fulfill a set of quality requirements as well as possible, given limited time and availability of experts. We show how this approach facilitates decision-making under severe (i.e., non-quantifiable) uncertainty. An example developed together with experts and users from the Danish Emergency Management Agency (DEMA) highlights the main characteristics of our approach. Finally, we summarise the main aspects and give an outline of open research questions.

ROBUST DECISION-MAKING IN STRATEGIC EMERGENCY MANAGEMENT: METHODS AND TOOLS

As the consequences of decisions in emergencies may have an important impact and the present uncertainties are considerable, decisions are required to be *robust* (Barthélemy, Bisdorff, & Coppin, 2002). Here, robustness refers to both the stability of the decision support system's results against minor changes in the parameters that characterise the actors' preferences, and against changes in the emergency situation or the perception thereof (within given limits). In the first case, the decision alternative chosen needs to be the preferred alternative within a range of parameters; in the second case, the decision must account for several possible developments within a given domain of possible situations.

While the first type of robustness can be warranted by performing sensitivity analyses (Geldermann et. al., 2009), *scenarios* can be employed to support the second robustness type. As there are a number of different definitions of the term scenario (Notten et al., 2003, it is important to note that we understand scenario as a plausible, internally fully coherent and consistent description of the situation and its future development (Schnaars, 1987) including both external factors beyond the control of the decision makers and the alternative implemented. The use of scenarios challenges existing mental frames and avoids the cognitive biases in the estimation of probabilities (Wright & Goodwin, 2009). Therefore, scenarios are particularly useful in cases of severe uncertainties, i.e., when the situation defies quantitative descriptions (Ben-Haim, 2000; Regan, 2005). Overall, scenarios support decision makers in

selecting an alternative that performs sufficiently well under a variety of possible developments. In this sense, scenarios support robust decision-making with respect to the stability against different developments (beyond minor perturbations) of the situation (Harries, 2003). To support decision makers who need to consider multiple goals and a set of scenarios, techniques from Multi-Criteria Decision Analysis (MCDA) can be adapted and used (Comes et al., 2009; Durbach & Stewart, 2003; Mahmoud et al., 2009; Montibeller, Gummer & Tumidei, 2006, van der Pas et al., 2010).

Although in medium and long term emergency management decisions do *not* need to be made immediately, time is restricted. Time criticality reduces the possibility of bringing together all experts involved in the scenario construction process in person as is usually done in discursive approaches to scenario planning (Schnaars, 1987). Furthermore, some experts may contribute to several decision problems or fulfill other tasks arising in the course of the emergency. Experts may not have time to participate in the entire scenario construction process. Classical expert systems attempt to deal with this problem by solving the decision problem autonomously, without the interference of the users, by using a (limited) model of the domain and a set of data or assumptions (Turban & Watkins, 1986). By excluding human experts, however, these systems require a vast continuously updated knowledge base covering all aspects of the available and feasible alternatives and all possible future developments (Dugdale, 1996; French, Maule, & Papamichail, 2009). In emergency management, this is clearly infeasible, since emergencies can be characterised as rare, often unexpected events (Cutter, 2003). Therefore, they defy (overly) standardised descriptions. The method for the construction of relevant scenarios presented in this paper is supported by an intelligent system enabling collaborative processing of information while taking into account the bounded availability of experts and the inadequacy of standardised descriptions. This approach combines the capabilities of multiple human experts and automated reasoning processes, each of which contributes specific expertise and processing resources to construct the scenarios. To organise and structure information processing and sharing, directed acyclic graphs (DAGs) that explicate cause-effect chains are used.

The robustness of alternatives can be investigated by evaluating the set of scenarios constructed for each decision alternative with techniques from MCDA and comparing the results. Furthermore, a deeper understanding of the circumstances under which a certain alternative performs particularly well (or badly) can be gained.

DEVELOPING THE DECISION MODEL

Decision Support Systems for Emergency Management

In emergency management, the decision-making task is usually modelled as a choice among a number of feasible alternatives based on a number of goals, making Multi-Attribute Decision Making (MADM) the MCDA technique of choice for our purposes (cf. Belton & Stewart, 2002). In MADM the basis for the evaluation of alternatives is a hierarchically structured attribute tree that is elicited from the decision makers (Keeney & Raiffa, 1976). The attribute tree shows how a strategic overall-objective is broken down first into less and less abstract sub-objectives *(criteria)* and finally into measurable *attributes,* taking the problem's framing from an initially vague and intuitive understanding to a more formal description that can be analysed mathematically (Stewart, 1992). Despite their advantages in reducing complexity and arriving at a common understanding of the problem, both deterministic and probabilistic MADM approaches have some drawbacks when applied under severe uncertainty. While the first

do not offer the possibility to account for uncertainty, the latter rely on probability distributions that are hard to define in emergencies, as these are rare events (Ben-Haim, 2000).

Scenario analysis is a well-established method for reasoning under severe uncertainty. By using scenarios as a basis for the decision support system, it is possible to overcome cognitive biases such as overconfidence and to integrate fundamental risks associated with certain developments of the situation that may be of very little probability into the reasoning framework (Schoemaker, 1993).

As a requirement for their acceptance, the scenarios constructed should show some qualitative properties (Heugens & van Oosterhout, 2001). A first quality requirement is *relevance* and *non-redundancy.* Each scenario should contain pieces of information that are decisive for the decision makers or have a (non-negligible) influence on one of the decisive variables. To avoid information overload, the scenarios should not contain redundant or irrelevant information. Furthermore, the set of scenarios should allow for exploring the realm of possible developments. Additionally, each scenario should be *plausible* (not going beyond the realm of possibility), *coherent* (having explicit logical connections explaining *how* the system evolves), and *consistent* (having no contradiction between its parts).

The task of constructing scenarios that respect these conditions can be difficult. In scenario planning, scenarios are developed in discursive procedures (Schnaars, 1987; Bañuls & Turoff, 2011), while formative scenario analysis (FSA) starts with identifying impact factors that influence the development of the situation (Scholz & Tietje, 2002). Both approaches do not explicitly include interdependencies *(how* variables are related to one another), thereby requiring substantial effort to ensure coherence and consistency. Furthermore, both methods require a substantial amount of time and the attendance of all experts involved to one or several interviews (Bañuls & Turoff, 2011).

Hence, the applicability of these techniques is limited when time is limited and the availability of experts is bounded.

In the next section, a novel approach to construct scenarios based on Causal Maps (CMs) is presented. This approach ensures plausibility, coherence and consistency as far as adequate given bounded availability of experts and limited time. Furthermore, it is shown that this approach facilitates *distributed reasoning* by implementing the underlying workflows within a distributed approach called Dynamic Process Integration Framework (DPIF).

Decision Model Structuring Using Causal Maps

The structuring of the decision problem is often characterised as one of the hardest, yet most crucial parts in developing decision support systems (Belton & Stewart, 2002). Originally, Causal Map (CMs) was developed as a problem structuring technique representing interlinked variables in a network (Montibeller & Belton, 2006). CMs depict variables that characterise the emergency as vertices. Directed edges describing cause-effect links connect vertices. The causality ensures that any CM is chronologically ordered: If a vertex i precedes a vertex j (represented graphically by an edge from i to j), the state of i influences the state of j causally. This implies that the state of i at time t influences the state of j at $t+\Delta$, $\Delta>0$. The temporal structure of the CM allows for the elimination of loops. Choosing the time steps appropriately, the CM can always be represented as a DAG.

A CM brings together several levels of a problem description. Considering only the vertices and their interdependences (represented by the edges) reveals the structure of the decision problem. This *structural level* shows the variables that have an impact on the decision and their interrelations. At the *functional level*, the relations or inference mechanisms between the variables, i.e., the way in

which each variable's possible value(s) influence its successors' values is analysed in more detail. In emergencies, these relations are often uncertain and captured in a framework allowing for reasoning under uncertainty. For instance, Bayesian Networks use (conditional) probability distributions to describe the variables' interdepencies (Russell & Norvig, 2003). Fuzzy Cognitive Maps (FCMs) rely on Fuzzy Logic and the respective inference mechanisms and fuzzyfication tools (Peña, Sossa & Gutiérrez, 2008). The CM's *informational level* captures the *results* associated to one instantiation of the system (Diehl & Haimes, 2004). In the framework presented in this paper, the value of a variable is not restricted to numbers. A value can, inter alia, be a map, a recorded speech, or text file according to what best fits the users' needs and is understood best by the further experts contributing to the scenario construction.

Distributed Problem Structuring Using the Dynamic Process Integration Framework (DPIF)

In structuring the decision problem, the first step is the elicitation of the attribute tree, where the decision makers define their objectives in terms of criteria and attributes (Keeney & Raiffa, 1976, cf. Figure 1). The attributes, which can be found on the lowest hierarchical level of the attribute tree, allow for *measuring* (or quantitatively estimating) the impact of the implementation of decision alternatives with respect to several objectives (Belton & Stewart, 2002). Attributes correspond to states in the physical world as do the variables in a CM. On this basis, we use the attributes as the intersection of the description of the situation and its development (represented in the CM) and the evaluation (represented in the attribute tree). The use of attributes facilitates filtering the variables in the CM (and ultimately, assessing the relevance of the scenarios constructed). *Relevant* variables have an impact on at least one attribute. All relevant variables should be contained in the

Figure 1. Attribute tree

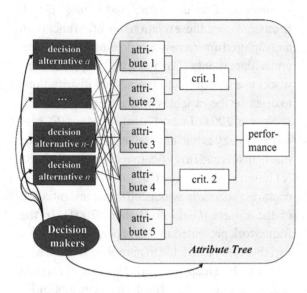

CM used to construct the scenarios for the decision problem at hand.

To discover the relevant variables and their dependencies (i.e., the vertices and edges in the CM), a distributed approach based on the resolution of task dependencies is used (cf. Comes, Hiete, Wijngaards, & Schultmann, 2010). The Dynamic Process Integration Framework (DPIF) lets experts (humans or automated systems) define their (reasoning) capabilities in terms of a task they can perform *(service)* and in terms of information this task requires. This task description reflects that an expert's output may rely on input he cannot determine autonomously. The presented system connects experts via *software agents* that are an interface between the expert(s) and the service-based discovery architecture the DPIF provides (Pavlin, Kamermans, & Scafes, 2010). Although it is necessary that the experts specify the *type* of information they need and can provide, the system allows for a flexible adaption of the formats used (e.g., images, spoken text or numbers).

If experts refer implicitly (in the case of humans) or explicitly (in the case of automated expert systems) to a local causal model that allows them

to provide their service, their reasoning processes are represented as local CMs (cf. Figure 2). These allow for mapping input from other experts to their own output captured in *local sink vertices*. A sink vertex of a network graph is a vertex without successors, see vertices with double borders in Figure 2. While locally, the output of each expert is captured in the sink vertex of its local CM, the sink vertex of the complete global CM merged with the attribute tree is the performance vertex. By iteratively combining the relevant local CMs, a global CM is developed in a top-down way (i.e., in the direction opposite to causation). In Figure 2, the black dashed edges show how the local CMs are combined at configuration time, whereas the red dotted edges correspond to the flow of information during computation. Strictly speaking, a system of collaborating experts arises combining their local knowledge such that the resulting system corresponds to a global CM. If all experts comply with the causal structure mentioned above, the resulting global CM can always be represented by a directed acyclic graph. In the following, it is assumed that this causality condition is fulfilled.

When a decision must be made, the configuration process starts with a request from the decision maker to perform an evaluation for a set of decision alternatives (in Figure 2 only shown for one alternative *a*). The request for attributes' scores *given each decision alternative* initiates the construction of the CM. In the first step, DPIF agents look for experts that can provide information about the attributes. When such experts are found, they refer to their local CMs and indicate that in order to supply information about the attribute they depend on further information (see Figure 2).

Resolving the first information requirement described above does not finish the map elicitation process. Figure 2 shows that all experts *A* to *E* need additional information to provide their services (namely, information about the variables *2* and *5–8*). Again, the DPIF manages the requests for information. The system iteratively configures and expands a distributed reasoning model, which

Figure 2. Constructing the global CM from local CMs

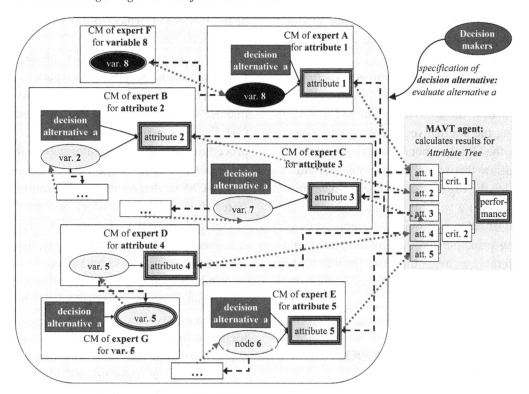

corresponds to the global CM This process finishes when the map covers all expertise needed to perform the scenario construction. This means that:

- All dependent vertices are *sufficiently connected*, i.e., for all local CMs where the sink vertex has predecessor vertices, experts capable of providing the required information are identified;
- Vertices without any predecessors are independent (depicted in black in Figure 2 e.g., variable 8), i.e., the respective experts can determine the state of the corresponding variables autonomously (e.g., via measurements).

All local CMs have a unique local sink vertex, which captures the information the respective experts are going to provide. The complete set of interconnected local CMs represents the *functional level* of the global CM, as it shows the expertise used to determine all relevant variables' states. To develop the corresponding CM on the *structural level,* the local CMs are merged. Individual experts are offered information which they have defined relevant for fulfilling their task. Any expert is not confronted with other information that is processed in the global CM. This mechanism ensures reduction of the problem of information overload.

This system allows for flexible reaction to the problem at hand, as the processing of information by each expert (i.e., the *reasoning*) is *not* standardised (Pavlin, Kamermans, & Scafes, 2010). While the DPIF connects experts based on service descriptions, each expert is free to choose the manner of reasoning (e.g., algorithms, heuristics, best practices). This property is of great importance in highly varying, dynamic and unpredictable situations such as emergencies. Particularly, it is possible that the expert adapts to the type and quality of information available (e.g., at first, information is uncertain, later it is confirmed). Yet, the structure of each expert's local CM (from the perspective of our model) usually remains unchanged in the course of the decision-making process.

The global CM shows how to go *structurally* (in the direction of causation) from all independent vertices to the attributes required for finally evaluating the alternatives. Contrarily to expert systems, the CM does not encode the knowledge regarding how this is achieved *functionally*. Rather, it specifies the *expert* responsible for determining the state of a variable and prescribes which input information is required. Several types of reasoning principles are hereby integrated into one map that can adapt their use flexibly to the problem at hand (cf. Comes, Hiete, Wijngaards, & Schultmann, in press).

CONSTRUCTING SCENARIOS USING CAUSAL MAPS

This section shows how we use CMs to construct decision-relevant scenarios. On basis of the concepts and terms developed previously, it is now possible to further clarify the term scenario. A scenario is one complete instantiation of the CM (on the informational level).

Scenario Construction

The scenario construction in the presented framework follows an iterative procedure. The process starts with decision makers to evaluate the performance of a finite set of decision alternatives $A = \{a_1, ..., a_N\}$, with the intention to decide on implementing one alternative for action. To this end, the decision makers request that consequences of each alternative under varying scenarios are determined. In the presented framework, these consequences are operationalised by a set of attribute scores for each alternative a_i in A. Then, all experts linked in the CM are activated and asked

to provide their specific service. The *configuration of the CM* is performed in a top-down manner, cf. black dashed lines in Figure 2.

In the *scenario construction* phase, content is added successively to the scenarios, by successively processing information in a bottom-up manner following the links established in the CM. Each expert uses his domain knowledge and procedures as well as the information about the values of the direct predecessor vertices in the global CM to determine possible and relevant states of the variable for which he agreed to provide results given the information he receives. In case this state is prone to uncertainty, the experts can specify several (relevant) values.

While the configuration of the CM starts with the attributes, the processing of information starts with the analysis of *independent vertices* in the CM (depicted in black, e.g., var. *8* in Figure 3). The experts providing information about the independent vertices assess the states of the corresponding variables (based on their local knowledge).

If there is uncertainty about the state of a variable, an expert can pass on several possible and relevant estimates for one variable he judges relevant. This information can be encoded as a set of numerical values, a number of maps, a set of text files etc. The only restriction to the type of information used is that all experts determining the values of the direct successor vertices must understand and be able to process it. To ensure scenario *consistency* all output variants must correspond to the input used to determine these estimates.

Figure 3 shows a part of the workflow where a set of scenarios arises. Assume that variable 8 is prone to uncertainty. The responsible expert F decides to transfer three possible states *(I, II and III)*. Then, expert A assessing attribute *1* must determine this attribute *under all states of* variable *8*. Expert *A* himself can determine a set of possible and relevant states for each of the states of variable *8*. In the example shown in Figure 3,

A passes on multiple possible states for each of the three possible states of variable *8,* namely, two states for *8=I* and *8=III,* and three states for *8=II.* The result is a total number of seven partial scenarios for attribute *1* under decision alternative *a.* The scenarios are called *partial* here as a complete scenario encompasses a description of all the variables in the CM, not only those relevant for the determination of just one attribute.

In brief, whenever there is uncertainty on how the emergency will develop the current partial scenarios branch into a number of 'new' (partial) scenarios. The arising set of scenarios can therefore be understood as a way of expressing uncertainty reflected in a range of possible and relevant values for each variable. The scenarios are completed iteratively, starting at the independent vertices and following the CM until the attributes are reached, by the development of consistent scenarios.

Properties of the Generated Set of Scenarios

Four requirements for the scenarios to provide a sound basis for the decision-making were identified: relevance, plausibility, consistency, and coherence. This section shows how far these requirements can be achieved by the scenario construction procedure presented.

First, this approach ensures the relevance of the scenario for the decision-making. The starting point of the CM configuration is the set of attributes. As the attributes are a means to operationalise the consequences of each scenario and as they are, ultimately, the basis for the evaluation, the scenarios contain all information relevant for the decision makers. Furthermore, as each expert is provided only with the information he himself judged relevant for his task, each variable within the scenario is itself relevant to determine the evaluation. In this manner, the scenarios constructed are relevant and sufficiently rich.

Figure 3. Scenario construction for alternative a

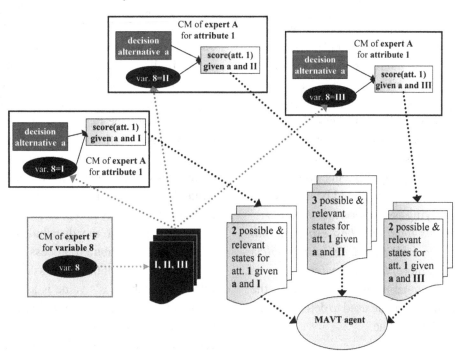

This approach ensures the *consistency* of each scenario from the beginning onwards, under the assumption that the local assessments of the experts reflect the best timely available procedures or techniques to determine the variable's value given their current knowledge. Furthermore, it is assumed that the experts are capable of specifying all pieces of information that are relevant for their task. Contrarily to Formative Scenario Analysis (FSA), where first a vast set of scenarios is created by combining all possible states of all variables arbitrarily, which is afterwards reduced by applying a pair-wise consistency assessment carried out by experts (Scholz & Tietje, 2002), in our approach consistency is ensured in a distributed manner by the experts providing each piece of information. Furthermore, each partial scenario (i.e., each intermediate step in the scenario construction phase) is itself consistent. The distributed implementation must properly combine the multiple experts' output into consistent individual scenarios. In this way, the number of scenarios is kept (much) smaller than in FSA.

The scenarios are *coherent*, as not only the states of all variables but also a description of the way in which they are interlinked is part of each scenario. Representing the scenarios as a network of causes and effects (or means and ends) allows for visualising these interdependencies.

Finally, the *plausibility* of each scenario is enhanced as our approach allows for understanding the underlying interdependencies between the variables, while integrating the best available expertise to assess their state(s). Overall, the fulfillment of the scenario quality requirements consistency, coherence and plausibility should be understood under the underlying assumption that *the experts' local assessments reflect the best expertise that is currently exploitable for the decision problem at hand*. Furthermore, the transparency can be augmented by annotating the vertices with the expert and/or the technique used to determine the respective variable's value.

Scenario Management

The number of scenarios constructed depends on the number of uncertain variables as well as on the number of values that are passed on per uncertain variable. Figure 4 shows an example. Information on the values of the variables *var 1* and *var 2* is needed to derive variable *var 3*'s value. For both variables *var 1* and *var 2*, the experts are uncertain about their values and provide m_1 and m_2 values of *var 1* and *var 2* respectively. These values are combined to new partial scenarios containing information on both variables value. As *var 1* and *var 2* are independent from each other, each value of *var 1* is combined with each value of *var 2* resulting in $m_1 \cdot m_2$ scenarios which are the basis to determine *var 3*. If *var 3*'s value is also prone to uncertainty, the number of scenarios increases further. This process may lead to a combinatorial explosion, with an immense number of scenarios being constructed.

As the number of scenarios depends on the number of variables prone to uncertainty and on the number of values per uncertain variable, there are two ways to control the number of scenarios: reducing the uncertainty about the emergency itself (i.e., reducing the number of uncertain variables) or restricting the number of values each expert is allowed to pass on. If it is impossible to reduce the number of uncertain variables (e.g., as time is limited and no further investigations can be performed), the maximum number of values that an expert is allowed to pass must be chosen carefully. Denote λ_j the maximum number of values allowed per uncertain variable *var j*. The value of λ_j can be understood as a compromise between uncertainty and ambiguity.

In the spirit of the distributed information processing, a distributed process to determine λ_j is proposed. Each of the experts depending on the information about x_j's value is asked how many values of x_j he can process within a certain time t_j. (This time t_j corresponds to the time each expert is assigned to perform his task. The longest (time-weighed) path through the network must have a length smaller than T, where T is the time available to make a decision.) Then, λ_j is the minimum of these values.

EVALUATION AND SCENARIO SELECTION

This approach is ultimately targeted at supporting decision makers who have to choose a decision alternative out of a finite set of options under severe uncertainty. To accomplish this aim, the CM is connected to the attribute tree (cf. Figure 5). For each decision alternative a_i $(i=1,...,n)$ an assessment of the attributes' scores is requested. Following the approach described previously, these scores are determined.

In case the states of all variables are deterministic, one set of attribute scores per decision alternative is assessed, i.e., one scenario per alternative and the usual MADM approach is followed

(cf. Belton & Stewart, 2002). If there is uncertainty, however, a set of scenarios $S(a_i)$ is created for each alternative a_i. The complete set of constructed scenarios is denoted Ω. When all attribute scores for all $s \in \Omega$ are determined, each scenario s is evaluated using the attribute tree and the decision makers' intra- and inter-criteria preferences, resulting in a performance $p(s)$ (Belton & Stewart, 2002). This technique allows for *comparing* the performances of each decision alternative under a variety of scenarios.

To facilitate decision-making encompassing all scenarios, a further aggregation-step is performed (cf. Figure 5). To this purpose, weights $w_j[s_j(a_i)]$ reflecting the *relative importance* of each scenario $s_j(a_i)$ are elicited from the decision makers, where $j=1,...,m$, $|S(a_i)|=m$ and $\sum_{j=1}^{m} w_j = 1$. These weights do not reflect the probability or likelihood of a scenario, rather they take into account the risk preferences of the decision makers. For a more detailed description on approaches to

Figure 4. Example of the combinatorial explosion of scenarios

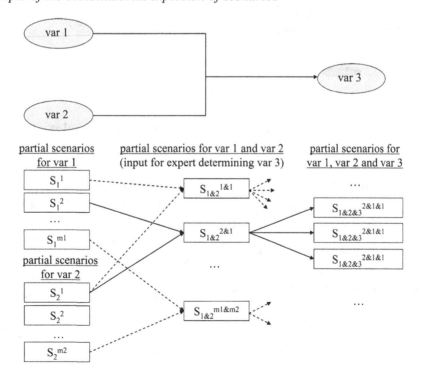

Figure 5. Robust decision support by aggregation of individual scenarios' results

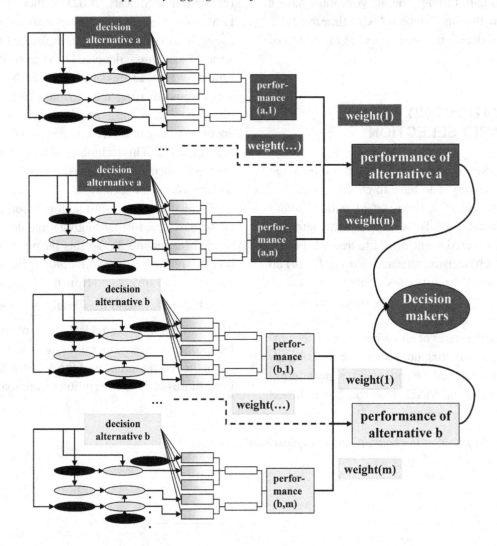

elicit these weights (e.g., on basis of the deviation from a desired value), see Comes, Hiete, and Schultmann (2010).

For each decision alternative a_i the performances of all scenarios $s_j(a_i)$ are aggregated as $p(a_i) = \sum_{j-1}^{m} w_j \ p(s_j(a_i))$ (cf. Figure 5). In this manner, *robust* decision-making, i.e., the choice of an alternative that performs sufficiently well for a set of scenarios (Ben-Haim, 2000; Comes, Hiete, & Schultmann, 2010) is supported. This approach can be particularly useful when Ω is large, or when there is overconfidence in a narrow range of scenarios, neglecting the significance of all other possible developments – a situation frequently encountered in emergency management (Wright & Goodwin, 2009).

To enhance acceptance and to facilitate consensus building on the preferences (e.g., as captured in scenario weights), it is not sufficient to provide the decision makers only with the total performance for each alternative. Rather, the result for the scenarios should be presented in more detail. A selection criterion is needed as, unfortunately, the cognitive capacity of decision makers is limited (ca. five scenarios at a time, cf. Godet,

1990). The integration of CM and attribute tree facilitates the use of the evaluation of scenarios to select for each decision alternative the scenarios for detailed analysis. This approach ensures that the chosen scenarios are the most distinct with respect to the decision makers' actual assessment. For each set $S(a_j)$, the scenarios with the worst, the best and the performance closest to the median are selected and presented to the decision makers in detail, adapting the usual choice of a pessimistic, an optimistic and a baseline scenario (Schnaars, 1987) to the aggregated results. This approach allows making the spread of possible evaluations for each alternative visible and facilitates robust decision-making, for example a minimum worst-case performance must be reached.

EMERGENCY MANAGEMENT EXAMPLE

This approach is illustrated by means of an emergency management example that has been developed together with the Danish Emergency Management Agency (DEMA).

Situation description: Two freight trains crash at the central train station in Odense, Denmark, causing the leakage of chlorine from a ruptured tank wagon. The Hazmat unit, a specialised unit from the nearest DEMA rescue centre, covers the leak and stabilises the situation temporarily. A permanent solution requires the chlorine to be transferred to another tank. This transfer is fraught with the risk of a further chlorine leakage creating a (possibly lethal) plume over the downwind area. In order to protect the population, a decision must be made regarding which preventive measure should be applied: *evacuation* of downwind areas or *sheltering* in house.

Problem Structuring

To develop the decision model for the example, an attribute tree is elicited which includes the cri-

teria *Health, Effort, Economic losses* and *Impact on society*. To illustrate our approach, we focus on the *Health* criterion, in particular on the attributes *Number of ill in hospital to be sheltered* and *Number of ill in hospital to be evacuated*, and first show how these attributes are determined.

The DPIF looks for experts whose services can provide scores for these attributes (e.g., a health expert). To determine the scores, the identified experts indicate the need for further information on the number of ill in the hospitals that are (potentially) exposed and on the alternative implemented. This is depicted in Figure 6, where both vertices are connected to the attributes.

The CM configuration process continues by determining the information required by the experts providing information about the *Decision alternative* and the *Number of ill in hospitals exposed*. For the former the information is provided by the decision makers, whereas for the latter an expert is consulted who in turn requires knowledge on the hospitals exposed. This is represented in Figure 6 as a further link connecting *Number of ill in hospital exposed* to *Hospitals exposed*. Continuing this process iteratively, the global CM expands until all independent vertices (represented in black) are reached. Figure 6 represents the partial CM that is developed by merging the involved experts' local CMs.

Scenario Construction

We now illustrate how scenarios are constructed for each decision alternative by processing information in a bottom-up manner following the causal links in the CM to determine the score of each variable in each path from an independent vertex to an attribute. We refer once more to the partial causal map in Figure 6. First, the scores for the independent vertices are determined. Assume that the relevant expert (the Local Incident Commander, LIC) is uncertain about the amount of chemical left in the tank. Consequently, he decides to pass on not a single but three estimates,

Figure 6. Partial CM to assess "Number of ill in hospital to be sheltered/evacuated". Grey dashed edges represent the way the vertices are connected during the configuration phase, red dotted edges indicate the flow of information.

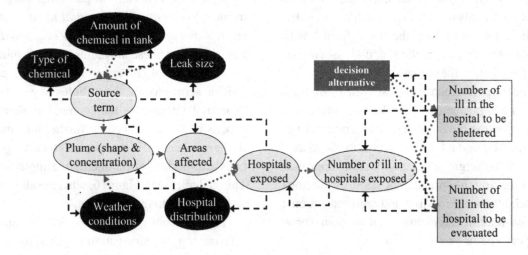

representing the cases that a *small* (10 t), a *medium* (30 t) or a *large* amount of chlorine (60 t) is left in the tank. A further expert (the Duty Hazmant Officer, DHO), who determines the source term, relies on this information. For each of the estimates the LIC passed on to him, the DHO develops the respective source term. This three-fold uncertainty regarding the amount of chlorine is passed on further (via the CM) to all vertices depending on the amount of chlorine in the tank. When further uncertainty arises, this is accommodated by passing on the respective number of possible states for each of the partial scenarios described by the predecessor vertices.

When distinct branches of the CM contribute to the determination of one variable and each of these includes one uncertain variable, it is important that they can be combined in a manner that ensures consistency. If *n* uncertain branches coming together in one vertex are independent (i.e., all paths do not have any vertices in common), combinations of all possible states for all *n* variables must be considered. In Figure 8, for example, the source term and the weather conditions are independent, so 2•3=6 partial scenarios

are considered in the assessment of "Plume" and "Areas affected".

It is important to notice that the number of decision-relevant scenarios does not necessarily increase following the CM. Figure 7 shows, for instance, that there are six distinct affected areas to be considered.

Despite the possibility of filtering decision-relevant scenarios, the total number of scenarios (when *all* attributes are considered) may still be very large due to the multiplicative combination of uncertain values. To control the number of scenarios in this example, the maximum number of values that each expert is allowed to pass on per set of scenarios $S(a_i)$ can be restricted. Assume, for instance that the expert calculating the *Number of ill in hospital exposed* states that he can only process two values (i.e., $\lambda_{\#\,ill\,in\,hosp.} = 2$, for all a_i). This means that, out of the six areas affected, the expert determining the *Hospitals exposed*HoHHhoHHhsafdsafd needs to choose the two values that he judges the most relevant. Here, only *one* hospital is situated next to the incident location. In some of the scenarios, this hospital is affected, while in other scenarios it remains unaffected.

Figure 7. Multiplicity of partial scenarios in the example

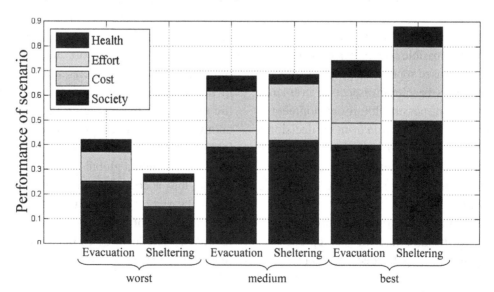

Therefore, he passes on both possible values: *Hospital A exposed* and *No hospital exposed.* Both values are representatives for a set of scenarios that share these values. This can be a simple way to keep the number of scenarios manageable, in case the number of (independent) uncertain variables and/or the number of values per uncertain variable are considerably large.

Evaluation of Decision Alternatives

When the CM is fully assessed for all decision alternatives with respect to *all* attributes, the attribute scores for each scenario are evaluated using MADM techniques. To overcome cognitive biases such as overconfidence and to avoid that the decision makers focus on a limited set of particularly impressing scenarios, an additional aggregation step is performed. To this end, scenario weights reflecting the relative importance

Figure 8. Evaluation of evacuation and sheltering

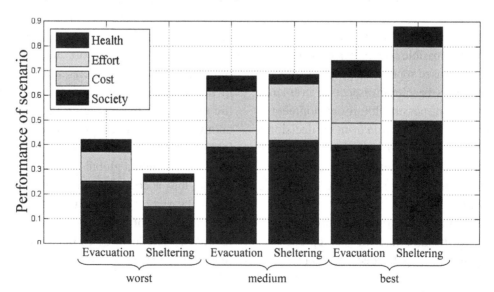

of each scenario are elicited from the decision makers. In this example, equal weights for the scenarios of each decision alternative were used. This results in a higher total performance for the sheltering alternative (0.66 for sheltering vs. 0.59 for evacuation).

In addition to these results, the decision makers are provided with stacked bar charts showing the performance of the worst, medium and best scenarios for both decision alternatives (see Figure 8, reading from left to right). This chart facilitates the assessment of robustness. Figure 8 shows, for example that the worst scenario for alternative evacuation results in a much better performance than the worst scenario for sheltering. A sensitivity analysis, varying the scenario weights, provides further support (cf. Comes et al., 2009).

CONCLUSION

This paper proposes a method that facilitates decision-making by considering scenarios in complex situations prone to severe uncertainty. The complexity of the decision problem stems from two sources: the modelling of the situation and the (potential) consequences of the implementation of different mitigation measures and the evaluation of the measures with respect to a set of objectives.

To reduce the complexity and uncertainty in determining the possible consequences of a decision, a scenario-based approach is proposed. Each scenario describes the relevant aspects of an emergency as well as one possible future development, given one alternative for action from the decision problem. Scenarios, being plausible, coherent, and consistent situation descriptions are easily understandable and help to overcome cognitive biases (Wright & Goodwin, 2009).

This paper introduces an approach to construct scenarios in a distributed manner. This approach is particularly useful, when the decision problem is so large and complex that it defies the use of standardised monolithic models. Using the attribute tree as a starting point for scenario construction ensures that the scenarios are sufficiently rich and provide all information necessary for the evaluation. By using the service-oriented architecture of the Distributed Process Integration Framework (DPIF), it is ensured that only information relevant for the decision at hand needs to be processed.

The presented approach facilitates distributed reasoning and involves human experts as well as automated systems. Each expert has the opportunity to freely choose his preferred methods to determine the state of each variable and to adapt these as the situation evolves (e.g., when previously lacking or uncertain information becomes known). Altogether, this approach results in overall distributed heterogeneous problem solving behaviour, enabling the system to adapt to a complex, dynamic, highly varying environment such as encountered in emergency management.

The sharing and processing of information is coordinated by having each expert specify which service can be provided and which information is relevant. Each expert's expertise is represented as a local Causal Map (CM) with the additional benefit that the expert is provided with all information he judged necessary for his task. Furthermore, an expert is *not* confronted with information irrelevant for performing a particular task, thus reducing information overload. As the scenarios developed in our framework are established by (human or artificial) experts, a major drawback in the use of computer-based systems, namely the problem of missing acceptance and trust in anonymous systems (Engelmann & Fiedrich, 2009), can be circumvented. The merging of the local models into a global CM makes it possible to take into account interdependencies between relevant variables. In this manner, the consistency and coherence of each scenario can be ensured to a reasonable level given that both availability of experts and time for the decision are bounded.

To come to a robust recommendation, a two-fold approach is proposed. First, a detailed analysis of the scenarios with the worst, the best and a medium performance for each decision alternative is provided. Thus, decision makers can gain deeper insights into the decision situation than those provided by standard methods from both SBR and MCDA. Second, an aggregated overall performance, which encompasses the evaluations of all scenarios for each decision alternative, allows for overcoming overconfidence in a small range of scenarios.

Our research aims at tailoring this approach so that it fits best the decision makers' needs. To improve scenario management and the control of the number of scenarios, scenario pruning and prioritisation mechanisms based on the difference of scenarios with respect to the attributes are investigated. Another research question concerns the coordination of multiple decision-making processes (performed in parallel on different hierarchical levels or sequentially). Ideally, the decision-making processes are coordinated such that an optimal set of decisions can be identified between all ongoing and future processes. Finally, the results must be presented in a transparent and easily understandable way. Additionally, a documentation of uncertainties related to each scenario must be developed and adequately presented.

ACKNOWLEDGMENT

This research has been conducted in the DIADEM project, funded by the European Union under the Information and Communication Technologies (ICT) theme of the 7th Framework Programme for R&D, ref. no: 224318, www.ist-diadem.eu. The authors wish to thank their project partners for the valuable comments.

REFERENCES

Barthélemy, J. P., Bisdorff, R., & Coppin, G. (2002). Human centered processes and decision support systems. *European Journal of Operational Research*, *136*(2), 233–252. doi:10.1016/S0377-2217(01)00112-6

Belton, V., & Stewart, T. (2002). *Multiple criteria decision analysis: An integrated approach*. Boston, MA: Kluwer Academic.

Ben-Haim, Y. (2000). Robust rationality and decisions under severe uncertainty. *Journal of the Franklin Institute*, *337*(2-3), 171–199. doi:10.1016/S0016-0032(00)00016-8

Bunn, D., & Salo, A. (1993). Forecasting with scenarios. *European Journal of Operational Research*, *68*(3), 291–303. doi:10.1016/0377-2217(93)90186-Q

Comes, T., Hiete, M., & Schultmann, F. (2010). A decision support system for multi-criteria decision problems under severe uncertainty in longer-term emergency management. In *Proceedings of the 25th Mini EURO Conference on Uncertainty and Robustness in Planning and Decision Making*, Coimbra, Portugal.

Comes, T., Hiete, M., Wijngaards, N., & Kempen, M. (2009). Integrating scenario-based reasoning into multi-criteria decision analysis. In *Proceedings of the 6th International Conference on Information Systems for Crisis Response and Management*, Gothenburg, Sweden.

Comes, T., Hiete, M., Wijngaards, N., & Schultmann, F. (2010). Enhancing robustness in multi-criteria decision-making: A scenario-based approach. In *Proceedings of the 2nd International Conference on Intelligent Networking and Collaborative Systems*, Thessaloniki, Greece.

Comes, T., Hiete, M., Wijngaards, N., & Schultmann, F. (in press). Decision maps: A framework for multi-criteria decision support under severe uncertainty. *Decision Support Systems*.

Cruz, A. M., Steinberg, L. J., & Vetere-Arellano, A. L. (2006). Emerging issues for Natech disaster risk management in Europe. *Journal of Risk Research*, *9*(5), 483–501. doi:10.1080/13669870600717657

Cutter, S. L. (2003). GI science, disasters, and emergency management. *Transactions in GIS*, *7*(4), 439–446. doi:10.1111/1467-9671.00157

Diehl, M., & Haimes, Y. (2004). Influence diagrams with multiple objectives and tradeoff analysis. *IEEE Transactions on Systems, Man, and Cybernetics. Part A, Systems and Humans*, *34*(3), 293–304. doi:10.1109/TSMCA.2003.822967

Dugdale, J. (1996). A cooperative problem-solver for investment management. *International Journal of Information Management*, *16*(2), 133–147. doi:10.1016/0268-4012(95)00074-7

Durbach, I., & Stewart, T. (2003). Integrating scenario planning and goal programming. *Journal of Multi-Criteria Decision Analysis*, *124-125*, 261–271. doi:10.1002/mcda.362

Engelmann, H., & Fiedrich, F. (2009). DMT-EOC – A combined system for the decision support and training of EOC members. In *Proceedings of the 6th International Conference on Information Systems for Crisis Response and Management*, Gothenburg, Sweden.

French, S., Maule, J., & Papamichail, N. (2009). *Decision behaviour, analysis and support*. Cambridge, UK: Cambridge University Press. doi:10.1017/CBO9780511609947

Geldermann, J., Bertsch, V., Treitz, M., French, S., Papamichail, K. N., & Hämäläinen, R. P. (2009). Multi-criteria decision support and evaluation of strategies for nuclear remediation management. *Omega*, *37*(1), 238–251. doi:10.1016/j.omega.2006.11.006

Godet, M. (1990). Integration of scenarios and strategic management: Using relevant, consistent and likely scenarios. *Futures*, *22*(7), 730–739. doi:10.1016/0016-3287(90)90029-H

Harries, C. (2003). Correspondence to what? Coherence to what? What is good scenario-based decision making? *Technological Forecasting and Social Change*, *70*(8), 797–817. doi:10.1016/S0040-1625(03)00023-4

Heugens, P., & van Oosterhout, J. (2001). To boldly go where no man has gone before: Integrating cognitive and physical features in scenario studies. *Futures*, *33*(10), 861–872. doi:10.1016/S0016-3287(01)00023-4

Keeney, R., & Raiffa, H. (1976). *Decisions with multiple objectives*. New York, NY: John Wiley & Sons.

Mahmoud, M., Liu, Y., Hartmann, H., Stewart, S., Wagener, T., & Semmens, D. (2009). A formal framework for scenario development in support of environmental decision-making. *Environmental Modelling & Software*, *24*(7), 798–808. doi:10.1016/j.envsoft.2008.11.010

Montibeller, G., & Belton, V. (2006). Causal maps and the evaluation of decision options - A review. *The Journal of the Operational Research Society*, *57*, 779–791. doi:10.1057/palgrave.jors.2602214

Montibeller, G., Gummer, H., & Tumidei, D. (2006). Combining scenario planning and multi-criteria decision analysis in practice. *Journal of Multi-Criteria Decision Analysis*, *14*(1-3), 5–20. doi:10.1002/mcda.403

Papamichail, K. N., & French, S. (2005). Design and evaluation of an intelligent decision support system for nuclear emergencies. *Decision Support Systems*, *41*(1), 84–111. doi:10.1016/j.dss.2004.04.014

Pavlin, G., Kamermans, M., & Scafes, M. (2010). Dynamic process integration framework: Toward efficient information processing in complex distributed systems. *Informatica*, *34*, 477–490.

Peña, A., Sossa, H., & Gutiérrez, A. (2008). Causal knowledge and reasoning by cognitive maps: Pursuing a holistic approach. *Expert Systems with Applications*, *35*(1-2), 2–18. doi:10.1016/j. eswa.2007.06.016

Ram, C., Montibeller, G., & Morton, A. (2010). Extending the use of scenario planning and MCDA for the evaluation of strategic options. *The Journal of the Operational Research Society*, *62*, 817–829. doi:10.1057/jors.2010.90

Regan, H. M., Ben-Haim, Y., Langford, B., Wilson, W. G., Lundberg, P., & Andelman, S. J. (2005). Robust decision-making under severe uncertainty for conservation management. *Ecological Applications*, *15*(4), 1471–1477. doi:10.1890/03-5419

Russell, S., & Norvig, P. (2003). *Artificial intelligence*. Upper Saddle River, NJ: Prentice Hall.

Schnaars, S. (1987). How to develop and use scenarios. *Long Range Planning*, *20*(1), 105–114. doi:10.1016/0024-6301(87)90038-0

Schoemaker, P. (1993). Multiple scenario development: Its conceptual and behavioral foundation. *Strategic Management Journal*, *14*(3), 193–213. doi:10.1002/smj.4250140304

Scholz, R., & Tietje, O. (Eds.). (2002). *Embedded case study methods. Integrating quantitative and qualitative knowledge*. Thousand Oaks, CA: Sage.

Turban, E., & Watkins, P. (1986). Integrating expert systems and decision support systems. *Management Information Systems Quarterly*, *10*(2), 121–136. doi:10.2307/249031

van der Pas, J., Walker, W., Marchau, V., Van Wee, G., & Agusdinata, D. (2010). Exploratory MCDA for handling deep uncertainties: the case of intelligent speed adaptation implementation. *Journal of Multi-Criteria Decision Analysis*, *17*(1-2), 1–23. doi:10.1002/mcda.450

van Notten, P. W. F., Rotmans, J., van Asselt, M. B. A., & Rothman, D. S. (2003). An updated scenario typology. *Futures*, *35*(5), 423–443. doi:10.1016/ S0016-3287(02)00090-3

Wright, G., & Goodwin, P. (2009). Decision making and planning under low levels of predictability: Enhancing the scenario method. *International Journal of Forecasting*, *25*(4), 813–825. doi:10.1016/j.ijforecast.2009.05.019

This work was previously published in the International Journal of Information Systems for Crisis Response and Management, Volume 3, Issue 4, edited by Murray E. Jennex and Bartel A. Van de Walle, pp. 17-35, copyright 2011 by IGI Publishing (an imprint of IGI Global).

Chapter 15
Exercise24:
Using Social Media for Crisis Response

Austin W. Howe
San Diego State University, USA

Murray E. Jennex
San Diego State University, USA

George H. Bressler
San Diego State University, USA

Eric G. Frost
San Diego State University, USA

ABSTRACT

Can populations self organize a crisis response? This is a field report on the first two efforts in a continuing series of exercises termed "Exercise24 or x24." The first Exercise24 focused on Southern California, while the second (24 Europe) focused on the Balkan area of Eastern Europe. These exercises attempted to demonstrate that self-organizing groups can form and respond to a crisis using low-cost social media and other emerging web technologies. Over 10,000 people participated in X24 while X24 Europe had over 49,000 participants. X24 involved people from 79 nations while X24 Europe officially included participants from at least 92 countries. Exercise24 was organized by a team of workers centered at the SDSU Viz Center including significant support from the US Navy as well as other military and Federal organizations. Dr. George Bressler, Adjunct Faculty member at the Viz Center led both efforts. Major efforts from senior professionals EUCOM and NORTHCOM contributed significantly to the preparation for and success of both X24 and especially X24 Europe. This paper presents lessons learned and other experiences gained through the coordination and performance of Exercise24.

DOI: 10.4018/978-1-4666-2788-8.ch015

INTRODUCTION

New social media technologies offer organizations increased agility, adaptivity, interoperability, efficiency, and effectiveness. Social media software can be used by civilians, non-governmental organizations (NGOs), and governments around the world for content creation, external collaboration, community building, and other applications. However, the ramifications of the proliferation of social media on national and international security exist for future operational challenges and obstacles as well as current traditional command and control systems of management for irregular, catastrophic, or disruptive events. Unfortunately, organizations are ill-prepared to move forward using such technologies, because of the lack of capabilities tests and exercises, not to mention the need for updated policies and funding. Failure to adopt and understand "fifth-generation command and control"—that is, the combination of technology and organizational response tools—may reduce an organization's relative capabilities over time. Globally, social software is being used effectively by businesses, individuals, activists, criminals, and terrorists. Those that harness this invaluable resource are the innovative leaders of the future; it is imperative that this particular aspect of up-and-coming technologies be utilized to its fullest in the response efforts of citizens and organizations to disaster events.

While GIS technologies are increasingly used for practical disaster event applications, research focused on overall issues regarding the use of GIS, crowdsourcing, and social media in disaster analysis is lacking. Currently, there are inadequate tools and a lack of knowledge about the current tools for the organizational decision maker, which ultimately hampers their situational awareness and inevitably leads to preventable losses of life and infrastructural damage. It is often that during or after a disaster, we learn that what we thought would work, in fact did not, because what should have been tested beforehand was not.

Disasters can be global emergencies that affect civilizations, economies, and public health. There is a need for an international approach in order that relief efforts are effective, no matter the complexity or size of an event. This paper is an action report on an effort by San Diego State University's Visualization Center, Google (and dozens of other companies), and the U.S. Navy to answer this call with the creation of Exercise24. The exercise tested how social media and crowdsourcing technologies could be implemented into humanitarian assistance and disaster cloud environments in an international multidisciplinary crisis simulation.

BACKGROUND

Cloud computing and social media have contributed to the wealth of sharing information across the globe, emergency managers, NGOs, and governments alike are seeking similar information management benefits for disaster relief and humanitarian assistance. Numerous emergency managers have formed in these social networking worlds such as LinkedIn, Facebook, MySpace, YouTube, Flickr, Twitter, etc. The capabilities of GIS in cloud environments have allowed for data collection, data processing, and data sharing across the emergency management field. For emergency managers, much of the data collection and processing can be performed as part of the preparation work. Data can be imported or created to reflect what might typically be required for a responding agency to operate effectively and communicate more openly with the counterparts involved in the emergency.

When an emergency strikes, the infrastructure of the cloud environment needs to be flexible enough to work in the field and to incorporate the multiple layers of additional data that will be collected and disseminated as part of the emergency response. In such a dynamic and challenging setting, the software technologies must be easy to handle and manage for those with limited knowledge of GIS or other software technologies. An overly complex software tool that requires additional specialists to run could result in response delays or a bottleneck situation just when the need for information becomes most critical.

The potential uses of cartography and forms of GIS technologies like Google Earth have brought the need to include not only official responders, but also members of the public (Palen et al., 2010). The first uses of mapping, in response to a crisis, date back to 1845, when London was reeling from a particularly virulent cholera outbreak. At its epicenter, five hundred people died in the space of ten days. By mapping the location of the infections, Dr. John Snow was able to provide evidence to local officials that the cause was not by noxious vapors, but by a local, contaminated water well. Once the well was closed, the spread of contagion rapidly abated (Gleicher & Hwang, 2010).

First responders are not, in practice, the trained professionals who are deployed to a scene in spite of the common use of that term for said professionals; they are instead the people from local and surrounding communities who provide first aid, transport victims to hospitals in their own cars, and begin search and rescue. An example of crisis response and social networking in the area of disaster can be seen following 9/11, when ferry captains and others self-organized to systematically evacuate people from Manhattan Island and relay information back to the Port Authority of New York (Gleicher & Hwang, 2010). Without the successful coordination of the ferry captains, thousands of individuals would have not been able to leave the critical areas where emergency response was most needed. Leveraging citizens as assets in Hurricane Katrina's aftermath allowed search and rescue GIS technologies to map mortality locations throughout the city.

A recent advent of crowdsourcing with social media in a terrorist event occurred during the attacks on Mumbai in 2008. The events of Mumbai unfolded online in real time, and the mainstream media (and in effect the world) got a first-person, eyewitness view of what was happening. The public was not the only one using the broadcasted media to figure out what was happening—the terrorists also used it to orchestrate their attacks and achieve the goal of creating fear in the world. "The terrorists were able to use sophisticated GPS navigation tools and detailed maps to sail from Karachi [in Pakistan] to Mumbai," said G. Parthasarathy, an internal security expert at the Center for Policy Research in New Delhi. "Our new reality of modern life is that the public also sent text messages to relatives trapped in hotels and used the Internet to try and fight back" (Wax, 2008).

This is the dilemma of social media when dealing with manmade disasters where both the creators and the responders are using the technology. Twitter streamed information and images during the terrorist event at such a rapid pace that mainstream media simply used footage without attribution and independent fact-checking. Hearsay and assumption also played a strong role in the information flow, and to some extent, "trust but verify" was suspended in favor of speed. While rapid, first-person intelligence via these new communications is valuable, there is the very real possibility of exploiting such streams to promote misinformation, particularly if decision makers do not understand the technology well (Dhanjani, 2008). From incidents like this, the very limited time that governments have to respond effectively in crises where social software is part of the information flow is becoming rapidly apparent (Dhanjani, 2008).

EXERCISE24

Exercise24 (X24) and X24-Europe, were two-day and three-day-long (respectively), international and multidisciplinary crisis simulations in which smart technologies were implemented to connect civilian and military organizations in efforts for humanitarian assistance and disaster relief. They explored how new social media components that support communication, logistics coordination, and disaster response affect mass populations and infrastructures. The idea was to see how national and international authorities, organizations, and citizens would react to such an event in Southern California utilizing these smart technologies.

The idea for X24 was originated by Dr. George Bressler of San Diego State University, SDSU. His vision had two basic outlooks. The first is that X24 is a comprehensive science-based holistic approach to emergency management which includes all-hazards, through all disciplines, that may occur at anytime. Secondly, the exercise was open-minded to new solutions and technologies and did not limit new ideas because "it is the way we have always done it." The goals for X24 and X24 Europe are to:

1. Demonstrate the use of no/low-cost, off-the-shelf social media, crowdsourcing, and collaboration web tools to gather, coordinate, and share actionable real-time information to build situational awareness to help victims of a natural disaster and help save lives;
2. Establish dialogue and build relationships between European partner nations, international organizations, and public/private partners regarding the use of online tools, make them more effective, and streamline cyber information sharing in preparation for the next real-world crisis;
3. Address the virtual flow of information and activities of international organizations during the first 180 days of a natural disaster; and
4. Encourage all formal and informal groups across the globe to actively participate or observe the exercise on-site in San Diego, CA or virtually on the web.

Design of Investigation

This is action research. The authors participated and reflected on the conduct of these exercises. X24 was first in September, 2010. Lessons learned from X24 were used to refine X24-Europe held in March, 2011.

DATA COLLECTION

Data collection occurred over the course of several months between August and October of 2010 and between March and April 2011. Data was collected using surveys and after action reports (AARs). Over 12,700 subjects participated September 24-26th of 2010 in X24, and over 49,000 subjects participated March 29-31, 2011 in X24-Europe. The data, in addition to being drawn from the subject population, is also drawn from the author's participation and observation as one of several leaders for the X24 team.

TOOLS AND TECHNOLOGIES

The following tools and technologies were utilized by participates in X24 as a means of collaboration, communication, logistics, and response for the exercise. The reason for utilizing these tools and technologies was to implement an easily accessible "Wiki" of applications at a low or no-cost value available to the majority of the world without social, economical, political, or organizational boundaries. 'Wiki' is a "Hawaiian word that means 'quick' and is used by the information systems community to refer to an open source, collaborative content management system (Raman et al., 2010). The applications used are illustrated by Figure 1.

Figure 1. Exercise24 tools and technologies diagram

EXERCISE CONTROL

To avoid confusion in the event of a real emergency the X24 team inside the Visualization Center at SDSUs campus were responsible for delivering all injects (updated scenario feeds) to the registered participants of the exercise. Individual injects were displayed on the X24 Sahana Map, the SAFER smart phone application, and sent to participant email addresses as they occur. A scrolling list of injects were also posted on the InRelief/Eushare websites. Non-registered participants and observers could view injects from the map and scrolling

list via the X24 public areas. There were no hidden injects and all injects included the statement "This is an exercise and not an actual event. If an actual event occurs, all registered participants will be notified 'This Is Not An Inject'...All exercise activities are complete. X24 participants and observers should contact their respective commands for further instructions. The Viz Center team will begin support of response efforts...This Is Not An Inject." Several media presses were released before the exercise and Twitter agreed to put out an advance message to alert social network users that any apocalyptic-sounding tweets are part of a test, not a real disaster. Each tweet included the words "Test. Not real," and included links to the X24 and InRelief /Eushare web sites.

DATA ANALYSIS

The data collected typically consists of situational awareness information, such as what roads were passable, maps of the area, weather data, and video/photo imagery of areas, as well as logistics data, such as what transport capabilities exist in the area, what materials/supplies were currently being shipped and when/where it would arrive. The gadgets or widgets used by X24's Dashboard used technologies that were made available for crowdsourcing via RSS, Atom, KML, and other feeds that could be readily disseminated into other commonly used websites as InRelief. The knowledge sharing through open source publishing, allowed users from the public to copy and redistribute the predestinated X24 Day 1 Injects and Day 2 Injects over cloud technologies such as Facebook, Twitter, and Global Talk. Social media companies like Buzz Manager then collected and tracked the progression of data related to X24 by individual Internet Protocol (IP) addresses from across the globe and gained an international overview of effectiveness of disseminating information through different means of social media.

X24

X24 Scenario

X24 utilized a scenario of an earthquake off the coast of Huntington Beach, California, USA generates a tsunami event in Baja, and a catastrophic subsurface and surface oil spill. A series of inland aftershocks result in reports of deaths and injuries, damage to the All American Canal, roadways, power lines, and other key resources and critical infrastructures in Southern California, USA and Northern Baja California, Mexico. A series of aftershocks, fires, loss of power, displaced populations, disease concerns, and other challenges continue throughout the exercise to facilitate participant objectives:

Objective One: Utilization of computing cloud to rapidly converge geographically dispersed global experts at the onset of a simulated international incident, deploy a foundation of guidance in concert with community leaders in a manner that empowers community members through education and smart technologies to support mitigation, response, recovery, and a resumption of societal normalcy at a level of functioning and order of magnitude higher than existed before.

Objective Two: Leverage smart phones, ultra-lights (United States), and unmanned air systems (Mexico) for rapid threat/damage assessment of a simulated seismic event that generates a significant oil spill off the coast of Southern California and Northern Baja California, as well as damage to critical infrastructure inland that necessitates mass sheltering of displaced community members.

Objective Three: Leverage the power of Non-Governmental Organizations (NGOs), faith-based groups, rapidly responding government and corporate groups, international groups, social networking communities (as occurred in Haiti), and other resilient networks to locate and nationally send aid to Southern California and Baja California.

X24 Performance

The scenario was initiated though injects (where an inject is an emergency announcement) on the InRelief website and by linking technology such as smart phones, back-end cloud computing and visualization, to help nurture the feeling of a real life event and time line. Though the information provided in the injects participants could simulate how they would respond, resources they would need, resources they could provide, and problem solve and create solutions for complex emergencies in difficult settings.

All injects that were placed into the scenario were specifically requested by participants and driven from experience, concern, and areas known to be problematic that individuals sought to work though. Once the injects were fused into a coherent, scientifically realistic, and globally impacting event they were mapped and posted. Images were also added to the injects to enhance the real life real time atmosphere. Although time compression was needed to meet all objectives the scenario provided a real life catastrophic event that allowed the global world to interact and prepare for the next disaster effectively.

Over 150 people physically attended the Viz Center at San Diego State University, which allowed them to observe and connect with over ten thousand people around the world using the Internet and social-networking tools via Google, Twitter, Facebook, and U-Stream. Furthermore, 79 different countries speaking 33 different languages located on every continents except Antarctica, participated virtually in the exercise. These visitors reached the X24 websites using mobile devices as well as personal computers, which further simulated a real life event and communication information flow. Statistics were collected using Buzz Manager. They included that 84% of the visitors to the X24 websites were first time visitors, not return or long-time users. The overall "Bounce Rate", the percentage of single-page visits or visits in which the person visited only the home page, was very small, 1.27%. Referrals from other sites, excluding Inrelief.org related sites, accounted for 74% of the visitors to the X24 websites. Significantly, referrals from CNN's website accounted for 50.5% of the non inrelief. org-related, suggesting that traditional media still has a role in a social media based emergency response. The Buzz Manager snapshot report indicates that visitors to the websites stayed long

enough to develop opinions regarding the conduct of the exercise.

The Exercise24 team was comprised of the InRelief.org staff who were funded by the United States Navy, the Viz Center staff, and undergraduate and graduate students from San Diego State University with varying majors.

X24 did not have a budget outside of existing salaries and systems. Essentially, X24 was functionally a zero cost simulation, which was the intended goal of the X24 team. By operating on a low or zero cost budget it simulated how it could be done in real life and function from the onset of an incident without needing to wait approval of funds from the private or public sectors.

X24 Observations

As previously stated, reaction data to X24 was collected from participants using surveys, AARs, and direct solicitation. Overall, the responses received from participants in X24 were optimistic and viewed social media is a valuable and powerful information tool in the future of emergency management. Participants were able to participate in new topics of discussion for how catastrophe training and exercises can be improved by utilizing available technologies and resources within an international, collaborative model of action. This research determined patterns in the AARs that contribute to a general consensus on improvements for X24 or similar social media exercises in the near future. These findings include:

- Social media is a valuable and powerful information and dissemination tool, but also has potential as a distractive force if data is not managed, analyzed, and acted upon in a methodical, planned manner.
- A hybrid of formal structure response capabilities combined with crowd sourced and informal self-activating capabilities appear to be the best sense of balance for disaster management and response.

- In the pre-exercise and exercise it was clear that training and practice using the technologies and common operational standards are necessary. When these the new technologies failed to work, participants reverted to more familiar technologies—in this case email, texting and cell phones.
- Sidebar conversations were highly beneficial, but knowledge dissemination was limited to participants located within these private clouds.
- Groups within the exercise desired a degree of anonymity and separation from other participants and observers of X24. The main reasons for this is their comfort levels or experience with exercises or groups wishing to experiment in private groups on information not pertaining to the general group exercise.
- Certain applications in video conferencing and chat (i.e., Global Talk and VSee) did not respond as efficiently in web browsers like Internet Explorer.
- The social media application GeoChat has several complications with sending and receiving user requests.
- There was confusion and a need for clarification about who the Point of Contact (POC) was for the various social media applications that were used during the exercise.
- Participants experienced with similar exercise trainings believed that the X24 injects were inclined to serve as an introduction to exercise environments instead of an actual, serious trial exercise, as per their expectations. Conversely, those less experienced with the social media or exercise environment tended to favor the visualization and informational approach of the disseminated injects.
- Participants often felt overwhelmed by the number of injects implemented in the short time frame of X24. However, individu- als expected the data being pushed to be tremendously overwhelming, particularly because the exercise did not portray a complete end-to-end response and recovery due to simulation constraints.
- There was an inconsistency regarding the number of preferred chat platforms for participants to use as many felt they were redundant and somewhat overwhelming. On the flipside, users acknowledged that while in a crisis, information should be replicated in a variety of locations since people may be using various social media tools to communicate and gather information. Several suggestions included a "one platform" tool that aggregates the relevant discussions together.

These observations were then used to refine the approach for implementation in X24-Europe.

X24-Europe

X24-Europe Scenario

The X24 Europe scenario involved a simulated seismic event that generated a tsunami in the Adriatic Sea and caused damage to key resources and critical infrastructure in the Balkans. Research determined that Balkans contain the most seismically active zones in Europe. Further research found that two universities, the Marine Biological Station, University of Ljubljana, Slovenia and the Bogaziçi University Kandilli Observatory and Research Institute, Istanbul-Turkey, had developed computer models of earthquake-generated tsunamis in the southern Adriatic Sea.

X24 Europe Objectives

- **Objective 1:** Demonstrate the use of no/low-cost, off-the-shelf social media, crowdsourcing and collaboration web tools to gather, coordinate, and share actionable real-time information to build situational awareness to help victims of a natural disaster and help save lives. This was achieved through:

- The use of online, no-cost/low-cost collaborative cloud computing tools to organize, plan, and coordinate a multi-national HADR exercise.
- The construction of http://x24.eu-share.org/ website that included Internet chat rooms for skill-set-specific groups (public health, logistics, partnerships) to collaborate.
- Application of Hootsuite.com as a social media portal for Twitter and Facebook. All injects, or events in the scenario, were transmitted by social media.
- Utilization of Crowdmap for social media aggregation and visualization.

- **Objective 2:** Foster collaborative dialogue by establishing and building relationships between European partner nations, international organizations, and public/private partners regarding the use of online tools, make them more effective, and streamline cyber information sharing in preparation for the next real-world crisis. This was achieved through:

  - The pre-event collaborative planning process required participant groups to meet via video teleconferences, teleconferences, shared documents stored in the cloud, webinars using tools like VSee, JustinTV, DVIDs, Google Docs, and Google Chat. This provided an opportunity to test the tools to be used during the exercise, dialogue with geographically separated partners, and build relationships.
  - Over 49,000 people from 92 nations were involved.
  - 78% of participants were from the targeted area of the Balkans.

- **Objective 3:** Address the virtual flow of information and activities between the various international organizations during the simulated first 180 days of a natural disaster. Humanitarian assistance and disaster relief experts were actively engaged in X24 Europe leading functional cluster groups, providing, which facilitated the transition of visualized situational awareness into actionable information for international organizations.

- **Objective 4:** Encourage all formal and informal groups across the globe to actively participate or observe the exercise on-site at one of the venues or virtually on the web. The majority of participation occurred in virtual venues, specifically social media. Social media proved to be a real-time method to enable geographically dispersed experts to empower responders in the simulated impacted area with their expertise. Social media also enabled those with aid to ask the right questions, which resulted in targeting the right-type of help to the right location.

X24-Europe Performance

X24 Europe commenced on March 29, 2011 with one day of "academics", consisting of presentations and demonstrations from exercise coordinators and public/private partners. While the exercise was hosted by SDSU Viz Center, the majority of presentations and demonstrations were broadcast from the remote sites. Social media video teleconferencing tools JustinTV and VSee were used to broadcast the mixture of live and taped presentations. JustinTV has an embedded chat room that allows the viewers to post questions and comments to the presenter. VSee is a multi-participant video-teleconferencing tool that allowed direct interaction with the widely separated exercise participants.

The next two days, 30-31 March 2011, focused on the exercise scenario and the flow of information/activities of international organizations in response to the virtual disasters. The exercise was broken up in four, four-hour phases that addressed specific time periods. The time periods covered were as follows:

Phase I: The first 48 hours after the initial disaster event;

Phase II: 2-15 days after the initial disaster event;

Phase III: 30-60 days after the initial disaster event; and

Phase IV: 90-180 days after the initial disaster event.

Crowdsourcing, and collaboration tools, information and storyboard injects were disseminated using Facebook and Twitter social networks. Responses and comments were received and collected from these social networks. The Ushahidi "crowdmap" tool was used to correlate and plot geo-locatable responses onto a map of the simulated disaster area. Periodic exercise updates were broadcast via JustinTV and emergent conferences were held using VSee.

Space was made on the X24 Europe website to allow the formation of "functional cluster groups." The purpose of the "functional cluster groups" was to create a space were persons and organizations with similar objectives could meet, exchange ideas, and discuss how to best respond to the HADR problems that may arise.

A small, 5-10 person team was stationed at the SDSU Viz Center to coordinate the exercise. Additionally, a five-person field team was sent to Montenegro and Croatia. The field team interacted with the local populace, visited locations that were impacted by virtual damage, and tested communications voice, text, chat, still imagery, and streaming imagery.

Finally, a virtual post-exercise review was conducted on March 31, 2011 to address lessons learned and comments from key participants.

During the two day exercise, 30,588 people (23,929 unique visitors) visited the X24 Europe website from 77 countries. When the duration is included to the week before and after the exercise, which includes the timing of media announcements, pre-exercise training, and post exercise interest, the website statistics are 49,822 visits (39,418 unique visitors) visited the X24 Europe website from 92 countries. A significant statistic derived from the Google Analytics data was more than 75% of the visitors to the X24 Europe site came from the Balkan region, which was the European area of focus for the exercise.

X24-Europe Observations

X24 Europe transcended all expectations in its ability to form a collaborative bridge between individuals, communities, and nations with over 49,000 participants from 92 nations that included two ambassadors, a US major general military officer, as well as representatives from US European Command (EUCOM), US Northern Command, US Transportation Command, Office of Navy Research, STAR-TIDES at the Center for Technology and National Security Policy from National Defense University, and many others. The focus area for the scenario was the Balkans region with a seismic event that generated a tsunami in the Adriatic Sea followed by aftershock damage inland. 78% of participants were from the Croatia, Macedonia, and Bosnia and Herzegovina. The United States was fourth on the list for number of participants with 3,419 and Serbia was fifth with 1,958. This is significant in consideration of the challenges experienced in this region.

This inclusive environment was facilitated by an open invitation for participation, which was hosted by the Immersive Visualization Center, San Diego State University. This openness challenged traditional exercise structures where events in a scenario, or injects, are associated with specific responses by participants with known capabilities. The complete range of skills and experience participated. This is very similar to actual humanitarian assistance and disaster relief operations. This requires a considerable degree of leadership agility on the part of the exercise controllers and evaluators. This agility enabled a unity of effort, when a unity of command was not possible.

X24 Europe provided an unprecedented opportunity to test, train, and explored leading-edge humanitarian assistance and disaster relief tools, technologies, and methodology with a global community of experts. This dramatically enhanced the knowledge base for all participants and observers. X24 Europe was as much an educational environment as it was an exercise. Injects were intentionally placed in the scenario to teach the global community the process for US Department of Defense Combatant Command (COCOM) humanitarian assistance and disaster relief capabilities to respond to impacted regions within a COCOM's Area of Responsibility. This unique opportunity to educate through social media would add considerable strategic and operational value to other COCOMs and US interests both within and outside its international borders.

X24 Europe provided a no-fault framework were government agencies, non-government organizations, academia, private industry, and volunteer groups could develop a working familiarity with the evolving tools, technologies, and methodology presently used by the open-source computing community to respond during humanitarian assistance and disaster relief crises. The integration of experienced military officers in the area of operations and logistics identified the improvement need for a time indicator addition to accompany geo-tagging. This will be a part of future analytic crowd mapping for significant events, since the majority of the processing for global disaster mapping occurs at the Immersive Visualization Center.

X24 Europe required no support funds outside of existing salaries and travel budgets. However, the complex nature of multi-nation exercise design, development, and implementation requires a small, full-time dedicated team that can scale as various skills are needed. The all-volunteer team at San Diego State University frequently relied on the gracious support of EUCOM representatives.

Future X24 events should focus on testing and teaching the strategic, operational, and tactical application for mission specific needs in the midst of a crisis. These include and are not limited to wildfire management, supply-chain security, mass migration, evacuations, and civil unrest. A cross-section of experts from an inter-agency environment should assess, develop, and test new policies and procedures for the integration of these systems.

OBSERVATIONS OF X24 TOOLS AND TECHNOLOGIES

X24 acted as an emergency management information system (EMIS) that supported and integrated all phases of emergency management and response into a multitude of simulated, real-life crisis situations. Emergency planners were able to actively engage different types of technologies and knowledge management interfaces to display and analyze the spatial relationships among possible event locations, shelters and other emergency management facilities and resources, transportation routes, and populations at risk. Participants acknowledged that response time and real-time data are important elements for an effective cloud EMIS, which enables emergency operators to accurately evaluate and quickly implement emergency response plans so as to reduce the risk to the affected population.

X24 utilized a combination of free social software with inexpensive mobile devices and donated computers to empower people to self-organize information-sharing networks that are not bound by federal, state, local, or other structural constraints. The use of social software on simple cell phones, computers, or personal digital assistants that incorporates geographical information is becoming ubiquitous. Internationally, empowerment through these tools and technologies have many opportunities for promoting collaboration and coordination during a global crisis, as well as offering security, building trust, and developing accountability.

X24 technologies and thought processes were analyzed using the Rayport and Heyword (2009) model with elements as follows:

1. **Universal connectivity:** Users must have near-ubiquitous access to the Internet.
2. **Open access:** Users must have fair, non-discriminatory access to the Internet.
3. **Reliability:** The cloud must function at levels equal to or better than current stand-alone systems.
4. **Interoperability and user choice:** Users must be able to move among cloud platforms
5. **Security:** Users' data must be safe.
6. **Privacy:** Users' right to their data must be clearly defined and protected.
7. **Economic value:** The cloud must deliver tangible savings and benefits.
8. **Sustainability:** The cloud must raise energy efficiency and reduce ecological impact.

The advantages and disadvantages in the X24 model as well as the reliability that end-users were able to engage as part of the three objectives set by X24 were also taken into consideration.

VIDEO AND VIDEO CONFERENCING

Video conferencing is a real-time and interactive tool used by a wide range of individuals for a variety of business, personal, and educational goals. It allows people to communicate via audio, video and computer technology across time zones and locations through a live connection that provides full-motion video images and high-quality audio. X24 used VSEE and UStream during the initial trainings and during the two-day exercise. Participants were able to view PowerPoint presentations broadcasted over the video conferencing software. UStream made the trainings available on InRelief after the secessions to allow users not present at the meetings to access the data at a later date. This was very beneficial to the international users who were asleep on the opposite end of the globe. VSEE and UStream video conferencing technologies delivered real-time injects and provided international users a means of virtually visualizing the presenters in the Visualization Center at SDSU.

While the video conferencing capabilities were easily disseminated across the web, participants often dealt with signal and noise complications. The signal was likely due to the individual's personal computer; meanwhile the noise component was often caused by additional participants in the Visualization Center carrying on side conversations. At times this was seen as time consuming and a major distraction from the exercise elements. A preventative measure in X24 Europe was to have pre-established leaders of an exercise completely separated from the general public. The use of real-time translators for presenters also complicated the listeners' ability to effectively hear and often prevented both languages from being understood over the speakerphone.

The benefits of video conferencing are numerous despite the minor complications during X24. Video conferencing saves travel time and money, it urges participants to reach decisions that may not come as easily in a face-to-face meeting, and it gives participants the chance to see others' body language and facial expressions, which are important factors for a sales or a board meeting. DeafLink, Inc., also a participating member of X24, incorporated video alerts to distribute test warning messages to those with hearing impairments. Including individuals with disabilities is crucial to understanding the needs of a community during an emergency response.

VIRTUAL CHAT

As chat applications have migrated to multiple platforms and morphed to include different speaker-audience relationships (one-to-one; one-to-many; many-to-many; known-to-known; known-to-unknown; unknown-to-unknown), they continue to figure centrally in evolving computer-mediated interactions. As more people adopt and maintain a digital presence, these ever-advancing forms of chat-based environments draw attention not only because of the synchronous and light-weight interactions they support, but also for the new information relationships they produce and the manner in which the media is adapted to suit technological constraints and social conditions (Starbird et al., 2010).

X24 offered a unique real-time chat-based software problem called "Global Talk" during the trainings and the main exercise for account users of InRelief. Global Talk's software was able to translate chat room environments into the user's native language without having to access additional software programs. This prevented delay or distraction when the information needed to be immediate and clear. The exercise offered nine languages as a trial, with more languages to be added in the future. While the accuracy of the program runs at about 80% efficiency in translating user's chats, that means there is another 20% of information that is not understood. However, Global Talk takes the next line of defense and allows participants to select the text of another user and highlight the text as an incorrect translation. This then allows for another user to correct the text or suggest the original author to rephrase their message. The most common reason for messages not correctly translating is the original author is using "slang" terminology or words that do not translate from one language to another.

Overall, the language translation tool could be improved and should likely be used by only a few individuals at one time to effectively communicate information in Esperanto. Due to time constraints, translation accuracy and the availability of a translator during a disaster is unlikely. Several AARs suggested the creation of a mobile application Wiki that could be implemented to communicate the essence of the information being discussed across language barriers. This ontology would have the ability to deliver the essence of the communication with high accuracy, thereby improving response efficacy. This tool was not used during X24-Europe for these reasons.

GeoChat, another open source communication technology, allowed team members to interact and maintain shared geospatial awareness of who is doing what and where—over any device, on any platform, over any network. X24 participants tested the system for SMS, email, and on the surface of a map in a web browser. Many complications occurred during the testing of GeoChat, i.e., backlogged text, interfaces not responding, and incorrect SMS messages being sent to users. For reasons undetermined the GeoChat technologies had previous successes in the Haiti earthquake but was limited in X24.

MOBILE APPLICATIONS

DefenCall was the only open-source cell phone application utilized during X24. The application serviced as a personal emergency response solution for smart phones. Currently, the application is only available through download at the Apple's Itunes store and is limited only to iPhones. The purpose of the application was to send SMS and email alerts to notify participant's first responders and X24 Visualization Center of personal emergency situations or disasters from strategic locations around the San Diego County (per their team assignment). The feedback from AAR included comments on the ease of use and accessibility of the application, the rapid sending/receiving of alerts, and the ability to send alerts via multiple communication devices, (i.e., cell phone, computers, land lines) and multiple mediums, (i.e., email, text, voice) at once. The testing of this application was limited to iPhone users and recommendations to encompass all types of smart phone technologies in future events are encouraged. Overall, the application proved to be a valuable asset in completing the goal of effectively disseminating information from X24 team members to participants and vice versa.

CARTOGRAPHY

The emergence of the Geospatial Web and particularly Web Mapping 2.0 has led to increases in geo-browsing activities (e.g., browsing through Google Maps or Google Earth). Web maps can function as an interface or index of additional information in a way that facilitates the up-to-date, dynamic, and interactive presentation and dissemination of geospatial data to many more users at a minimal cost. Web maps also allow users to explore and find answers to location-specific questions as opposed to mainstream media's broad reporting. Google Maps, Sahana, and OpenStreetMap were used for geotagging—a process of tagging images and injects from X24 to various open layers, in the form of geospatial metadata, where users can find a variety of location-specific information. By making it possible to integrate different types of data from diverse sources, collaborative post-disaster efforts were able to strengthen analytical capabilities and decision making for disaster response (Maiyo et al., 2010). The current architecture for these programs uses remote user profiles and allows dissemination of post-disaster damage maps without any major constraints. Collaboration gives emergency management organizations a pool of expertise far larger than one organization itself can provide (Maiyo et al., 2010).

Sahana offered X24 a multitude of software programs including web based collaboration tools that addressed common coordination problems during a disaster. These include finding missing people, managing aid, managing volunteers, and tracking camps effectively between Government groups, NGOs, and the victims themselves. From the Haiti earthquake to the Pakistan floods, Sahana has had several successes in utilizing their software programs during a real-time disaster.

People Finder, Location Registry, and Request Management by Sahana all proved to be valuable tools during X24. The ease and simplicity of the programs meant they were highly accepted and utilized by participants. OpenStreetMaps, another successful tool for X24, provided a Wiki-style map of any location in the world that users could edit. Walking papers allowed users the ability to select an area on a map where OpenStreetMap's software had coded the data and permitted users to easily create coded maps for printing. Once the walking papers had been updated with written data, participants digitally scanned their map into the cloud software of OpenStreetMap. The software allowed layered data to be complied digitally for other users who were remote so that they could gain knowledge of the situations on the ground. After a major disaster, situational awareness is key in determining which roads are accessible, what buildings have suffered critical infrastructure damage, or where the nearest shelter is located.

The data produced by Sahana and OpenStreet-Maps provided several solutions to problems brought about by major disasters, both during the event itself, and in the immediate aftermath. As reported by the US National Research Council, "data and tools should be essential parts of all aspects of emergency management—from planning for future events, through response and recovery, to the mitigation of future events. While these tools were beneficial to the survey respondents who had utilized it, only a few X24 participants tested the technologies firsthand due to time constraints of the exercise.

SOCIAL MEDIA

With over 500+ million active users with a Facebook account and another 100+ million followers on Twitter, it is clear why X24 team members decided to select these two social media outlets to track the flow of data being disseminated across the globe. Over 90% of the participants acknowledged to having either a Facebook or Twitter account. Injects were communicated over both social media components by Buzz Manager. Buzz Manager's purpose was to uncover patterns of public use of social media as people learned about X24 and when they shared the disaster information. The study included tracking public perception about the gravity of the event, quality of response, info sharing, and expectations. After tracking the information, Buzz Manager gives the response a "Buzz Rating" which quantifies volume and impact of conversations.

Social media is an excellent tool for reaching out to a large, dispersed audience. However, respondents of X24 found that the amount of unverifiable information in a real-time crisis is generally overwhelming within social media channels. In addition, users felt there was a lot of interesting information, but the social media aspect was un-actionable. In other words, some participants felt their organizational protocols would not allow them to move forward based solely on civilians "tweeting" or "facebooking" alone. X24-Europe addressed this issue with experts through registration, but solutions have yet to be identified for affected populations. There will need to be a considerable amount of research on ways to improve the perceived credibility of messages from the public to responders and vice versa. 65% of respondents felt social media originating from a general member of the public is likely to be inaccurate. This study suggests reliability in social media will need to originate from the accounts of X24, or similar emergency management agencies pushing the message outward while cautiously handling any reports coming inward.

CONCLUSION

Jennex (2010) postulated the use of knowledge management strategy as a method for incorporating social media, cloud computing, and web 2.0 technologies into organizational crisis response and crisis response systems. X24 and X24-Europe utilized knowledge and lessons learned to refine the use of these technologies into crisis response, this is a knowledge management approach. However, X24 is really about demonstrating how these technologies can be used and less about incorporating them into a specific crisis response system. It is left to future research to fully incorporate these technologies into organizational crisis response. However, X24 has been very useful in demonstrating the ability of these technologies to facilitate crisis response collaboration on a massive scale; something that hadn't been done prior to X24. This paper is a action report on the X24 exercise. Its goal is to make the crisis response community aware of capabilities they may not be aware of and of an exercise approach that works for involving thousands of experts. The value of this paper is in the experience that is being reported.

Additionally, X24 utilized the "democratization of technology," which brought about open and inexpensive cloud computing technologies, has given individuals caught in the midst of an emergency situation the ability to decide their own course of action through crowdsourcing readily available information. Citizens no longer have to be empty vessels waiting to receive unidirectional news from a government press release, the 6:00PM network news, or the morning edition of USA Today. Instead, they serve as information gatherers and receivers, as direct links between the multitude of resources that need to be utilized during and after a natural or manmade emergency situation. Given that there appears to be an increasing upward trend in natural disasters as well as an ongoing global economic downturn, demonstrating the viability of low cost crisis response solutions is valuable to victims, responder organizations, and governments.

Finally, it is important that information sharing in a crisis occurs between individuals within government, between government employees and communities of interest, between researchers and government data, between the government and its citizens, and between governments of different countries. We need to encourage user input. Efforts must be made to leverage private, public, and university resources in aiding a community-based social knowledge network. There are only a few existing tools, policies, and trainings on the potentials of social media in emergency management. We must have comprehensive emergency-response plans and funding opportunities for groups to continue to research and test the tools and technologies available to figure out what works before a tragedy, not only after it has occurred.

The challenge for future research is to ensure that planning measures are universally designed for all, are useful in the community, and promote equal access, dignity, choice, and security in response and recovery. Millions of people in developing nations are still struggling with basic technologies, or just, the lack, in numbers of available smart-technologies. Citizens must be able to believe the warnings are credible and from multiple reliable resources. Our tools and technologies utilized in this paper not only need to be perfected, but made absolutely available to as many individuals on the planet in order to be an effective means of disaster preparation, response and relief.

Agencies are inevitably stretched thin during a natural disaster or manmade emergency, especially one that threatens a large community with the loss of life and property. Agencies have limited staff and limited abilities to acquire and synthesize the geographic information that is vital to effective response (Goodchild & Glennon, 2010). On the other hand, X24, X24-Europe and its cloud computing technologies equip the average citizen with the power of observation, and then empower them with the ability to register those observations, transmit that information through the cloud computing tools, and to finally synthesize the information into readily understood situational awareness models.

X24 and X24 Europe demonstrated that volunteer NGOs, partner nations, and interested individuals are willing to work together and share critical information during a time of crisis. More importantly, by using no-cost, open-source tools and technologies, volunteer participants felt secure in sharing information. This can be attributed to the fact that the US government or military did not control the applications or infrastructure on which the tools were hosted. NGOs in particular have historically been skeptical of using tools hosted on government owned infrastructure for fear that their personal information may be captured, tracked, etc., by intelligence entities. Even though this is understandably false, the fear and perception are real and any proposed military or other government info sharing solution needs to understand this basic constraint. No personal information was collected on any participant. The focus of information sharing was in regards to situational awareness of the simulated disaster.

Any interesting observation during the preparation of this paper was the occurrence of the San Diego Blackout on September 8, 2011. This was a total blackout that also affected Baja Mexico and parts of Orange and Imperial counties in Southern California. The interesting observation is that even in a total blackout situation, affected population was still using twitter, Facebook, cell phones, and smart phones to communicate and plan crisis response. It is too early to truly understand the impact of social media on this event but initial observations show that social media is viable even in a total blackout event.

AREAS FOR FUTURE STUDY

Several good research questions have been identified through X24 and X24-Europe (Note that in the following "national," "federal," and other generic terms refer to any nation or government and also include multinational concerns):

- When dealing with national security, a host of factors encourage "stove piped" communications that make interagency coloration difficult. Extraordinary changes in law, organization, and processes will need to be actualized in order to clarify roles and responsibilities and improve relations in a crowdsourcing environment.
- There will be significant legal questions raised and precedents set by governmental use of cloud computing, and legislation addressing both IT and business needs and consumer fears and protections will need to be a major focus for future research.
- Pre-existing contracts and plans between federal (and possibly state) agencies should exist prior to disaster events. These contracts should clearly state requirements regarding the imagery, its collection time/processing, and delivery expectations. The need for such contracts is not a new concept for disaster event planning, but these may not be pursued for a variety of reasons.
- Security, accountability, privacy, and other concerns often drive national security institutions to limit the use of open tools such as social software, whether on the open web or behind government information system firewalls. Information security concerns are very serious and must be addressed, but understanding our restrictions may diminish our ability to effectively communicate as well.
- During the training session, a number of participants used Skype to collaborate. They naturally turned this into a Social Media Emergency Operations Center (EOC) for volunteer technical communities. This type of activity was found

to be necessary to best capitalize on the crowdsourcing.

- With the vast amount of acronyms in emergency management, it is important to clarify which letters mean what. It would also be beneficial all players, if participants would utilize the same symbols for mapping.
- Information about who the participants are and what they bring to the exercise should be collected well before the exercise date, as well as information regarding how can their tool or service can be combined with others to create innovative solutions. This will enable better collaboration.
- The realism and utility of exercises could be enhanced through the inclusion of "affecteds" in addition to "responders". Past experience has demonstrated that the persons on the ground who are affected by the disaster are the catalyst that drives social media takeoff immediately after the event. This has typically involved use of a wiki as a coordinating force, which by its very nature allows for expansion, rectification and the ability to branch-off as the response progresses.
- Social media can tend to lead to mass hysteria and the free flow of misinformation, and potentially disinformation. This can be minimized if organizations that are distributing accurate information (i.e., Red Cross) are sure to label and site their information. This will help to ensure the general public's trust and reliance on information that comes from credible sources, and will encourage the public to access these sources first.

Additionally, further analysis of the data gathered during X24 and X24-Europe needs to be performed. This will be done in a future article that identifies an analysis framework and then views the data through this research lens.

A final area for future research is the performance of more X24 exercises. Currently X24-Mexico is tentatively planned for February, 2012. This exercise has the goals of:

1. Demonstrate the use of no/low-cost, off-the-shelf social media, crowdsourcing, and collaboration web tools to gather, coordinate, and share actionable real-time information to build situational awareness to help victims of a natural disaster and help save lives;

2. Establish dialogue and build relationships between all partner nations, international organizations, and public/private partners regarding the use of online tools;

3. Test online tools to measure effectiveness, and streamline cyber information sharing in preparation for the next real-world crisis;

4. Address the virtual flow of information and activities of international organizations during a natural disaster and a terror attack utilizing biological weapons of mass destruction/effect; and

5. Encourage all formal and informal groups across the globe to actively participate or observe the exercise.

ACKNOWLEDGMENT

The authors wish to acknowledge the many persons and organizations involved in holding the X24 exercises. First are the adjunct faculty, researchers, and students in SDSU's Homeland Security Program. We also would like to acknowledge some key people: Doug Wied, Chris Maxin, and Mike Fernandez were deeply involved in leading the exercises working under Dr. George Bressler's leadership; Steve Birch, Mike Hennig, and Eric Ackerman were deeply involved in the technical aspects of supporting the exercises. Additionally several companies and organizations contributed to the exercises including Google, Twitter, Buzz Manager, and InRelief, Finally, we'd like to thank all the individuals and organizations that participated in the exercise.

REFERENCES

Dhanjani, N. (2008, December 18). How terrorists may abuse micro-blogging channels like Twitter. *Nitesh Dhanjani Blog*. Retrieved October 1, 2010, from http://www.dhanjani.com/blog/2008/12/how-terrorists-may-abuse-microblogging-channels-like-twitter.html

Gleicher, N. J., & Hwang, J. S. (2010). Geospatial information services: Balancing privacy and innovation. In *Proceedings of the AAAI Spring Symposium Series* (pp. 75-81).

Goodchild, M. F., & Glennon, J. A. (2010). Crowdsourcing geographic information for disaster response: a research frontier. *International Journal of Digital Earth*, *3*(3), 231–241. doi:10.1080/17538941003759255

Jennex, M. E. (2010). Implementing social media in crisis response using knowledge management. *International Journal of Information Systems for Crisis Response and Management*, *2*(4), 20–32.

Maiyo, L., Kerle, N., & Köbben, B. (2010). Collaborative post-disaster mapping via geo web services. In Konecny, M., Ziatanova, S., & Bandrova, T. L. (Eds.), *Geographic information and cartography for risk and crisis management* (pp. 221–232). New York, NY: Springer. doi:10.1007/978-3-642-03442-8_15

Palen, L., Anderson, K. M., Mark, G., Martin, J., Sicker, D., Palmer, M., & Grunwald, D. (2010, April 13-16). A vision for technology-mediated support for public participation & assistance in mass emergencies & disasters. In *Proceedings of the ACM-BCS Conference on Visions of Computer Science*, Edinburgh, UK (pp. 1-12). Retrieved September 11, 2010, from http://www.cs.colorado.edu/~palen/computingvisionspaper.pdf

Raman, M., Ryan, T., Jennex, M. E., & Olfman, L. (2010). Wiki technology and emergency response: An action research study. *International Journal of Information Systems for Crisis Response and Management*, 2(1), 49–69. doi:10.4018/jiscrm.2010120405

Rayport, J., & Heyward, A. (2009). Envisioning the cloud: The next computing paradigm. *Marketspace*. Retrieved September 20, 2010, from http://www. marketspaceadvisory.com/cloud/Envisioning-the-Cloud.pdf

Starbird, K., Palen, L., Hughes, A. L., & Vieweg, S. (2010, February 6-10). Chatter on The Red: What hazards threat reveals about the social life of microblogged information. In *Proceedings of the ACM Conference on Computer Supported Cooperative Work*, Savannah, GA (pp. 241-250).

Wax, E. (2008, December 3). Gunmen used technology as a tactical tool. *Washington Post*. Retrieved October 10, 2010, from http://www.washingtonpost.com/wp-dyn/content/article/2008/12/02/AR2008120203519.html

This work was previously published in the International Journal of Information Systems for Crisis Response and Management, Volume 3, Issue 4, edited by Murray E. Jennex and Bartel A. Van de Walle, pp. 36-54, copyright 2011 by IGI Publishing (an imprint of IGI Global).

Chapter 16
Ubiquitous Computing for Personalized Decision Support in Emergency

Alexander Smirnov
St. Petersburg Institute for Informatics & Automation of the Russian Academy of Sciences, Russia

Tatiana Levashova
St. Petersburg Institute for Informatics & Automation of the Russian Academy of Sciences, Russia

Nikolay Shilov
St. Petersburg Institute for Informatics & Automation of the Russian Academy of Sciences, Russia

Alexey Kashevnik
St. Petersburg Institute for Informatics & Automation of the Russian Academy of Sciences, Russia

ABSTRACT

Ubiquitous computing opens new possibilities in various aspects of human activities. The paper proposes an approach to emergency situation response that benefits from the ubiquitous computing. The approach is based on utilizing profiles to facilitate the coordination of the activities of the emergency response operation members. The major approach underlying idea is to represent the operation members jointly with information sources as a network of services that can be configured via negotiation of participating parties. Such elements as profile structure, role-based emergency response, negotiation scenarios, and negotiation protocols are described in detail.

INTRODUCTION

Critical aspects of decision support for emergency situation management incorporate managing and controlling sources of information, processing real-time or near real-time streams of events,

representing and integrating low-level events and higher level concepts, multi-source information fusion, information representation that maximizes human comprehension, reasoning on what is happening and what is important (Jakobson et al., 2005; Scott & Rogova, 2004; Smirnov et al., 2007).

Although many research efforts are aiming at investigating the issues above, an efficient complex

DOI: 10.4018/978-1-4666-2788-8.ch016

solution is still required in this field. The paper proposes an innovative approach based on emerging information technologies to contribute into emergency decision support. The present research considers the emergency situation management that incorporates the following types of operations: first aid, emergency control, and evacuation. These operations can involve autonomous entities like public/governmental organizations, different private organizations and volunteers (referred to as operation members). Particularly, an event of fire is considered as an example of the emergency situation that requires the emergency response operation.

Organization of collaborative environments out of autonomous entities is a focus of approaches aimed at building context aware decision support systems (e.g., Kwon et al., 2005; Burstein et al., 2009), self-optimization and self-configuration in wireless networks (Cordis, 2008), organization of context-aware cooperative networks (Ambient Networks, 2006) and collaborative context-aware service platforms (Ejigu et al., 2008), etc.

This work proposes an approach to developing a decision support system intended for situation management. The approach benefits from certain features of ubiquitous computing and incorporates technologies of context management, intelligent agents, Web-services and profiling. Ubiquitous computing (ubicomp) is a post-desktop model of human-computer interaction that thoroughly integrates information processing into everyday objects and activities. In the course of ordinary activities, someone "using" ubiquitous computing engages many computational devices and systems simultaneously, and may not necessarily even be aware that they are doing so (Wikipedia, http://www.wikipedia.org).

The technology of Web-services is used as a technology allowing the heterogeneous resources (operation members and information sources) to cooperate for a common purpose. This technology defines formal interface agreement (Austin, Barbir, Ferris, & Garg, 2004), however, does not address the semantics of those interfaces. An agent-based service model is used to provide the Web-services with semantics and to turn them into active collaborative components. In different approaches integration of intelligent agents and Web-services has been applied for various purposes, for instance, it served as a basis for distributed service discovery and negotiation system in B2B (Lau, 2007; Moradian, 2010) describes a way to protect sensitive business information against being disclosed, modified and lost. In the presented approach the agents "activate" Web-services when required and make Web-service descriptions sharable via ontology.

In real life situations consideration often must be given to their continuous changes including people movement, traffic situation (traffic jams, closed roads, etc.) that makes the problem quite complex and requires its real-time solving. For this purpose the approach utilizes the technology of context management; at that, the context is constantly updated to provide up-to-date information for situation management.

Application of the profiling technology significantly facilitates the emergency situation management. Dynamic profiles are considered, for instance, in Carillo-Ramos et al. (2007) and Kirsch-Pinheiro et al. (2006). The authors of Thomsen et al. (2009) propose dynamic user profiles that reflect the current situation of the user (e.g., his/her location). In the presented approach the profile is considered as an information source for forming the situation context. The operation member profiles contain such information as available transportation, current geographical coordinates, competencies, preferences, as well as roles of the response operation members. Competencies are described by operation members' capabilities, capacities, price-list in case of implementation by a private organization, and agility. The preferences determine the constraints preferable for the operation members.

The rest of the work is structured as follows. First, the motivating scenario is described. Then major principles of the proposed approach and main basic approach components like the profile structure, role-based situations in emergency, negotiation scenarios, and negotiation protocols are discussed successively; most important results are summarized in the conclusion.

Motivating Scenario

Let us consider the following scenario. All of a sudden a fire has started in some area and was detected by smart fire sensors. Each person in-

volved in the fire situation is supposed to have a mobile device (e.g., a mobile phone) that can measure impact on the owner (e.g., with a help of G-sensor) and register his/her current location. The emergency professionals, volunteers or police who can render assistance in the event of fire and better explain the situation to the specialists can be near the emergency site. These people are also considered to have mobile devices, which would inform them about the needed help. The proposed case study scenario is described below and represented in Figure 1.

Figure 1. Case study scenario

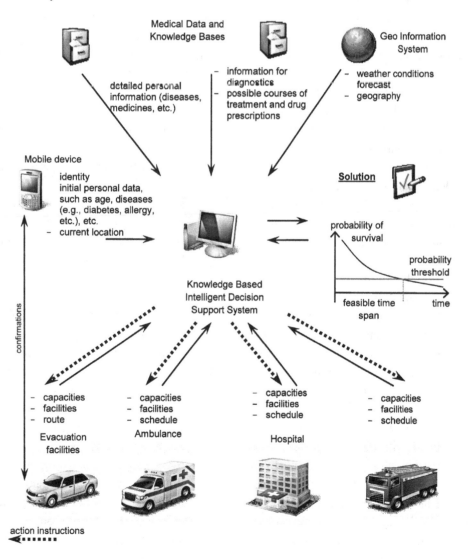

The goal of the approach is to produce efficient plans of actions both for people at the emergency site and the rescue facilities (ambulances, fire fighters, etc.) based on available information from different sources and benefiting from the features of the ubiquitous computing. The efficiency is estimated via a multicriterion function accounting for the time taken for the situation response and associated costs.

The information about people in emergency (their locations, possible injuries, some specific data (e.g., diseases such as diabetes or allergy, etc.) is stored in their mobile devices. This information and the information about the existing local infrastructure (available hospitals, road network, etc.) are transferred to the decision support system. Besides the listed information, the system acquires additional information and knowledge required for the problem solving. Some of this information and knowledge are acquired from available sources such as medical databases, weather conditions' sensors and Internet-sites, weather forecasts' sources, etc. The information from the sources is extracted in a particular context (incident location, type of incident, etc.). Other information is provided by the operation members: ambulances, hospital and evacuation facilities provider(s) that supply to the system actual information about their facilities as well as current and possible future capacities.

Most of the information supplied to the system can be of certain probability or uncertain. The objective here is to reduce the situation response time and, as a result, to increase the probability of the victims' survival (see the graph in Figure 1).

Taking into account the current locations, competences, capabilities and current conditions of people at the emergency site the decision support system might instruct them and the operation members through their mobile devices to perform certain actions e.g., doctors can be instructed to render the first aid and to explain the situation and conditions of the injured people to medical brigades that would arrive in a certain time.

APPROACH

Figure 2 represents the generic scheme of the approach aimed at the emergency situation management. The main idea behind the approach is to represent the operation members and information sources by sets of services they provide. This allows for replacing the configuration of the emergency situation response facilities by the distributed services. To resolve the problem of semantic interoperability between the services they are represented by Web-services using a common vocabulary supported by an application ontology (AO) of the emergency management domain (Smirnov et al., 2009). This ontology consists of 7 taxonomy levels, more than 600 classes, 160 attributes, 40 hierarchical constraints, 50 associative constraints and 30 functional constraints.

At the first stage of the research the main ideas the approach is based on were formulated:

1. A common shared AO serves for terminology unification. Each service has a fragment of this ontology corresponding to its capabilities/responsibilities. This fragment is synchronized automatically when necessary (not during the response operation).
2. An emergency situation is represented by two levels. At the first level it is represented by an *abstract context* that is non-instantiated ontology knowledge relevant to the emergency situation. At the second level it is represented by an *operational context* that is an instantiated abstract context.
3. The abstract context is instantiated by a *network of Web-services*.
4. Emergency response plans are generated using the *constraint satisfaction technology*.
5. Each operation member is represented by a *profile* describing its capabilities. Information from the operation member's profile is used for determining if this member is capable of carrying out a specified task and, hence, can be chosen as a team member.

Figure 2. Generic scheme of the approach

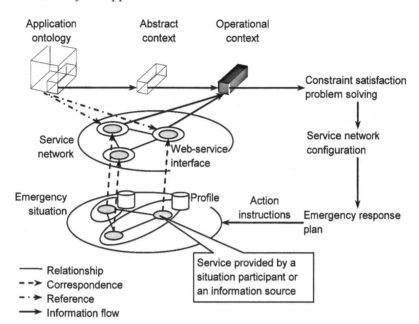

6. Web-service standards are used for the interactions. External sources (e.g., medical databases, transport availability sources, weather forecasts' sources) should also support these standards and the terminology defined by AO. This is achieved by developing wrapping services for each particular source.

7. Each service is assigned an intelligent agent that represents it (together they will be called *"agent-based service"*). The agent collects information required for situational understanding by the service and *negotiates* with other agents for creating ad-hoc action plans. The agent has predefined rules to be followed during the negotiation processes. These rules depend on the role of the appropriate member.

The main elements of the approach, namely the profile structure, the role-based situations in emergency, the negotiation scenarios, and the negotiation protocol are described in detail.

Operation Member Profile

The structure of the operation member profile is given in Figure 3. The operation member profile represents the following types of information about the member: *General Information, Operation Member Information, Instruction History, Operation Member Preferences.*

The *General Information* part describes general information about an operation member organisation. It contains a name of organisation, its identifier in the system, date of its foundation, and URL of the organization's Web page.

Operation Member Information is a set of tuples describing information about an operation member. Each tuple possesses the following properties:

- **Member Name**: The name of an operation member;
- **Location:** Current geographical location of a member, it can be taken into account for estimating a rapidity and quality of acting instruction performance in a particular

Figure 3. Operation member competence profile

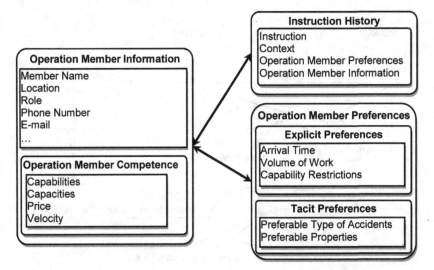

situation; this property is used by GIS for generating the map of the region representing the situation, road network, operation members, and hospitals.

- **Role:** The role an operation member fulfills in the emergency situation; the roles of decision maker, fire-fighter, emergency doctor, leader of a team, car driver, victim, etc. can be pointed out;
- **Phone Number, E-mail:** Contact information;
- Other characteristics of an operation member (the time zone, the list if languages, etc.).

Operation Member Competencies includes the following properties:

- **Capabilities:** Determine types of operations that operation member can implement, the capabilities are described by a list of corresponding classes (describing possible activities) from AO;
- **Capacities:** Determine capacity of an operation member (in case of evacuation this attribute determines how many people this operation member can evacuate);

- **Prices:** Determine prices of member's activities in case of implementation by a private organization;
- **Velocity:** Determines velocity of operation member's activities.

Instruction History is also a set of tuples. Each tuple possesses the following properties:

- **Instruction:** An acting instruction sent to a member;
- **Context:** A situation context used to analyze performance of a member (other members can see solutions generated in particular situations) and to identify detectable member preferences;
- **Operation Member Preferences:** Member preferences at the moment of an acting instruction sending. They contain a snapshot of all the properties of the category "Operation Member Preferences";
- **Operation Member Information:** Specific information about a member at the moment of an acting instruction sending. It contains a snapshot of all the properties of the category "Operation Member Information".

The *Operation Member Preferences* part consists of Explicit Preferences and Tacit Preferences. Explicit Preferences describe member preferences that are manually introduced by a member. These preferences are used for choosing a member for a particular situation, and contain the member preference for arrival time, volume of work, and capability constraints. The latter stores several capabilities and logical restrictions from a list of all the capabilities for the domain. *Tacit Preferences* describe automatically detectable member preferences.

Role-Based Emergency Response

The decision support system presents the emergency response plans to the operation members in accordance with the roles these members fulfill in the emergency situation. The plans are designed based on the information from the situation context acquired from external information sources (e.g., medical databases, weather forecasts, etc.) and operational members' profiles. An emergency response plan is a set of operation members with transportation routes for the mobile responders or with required helping services for the organizations, e.g., the decision maker must be aware of complete information about the emergency situation. The emergency response plan presented to the person fulfilling the role of *decision maker* represents all the operational members along with corresponding transportation routes (see Figure 4). The operation members do not need to have the full picture of the situation. For them a slice of the common plan is presented. This slice represents information and parts of the common plan that are relevant to the actions expected from the particular operation member. Figure 5 shows part of the response plan presented to the operation member fulfilling the role of *leader of an emergency team* going by ambulance. This plan is displayed on the smart phone of the leader.

In emergency situations roles may change. Roles interplay is illustrated by their change during evacuation of people from the dangerous area. The procedure of the evacuation is supported by the ridesharing technology. The technology can be applied to uninjured people that do not need any assistance.

Persons who need to be evacuated fulfill the role of *potential victims*. Through the user interface in their mobile devices they invoke agents implementing the ridesharing technology functions. The profiles of these persons contain real-time information about their locations. The persons enter the locations they would like to be conveyed. The agents search for cars going to or by the same or close destinations that the persons would like to be. They search the cars among the vehicles passing the person locations. Information about the destinations that the car drivers are going to is read from the navigators that the drivers use or from the drivers' profiles. The profiles store periodic routes of the drivers and cars' properties as number of passenger seats, availabilities of baby car seats, etc. (see Figure 6).

The agents send the information about the person locations and destinations, the locations and destinations of the found cars, and the cars' properties to the decision support system. The system generates a set of feasible routes for person transportations and proposes efficient ones. The criteria of the efficiency are minimum time of evacuation and maximum evacuation capacity. As well the system determines points where the drivers are expected to pick up the passenger(s). The decision maker comes to an agreement with the drivers included in the plan and the persons to be evacuated. From this point the car drivers change the role of *driver* to *evacuation driver*; the persons evacuated change the role of *potential victim* to *passenger*.

Examples of routes generated for a driver and a passenger and displayed on the mobile devices of persons fulfilling the roles of *evacuation driver* and *passenger* are given in Figure 7 and Figure 8. The encircled car in the figures shows the site where the driver is offered to pick up the passenger.

Figure 4. Emergency response plan: decision maker view

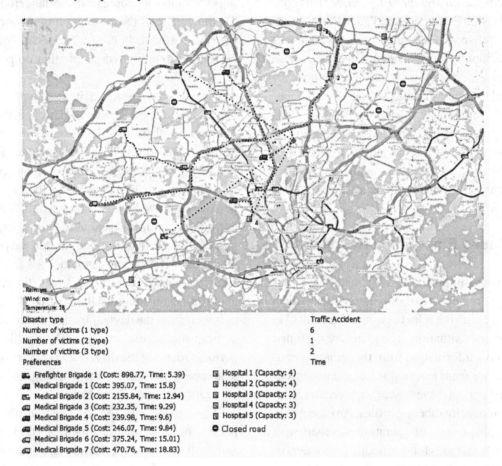

Rain: yes
Wind: no
Temperature: 18

Disaster type		Traffic Accident
Number of victims (1 type)		6
Number of victims (2 type)		1
Number of victims (3 type)		2
Preferences		Time

Firefighter Brigade 1 (Cost: 898.77, Time: 5.39)   Hospital 1 (Capacity: 4)
Medical Brigade 1 (Cost: 395.07, Time: 15.8)   Hospital 2 (Capacity: 4)
Medical Brigade 2 (Cost: 2155.84, Time: 12.94)   Hospital 3 (Capacity: 2)
Medical Brigade 3 (Cost: 232.35, Time: 9.29)   Hospital 4 (Capacity: 3)
Medical Brigade 4 (Cost: 239.98, Time: 9.6)   Hospital 5 (Capacity: 3)
Medical Brigade 5 (Cost: 246.07, Time: 9.84)   Closed road
Medical Brigade 6 (Cost: 375.24, Time: 15.01)
Medical Brigade 7 (Cost: 470.76, Time: 18.83)

Negotiation Scenarios

Figure 9 shows a scenario demonstrating negotiations between agents representing services provided by the emergency response operation members at decisions making according to the plan of the response actions. In the figure "Decision maker" means an agent that represents the services provided by the person fulfilling the role of decision maker, "Emergency responder" means an agent that represents the services provided by the emergency responder.

The decision making procedure starts after a set of feasible emergency response plans is generated and an efficient plan is selected from this set. An efficient plan is determined based on criteria of minimal time and cost of transportation of all

the victims to hospitals, and minimal number of mobile teams involved in the response operation.

The decision maker submits the efficient plan to the operation members. The operation members either approve the plan or decline it. In the latter case another set of plans have to be generated or the selected plan have to be adjusted (so that the potential participant who refused to act according to the plan does not appear in the adjusted plan) and submitted to approval. As soon as all the operation members have confirmed the plan they are in, the decision maker notifies the operation members to put the confirmed plan into actions.

Figure 10 shows agent interactions when all the operation members agree to participate in the response actions according to the plan selected by the decision maker (in the figure the

Figure 5. Emergency response plan: view for a leader of emergency team

operation members are represented by vehicles that they use – ambulance, fire truck, and rescue helicopter). It is seen that the decision maker sends simultaneous messages to all the operation

members with the plan for each member, waits for their replays on plan acceptance (Ready), and sends them simultaneous messages to take the response actions (Start).

The plan adjustment consists in a redistribution of the actions among operation members that are contained in the set of feasible plans. If such a distribution does not lead to a considerable loss of time (particularly, the estimated time of the transportation of injured people to hospitals does not exceed "The Golden Hour"), then the adjusted plan is submitted to the renewed set of operation members for approval. If a distribution is not possible or leads to loss of response time a new set of plans has to be generated.

Figure 11 demonstrates agent interactions in case when all ambulances selected for the response actions are not ready to participate, and the plan selected by the decision maker cannot be adjusted. Two ambulances (Ambulance 1 and Ambulance 2) replay "Not ready" to the messages of the decision maker. This replay is accompanied with the messages to the decision maker and the decision

Figure 6. Driver's profile

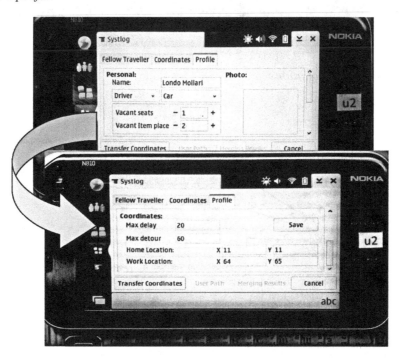

Figure 7. Ridesharing route: driver's view

Figure 8. Ridesharing route: passenger's view

Figure 9. Decision making negotiations

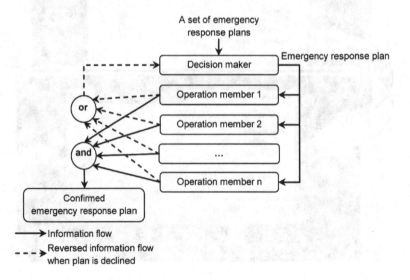

Figure 10. Emergency responders accept emergency response plan

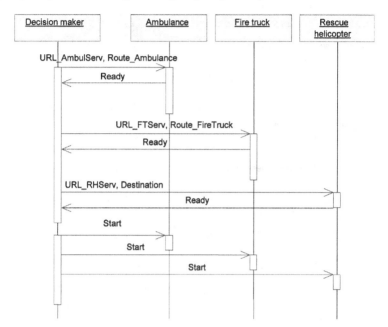

support system (DSS) with the reasons of their refusals. Examples of such reasons can be: the road is destroyed, the ambulance is blocked, etc.

The decision maker forwards the messages with the reasons of their refusals to DSS. This increases the chances that DSS will receive information that it was unaware of up to this moment. As well the decision maker sends to DSS the message about excluding the two ambulances from the list of available emergency responders.

DSS corrects the operational context according to the information contained in the reasons, generates a new set of plans, and sends it to the decision maker.

Negotiation Protocol

In order to choose a protocol for negotiation between agents representing services provided by the emergency response operation members, the main features have been formulated as follows:

1. **Contribution:** The agents have to cooperate with each other to make the best contribution into the overall system's benefit – not into the agents' benefits;

2. **Task performance:** The main goal is to complete the task performance – not to get profit out of it;

3. **Mediating:** The agents *operate* in a decentralized community, however in all the negotiation processes there is an agent managing the negotiation process and making a final decision;

4. **Trust:** Since the agents *participate* in the same situation they have to completely trust each other;

5. **Common terms:** The agents *are* supposed to use common terms for communication (based on the dictionary of the common AO).

The protocols analyzed in order to choose one most suitable include voting, bargaining, auctions, general equilibrium market mechanisms, coalition games, and constraint networks (Weiss, 2000). Based on the analysis of these protocols and the above requirements to them, the contract

Figure 11. Plan regeneration

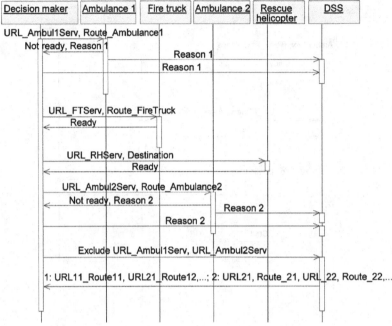

net protocol (CNP) was chosen as a basis for the negotiation model in the approach. As it can be seen from Table 1 this protocol meets most of the requirements.

CNP is one of basic coordination strategies between agents in multi-agent systems. It was originally introduced by Davis and Smith (1983). The main features of this protocol are (1) *managers* (*initiators* in FIPA) divide tasks, (2) *contractors* (*participants* in FIPA) bid, (3) manager makes contract for lowest bid, (4) there is no negotiation of bids. The UML sequence diagram of FIPA-based contract net protocol is presented in Figure

12. Since CNP is a basic protocol any particular multi-agent system requires some modifications for CNP to be implemented (Payne et al., 2002). Table 2 describes the modifications made for original CNP.

To compare the results of the conventional CNP and constraint-based CNP the following example is considered. Configuration agent (traffic accident response manager, CA) needs to involve three operation members (accident investigators that are supposed to work together) represented by the contractor wrapper agents (W1, W2, and W3), with the time and cost of help be-

Table 1. Comparison of negotiation protocols

Protocols Criteria	Voting	Bargaining	Auctions	General Equilibrium Market Mechanisms	Coalition Games	Contract Nets
Contribution	☑	☑	☐	☐	☐ / ☑	☑
Task Performance	☐ / ☑	☑	☐	☐	☐	☐ / ☑
Mediating	☐	☐	☑	☐	☐	☑
Trust	☑	☑	☑	☑	☐	☑
Common Terms	☑	☑	☑	☑	☑	☑

Figure 12. Example of the constraint-based iterative negotiation

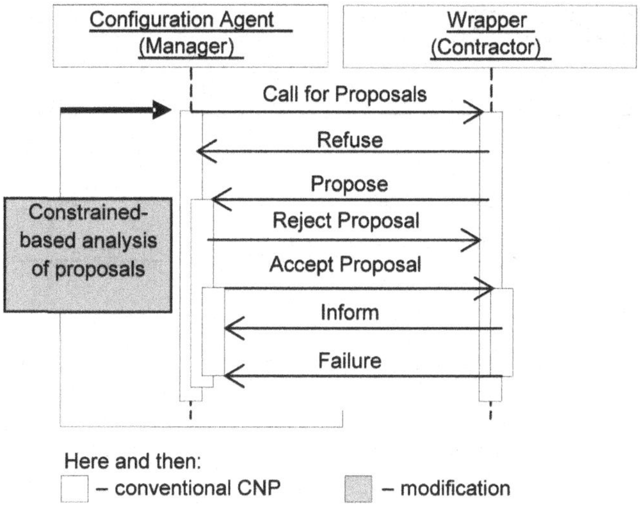

ing minimal. These criteria (time and costs) have been chosen as demonstration criteria. The set of criteria can be extended with other parameters. It is also preferable for the Configuration agent to choose smaller costs if the time is the same:

CA: time → min, costs → min

The Wrappers can make different offers such that the costs inversely depend on the time of help. This dependency is described by a table function given below (the cost is measured in abstract units):

W1: 3.0 min/15

W2: 1.5 min/20; 2.5 min/10; 4.5 min/5
W3: 5.0 min/25; 6.0 min/15; 7.0 min/10

The resulting time and costs are calculated as follows:

$$time = \max(time_{W1}, time_{W2}, time_{W3})$$

$$costs = \text{sum}(costs_{W1}, costs_{W2}, costs_{W3})$$

The comparison of the two scenarios is presented in Table 3. The left column describes negotiation performed in accordance with the conventional CNP (see Figure 13). The right column

Table 2. Changes in features of the conventional CNP

Protocols Feature	Conventional CNP	Modified CNP
Iterative Negotiation	-	The negotiation process can be repeated several times until acceptable solution is achieved
Conformation	-	Concurrent conformation between manager and contractors
Available Messages	Fixed set of 8 Messages (see Figure 12)	Flexible set: new specific messages and corresponding to FIPA *Request* and *Confirm* communicative acts, and message *Clone* not corresponding to any FIPA communicative act are included
Participants Roles	Manager and Contractors	Manager and two types of contractors: (1) "classic" contractors negotiating proposals, and (2) auxiliary service providers not negotiating but performing operations required for decision support system functioning (e.g., AO modification, user interfacing, etc.)
Role Changing	-	Agents can change their roles during a scenario

Figure 13. Experimentation: scenario 1

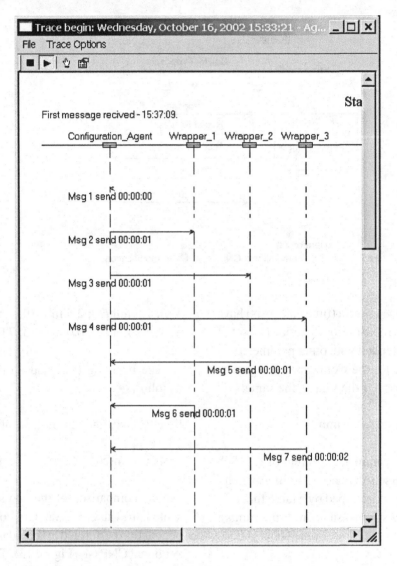

describes negotiation performed in accordance with the modified CNP; it contains concurrent conformation and iterative negotiation (see Figure 14). As it can be seen, the constraint-based CNP proposed here allows achieving a better solution with reduced cost. The messages (Msg 1, Msg 2, etc.) are indicated in both the table and the figures. Msg 1 in the figure is an auxiliary message that is used to start the negotiation process. The sequence of the replies arriving from wrapper agents (e.g., wrapper 2 answers faster than wrapper 1) is more or less random and should not be paid attention.

CONCLUSION

The paper proposes an approach to emergency situation response that benefits of the features of ubiquitous computing. It is based on utilizing profiles of operation members in the emergency situation for coordination of their activities. The major idea of the approach is to represent the operation members together with information sources as a network of services that can be configured via negotiation of participating parties.

Table 3. Comparison of the conventional and modified CNP

Conventional CNP	Modified CNP
The configuration agent sends calls for proposals to all the wrappers concurrently. Besides description of the task to be performed each call contains additional constraints. In this case these constraints will contain the following: Msg2, Msg 3, Msg4: time → min	
The offers from wrappers will contain the following: Msg6: W1: 3.0 min/15 Msg5: W2: 1.5 min/20 Msg7: W3: 5.0 min/25	
At this stage the work of the conventional CNP ends.	From this stage the second iteration of the modified CNP starts. After this the Configurator analyses the results and sends new calls to the Wrappers 1 and 2: Msg8, Msg9: time ≤ 5.0 AND costs → min
	The Wrappers reply as follows: Msg10: W1: 3.0 min/15 Msg11: W2: 4.5 min/5
The result will be 5.0 min and 60 cost units: W1: 3.0 min/15 W2: 1.5 min/20 W3: 5.0 min/25	The result is 5.0 min and 45 cost units: W1: 3.0 min/15 W2: 4.5 min/5 W3: 5.0 min/25

The profiles describe both static and dynamic aspects of the operation members such as their competences, locations, or roles they fulfill in the emergency situation and store the history of the acting instructions sending to them and their responses. The proposed structure of the profiles makes it possible to perform revealing of tacit preferences and competences. The algorithm of ontology-based clasterization for revealing the tacit preferences was presented in Smirnov et al. (2005). As well the profiles' structure allows the decision support system to present the emergency response plan to the operation members in accordance with the roles these members fulfill at a given moment of time.

The negotiation scenarios discussed in the paper allowed the authors to formulate main specifics of the negotiation in emergency. Taking into account these specifics the contract net protocol has been chosen as the basis for the negotiation between the operation members. At that, the conventional contract net protocol has been modified. Based on the comparative analysis of the conventional contract net protocol and the modified the conclusion can be made that the modified protocol allows for obtaining more efficient negotiation results than the conventional one.

ACKNOWLEDGMENT

The paper is due to the research carried out as a part of the project funded by grants 09-07-00436, 09-07-00066, 10-07-00368, 10-08-90027, 11-07-00045, and 11-07-00058 of the Russian Foundation for Basic Research, and project 213 of the Russian Academy of Sciences.

Figure 14. Experimentation: scenario 2

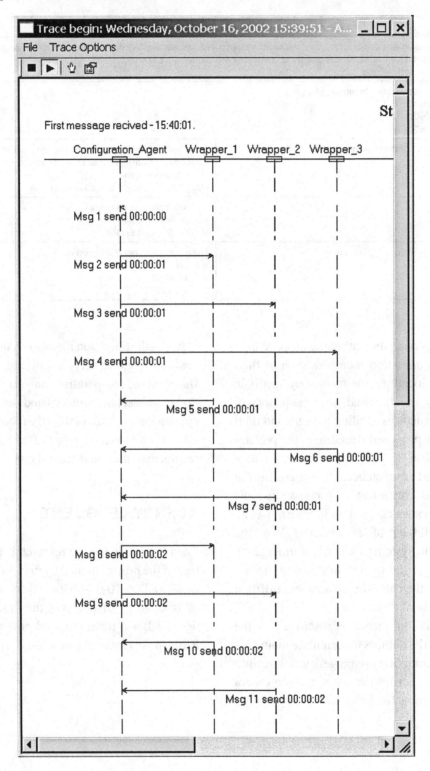

REFERENCES

Ambient Networks Project. (2006). *Mobile and wireless systems and platforms beyond 3G* (Tech. Rep. No. IST-2004-2.4.5). Retrieved from http://www.ambient-networks.org/

Austin, D., Barbir, A., Ferris, C., & Garg, S. (Eds.). (W3C, 2004) Austin (2004). *Web services architecture requirements*. Retrieved from http://www.w3.org/TR/2004/NOTE-wsa-reqs-20040211/

Burstein, F., Zaslavsky, A., & Arora, N. (2005). Context-aware mobile agents for decision-making support in healthcare emergency applications. In *Proceedings of the Workshop on Contextual Modelling and Decision Support* (p. 144).

Carillo-Ramos, A., Villanova-Oliver, M., Gensel, J., & Martin, H. (2007). Contextual user profile for adapting information in nomadic environments. In M. Weske, M.-S. Hacid, & C. Godart (Eds.), *Proceedings of the 8th International Conference on Web Information Systems Engineering* (LNCS 4832, pp. 337-349).

Cordis. (2008). *Self-optimisation and self-configuration in wireless networks (SOCRATES). The network of the future* (ICT-2007.1.1.) Retrieved from http://cordis.europa.eu/

Davis, R., & Smith, R. G. (1983). Negotiation as a metaphor for distributed problem solving. *Artificial Intelligence, 20*(1), 63-109.

Ejigu, D., Scuturici, M., & Brunie, L. (2008). Hybrid approach to collaborative context-aware service platform for pervasive computing. *Journal of Computers, 16*(1), 40-50.

Jakobson, G., Kokar, M. M., Lewis, L., Buford, J., & Matheus, C. J. (2005). Overview of situation management at SIMA 2005. In *Proceedings of the Military Communications Workshop on Situation Management* (pp. 17-20).

Kirsch-Pinheiro, M., Villanova-Oliver, M., Gensel, J., & Martin, H. (2006, June). A personalized and context-aware adaptation process for web-based groupware systems. In *Proceedings of the 4th International Workshop on Ubiquitous Mobile Information and Collaboration Systems* (pp. 884-898).

Kwon, O., Yoo, K., & Suh, E. (2005). UbiDSS: A proactive intelligent decision support system as an expert system deploying ubiquitous computing technologies. *Expert System Applications, 28*(1), 149-161.

Lau, R. Y. K. (2007). Towards a web services and intelligent agents-based negotiation system for B2B eCommerce. *Electronic Commerce Research and Applications, 6*(3), 260-273.

Moradian, E. (2010). Integrating web services and intelligent agents in supply chain for securing sensitive messages. In I. Lovrek, R. J. Howlett, & L. C. Jain (Eds.), *Proceedings of the 12th International Conference on Knowledge-based Intelligent Information and Engineering Systems* (LNCS 5179, pp. 771-778).

Payne, T., Singh, R., & Sycara, K. (2002) Communicating agents in open multi-agent systems. In *Proceedings of the First GSFC/JPL Workshop on Radical Agent Concepts*.

Scott, P., & Rogova, G. (2004, June 28-July 1) Crisis management in a data fusion synthetic task environment. In *Proceedings of the 7th Conference on Multisource Information Fusion*, Stockholm, Sweden (pp. 330-337).

Smirnov, A., Pashkin, M., Chilov, N., Levashova, T., Krizhanovsky, A., & Kashevnik, A. (2005). Ontology-based users and requests clustering in customer service management system. In V. Gorodetsky, J. Liu, & V. Skormin (Eds.), *Proceedings of the International Workshop on Autonomous Intelligent Systems: Agents and Data Mining* (LNCS 3505, pp. 231-246).

Smirnov, A., Kashevnik, A., Levashova, T. Pashkin, M., & Shilov, N. (2007). Situation modeling in decision support systems. In *Proceedings of the International Conference on Integration of Knowledge Intensive Multi-Agent Systems: Modeling, Evolution and Engineering* (pp. 34-39).

Smirnov, A., Levashova, T., Krizhanovsky, A., Shilov, N., & Kashevnik, A. (2009). Self-organizing resource network for traffic accident response. In *Proceedings of the International Conference on Information Systems for Crisis Response and Management*.

Thomsen, J., Vanrompay, Y., & Berbers, Y. (2009). Evolution of context-aware user profiles. In *Proceedings of the International Conference on Ultra Modern Telecommunications & Workshops* (pp. 1-6).

Weiss, G. (2000). Introduction. In G. Weiss (Ed.), *Multiagent systems: A modern approach to distributed artificial intelligence*. Cambridge, MA: MIT Press.

This work was previously published in the International Journal of Information Systems for Crisis Response and Management, Volume 3, Issue 4, edited by Murray E. Jennex and Bartel A. Van de Walle, pp. 55-72, copyright 2011 by IGI Publishing (an imprint of IGI Global).

Compilation of References

ActionAid. (2006). *Tsunami Response: Human Rights Assessment.* Retrieved September 27, 2010, from http://www.actionaid.org/docs/tsunami_human_rights.pdf

Adams, M. J., Tenney, Y. J., & Pew, R. W. (1995). Situation awareness and cognitive management of complex systems. *Human Factors, 37*(1), 85–104. doi:10.1518/001872095779049462

Aedo, I., Díaz, P., Carroll, J. M., Convertino, G., & Rosson, M. D. (2010). End-user oriented strategies to facilitate multi-organizational adoption of emergency management information systems. *Information Processing & Management, 46*(1), 11–21. doi:10.1016/j.ipm.2009.07.002

Aedo, I., Díaz, P., & Díez, D. (2009). Cooperation amongst autonomous governmental agencies using the SIGAME web information system. In *Proceedings of the EGOV Conference.*

Agostini, T., & Bruno, N. (1996). Lightness contrast in CRT and paper-and-illuminant displays. *Perception & Psychophysics, 58*(2), 250–258. doi:10.3758/BF03211878

Alani, H., Jones, C. B., & Tudhope, D. (2001). Voronoi-based region approximation for geographical information retrieval with gazetteers. *International Journal of Geographical Information Science, 15*(4), 287–306. doi:10.1080/13658810110038942

Ambient Networks Project. (2006). *Mobile and wireless systems and platforms beyond 3G* (Tech. Rep. No. IST-2004-2.4.5). Retrieved from http://www.ambient-networks.org/

Amitay, E. Har'EI, N., Sivan, R., & Soffer, A. (2004). Web-a-where: Geotagging web content. In *Proceedings of the 27th Annual International ACM SIGIR Conference on Research and Development in Information Retrieval.*

Andersen, H. B., Garde, H., & Andersen, V. (1998). MMS: An electronic message management system for emergency response. *IEEE Transactions on Engineering Management, 45*(2), 132–140. doi:10.1109/17.669758

Andrienko, N., & Andrienko, G. (2005). A concept of an intelligent decision support for crisis management in the OASIS project. In van Oosterom, P., Zlatanova, S., & Fendel, E. M. (Eds.), *Geo-information for disaster management* (pp. 669–682). Berlin, Germany: Springer-Verlag. doi:10.1007/3-540-27468-5_48

Andrienko, N., & Andrienko, G. (2007). Intelligent visualisation and information presentation for civil crisis management. *Transactions in GIS, 11*(6), 889–909. doi:10.1111/j.1467-9671.2007.01078.x

Applied Technology Council (ATC). (2001). ATC-20 Procedures used to evaluate Damaged Buildings near World Trade Center. *ATC News Bulletin.* Retrieved from http://www.atcouncil.org/pdfs/1101news.pdf

Auf der Heide, E. (1989). *Disaster Response: Principles of Preparation and Coordination.* St. Louis, MO: CV Mosby.

Austin, D., Barbir, A., Ferris, C., & Garg, S. (Eds.). (W3C, 2004) Austin (2004). *Web services architecture requirements.* Retrieved from http://www.w3.org/TR/2004/NOTE-wsa-reqs-20040211/

Aziz, Z., Peña-Mora, F., Chen, A. Y., & Lantz, T. (2009). Supporting Urban Emergency Response and Recovery Using RFID-Based Building Assessment. *Disaster Prevention and Management: An International Journal, 18*(1), 35–48. doi:10.1108/09653560910938538

Bali, R. K., Wickramasinghe, N., & Lehaney, B. (2009). *Knowledge management primer.* New York, NY: Routledge.

Bardini, T. (1997). Bridging the gulfs: From hypertext to cyberspace. *Journal of Computer-Mediated Communication, 3*(2).

Barthélemy, J. P., Bisdorff, R., & Coppin, G. (2002). Human centered processes and decision support systems. *European Journal of Operational Research, 136*(2), 233–252. doi:10.1016/S0377-2217(01)00112-6

Baskerville, R. (1991). Risk analysis: An interpretive feasibility tool in justifying information systems security. *European Journal of Information Systems, 1*, 121–130. doi:10.1057/ejis.1991.20

Battista, D., Graziano, D., Franchi, V., Russo, A., de Leoni, M., & Mecella, M. (2009). *A Web Service-based Process-aware Information System for Smart Devices.* Retrieved from http://padis2.uniroma1.it:81/ojs/index.php/DIS_TechnicalReports/issue/view/198

Beizer, D. (2009, February 23). USA.gov Will Move to Cloud Computing. *Federal Computer Week.* Retrieved September 11, 2010, from http://fcw.com/articles/2009/02/23/usagov-moves-to-the-cloud.aspx

Belton, V., & Stewart, T. (2002). *Multiple criteria decision analysis: An integrated approach.* Boston, MA: Kluwer Academic.

Ben-Haim, Y. (2000). Robust rationality and decisions under severe uncertainty. *Journal of the Franklin Institute, 337*(2-3), 171–199. doi:10.1016/S0016-0032(00)00016-8

Benner, T., Binder, A., & Rotmann, P. (2007). *Learning to build peace? United Nations peacebuilding and organizational learning: Developing a research framework.* Berlin, Germany: Global Public Policy Institute.

Bentley, J. L. (1975). Multidimensional binary search trees used for associative searching. *Communications of the ACM, 18*(9), 509–517. doi:10.1145/361002.361007

Berkowitz, S. D. (1982). *An introduction to structural analysis: The network approach to social research.* Toronto, ON, Canada: Butterworths.

Berner, E. S. (2006). *Clinical decision support systems: Theory and practice.* Berlin, Germany: Springer-Verlag.

Bertelli, G., de Leoni, M., Mecella, M., & Dean, J. (2008). Mobile Ad hoc Networks for Collaborative and Mission-Critical Mobile Scenarios: a Practical Study. In *Proceedings of the 17th IEEE International Workshops on Enabling Technologies: Infrastructure for Collaboration Enterprises* (pp. 147-152).

Bharosa, N., Appelman, J., & de Bruin, P. (2007). Integrating technology in crisis response using an information manager: First lessons learned from field exercises in the Port of Rotterdam. In *Proceedings of the International Conference on Information Systems for Crisis Response and Management,* Brussels, Belgium.

Biddick, M. (2010, August 6). In Government, Private Clouds May Trump Google Apps. *Information Week.* Retrieved October 5, 2010, from http://www.privatecloud.com/2010/09/17/in-government-private-clouds-may-trump-google-apps-2

Biewener, A., Aschenbrenner, U., Rammelt, S., Grass, R., & Zwipp, H. (2004). Impact of helicopter transport and hospital level on mortality of polytrauma patients. *The Journal of Trauma Injury Infection and Critical Care, 56*(1), 94–98. doi:10.1097/01.TA.0000061883.92194.50

Bilham, R. (2010). Lessons from the Haiti earthquake. *Nature, 463,* 878–879. doi:10.1038/463878a

Bishop, M. (2003). *Computer security: Art and science.* Reading, MA: Addison-Wesley.

Bissell, A. B., Pinet, L., Nelson, M., & Levy, M. (2004). Evidence of the Effectiveness of Health Sector Preparedness in Disaster Response. *Family & Community Health, 27*(3), 193–203.

Boersma, K., Groenewegen, P., & Wagenaar, P. (2009). Emergency response rooms in action: An ethnographic case study in Amsterdam. In *Proceedings of the International Conference on Information Systems for Crisis Response and Management,* Gothenburg, Sweden.

Bortenschlager, M., Goell, N., Haid, E., Rieser, H., & Steinmann, R. (2008). GeoCollaboration-Location-based Collaboration in Emergency Scenarios. In *Proceedings of the 17th IEEE International Workshops on Enabling Technologies: Infrastructure for Collaborative Enterprises.*

Bressler, G. (2010). *Exercise24* [Lecture notes]. San Diego, CA: San Diego State University.

Brooks, C. H., & Montanez, N. (2006). Improved annotation of the blogosphere via autotagging and hierarchical clustering. In *Proceedings of the 15ᵗʰ International World Wide Web Conference*.

Brotcorne, L., Laporte, G., & Semet, F. (2003). Ambulance location and relocation models. *European Journal of Operational Research*, *147*(3), 451–463. doi:10.1016/S0377-2217(02)00364-8

Brown, S. L., & Eisenhardt, K. M. (1997). The art of continuous change: Linking complexity theory and time-paced evolution in relentlessly shifting organizations. *Administrative Science Quarterly*, *42*(1), 1–34. doi:10.2307/2393807

Browne, P. (2009). *Jboss drools business rules*. Birmingham, AL: Packt Publishing.

Bunn, D., & Salo, A. (1993). Forecasting with scenarios. *European Journal of Operational Research*, *68*(3), 291–303. doi:10.1016/0377-2217(93)90186-Q

Burstein, F., Zaslavsky, A., & Arora, N. (2005). Context-aware mobile agents for decision-making support in healthcare emergency applications. In *Proceedings of the Workshop on Contextual Modelling and Decision Support* (p. 144).

Bush, V. (1945). As we may think. *Atlantic Monthly*, *176*(1), 101–108.

Cai, G. (2005). Extending distributed GIS to support geo-collaborative crisis management. *Geographic Information Sciences, 11*(1).

Cai, G., MacEachren, A. M., Sharma, R., Brewer, I., Fuhrmann, S., & McNeese, M. (2005). Enabling geo-collaborative crisis management through advanced geoinformation technologies. In *Proceedings of the National Conference on Digital Government Research* (pp. 227-228).

Callon, M. (1986). Some elements of a sociology of translation: Domestication of the scallops and the fishermen in St. Brieuc's bay. In Law, J. (Ed.), *Power, action and belief: A new sociology of knowledge?* (pp. 196–219). London, UK: Routledge & Kegan Paul.

Callon, M. (1991). Techno-economic networks and irreversibility. In Law, J. (Ed.), *A sociology of monsters: Essays on power, technology and domination* (pp. 132–161). London, UK: Routledge.

Campbell, B. (2010). *Adapting simulation environments for emergency response planning and training*. Unpublished doctoral dissertation, University of Washington, Seattle, WA.

Campbell, B., Mete, O., Furness, T., Weghorst, S., & Zabinsky, Z. (2008, May 12-13). Emergency response planning and training through interactive simulation and visualization with decision support. In *Proceedings of the IEEE International Conference on Technologies for Homeland Security*, Waltham, MA (pp. 176-180).

Campbell, B., & Schroder, K. (2009). Training for emergency response with RimSim:Response! In *Proceedings of the Defense, Security, and Sensing Conference*, Orlando, FL.

Campbell, B., Schroder, K., & Weaver, C. (2010). RimSim visualization: An interactive tool for post-event sense making of a first response effort. In *Proceedings of the 7th International Conference on Information Systems for Crisis Response and Management*, Seattle, WA.

Canós, J. H., Alonso, G., & Jaen, J. (2004). A multimedia approach to the efficient implementation and use of emergency plans. *IEEE MultiMedia*, *11*(3), 106–110. doi:10.1109/MMUL.2004.2

Canós, J. H., Penadés, M. C., Solís, C., Borges, M. R. S., & Llavador, M. (2010). Using spatial hypertext to visualize composite knowledge in emergency responses. In *Proceedings of the 7th International ISCRAM Conference*, Seattle, WA.

Carillo-Ramos, A., Villanova-Oliver, M., Gensel, J., & Martin, H. (2007). Contextual user profile for adapting information in nomadic environments. In M. Weske, M.-S. Hacid, & C. Godart (Eds.), *Proceedings of the 8ᵗʰ International Conference on Web Information Systems Engineering* (LNCS 4832, pp. 337-349).

Carlsson, S. A. (2001). *Knowledge management in network contexts*. Paper presented at the the 9th European Conference on Information Systems.

Carroll, J. M., Neale, D. C., Isenhour, P. L., Rosson, M. B., & McCrickard, D. S. (2003). Notification and awareness: Synchronizing task-oriented collaborative activity. *International Journal of Human-Computer Studies*, *58*, 605–632. doi:10.1016/S1071-5819(03)00024-7

Carver, L., & Turoff, M. (2007). Human computer interaction: The human and computer as a team in emergency management information systems. *Communications of the ACM, 50*(3), 33–38. doi:10.1145/1226736.1226761

Catarci, T., Leoni, M. D., Rosa, F. D., Mecella, M., Poggi, A., Dustdar, S., et al. (2007). The WORKPAD P2P service-oriented infrastructure for emergency management enabling technologies: Infrastructure for collaborative enterprises. In *Proceedings of the 16th IEEE International Workshops on Enabling Technologies: Infrastructure for Collaborative Enterprises* (pp. 147-152).

Cazals, F., Giesen, J., Pauly, M., & Zomorodian, A. (2005). Conformal alpha shapes. In *Proceedings of the Second Symposium on Point Based Graphics* (pp. 55-61).

Center for Technology in Government. (2004). *Learning from the crisis: Lessons from the World Trade Center response*. Retrieved from http://www.ctg.albany.edu/publications/reports/wtc_symposium

Centre for Research on the Epidemiology of Disasters (CRED). (2009). *2008 Disasters in Numbers*. Retrieved from http://www.reliefweb.int/rw/rwb.nsf/db900SID/LSGZ-7NJKJV?OpenDocument

Ceri, S., Fraternali, P., & Bongio, A. (2000). Web modeling language (WebML): A modeling language for designing web sites. *Computer Networks, 33*(1-6), 137-157.

Chang, R., Ziamkiewicz, C., Green, T. M., & Ribarsky, W. (2009). Defining insight for visual analytics. *IEEE Computer Graphics and Applications, 29*(2), 14–17. doi:10.1109/MCG.2009.22

Chen, A. Y., Peña-Mora, F., Mehta, S. J., Plans, A. P., Brauer, B. R., Foltz, S., & Nacheman, S. (2010, May 2-5). A GIS Approach to Equipment Allocation for Structural Stabilization and Civilian Rescue. In *Proceedings of the 7th International Conference on Information Systems for Crisis Response and Management (ISCRAM)*, Seattle, WA.

Chen, A. Y., Peña-Mora, F., & Ouyang, Y. (in press). A Collaborative GIS Framework to Support Equipment Distribution for Civil Engineering Disaster Response Operations. *Automation in Construction*.

Chen, R., Sharman, R., Rao, H. R., & Upadhyaya, S. J. (2008). Coordination in emergency response management. *Communications of the ACM, 51*(5), 66–73. doi:10.1145/1342327.1342340

Chiu, D., Li, Q., & Karlapalem, K. (2000). A logical framework for exception handling in ADOME workflow management system. In *Proceedings of 12th International Conference on Advanced Information Systems Engineering* (pp. 110-125).

Cholewo, T. J., & Love, S. (1999). Gamut boundary determination using alpha-shapes. In *Proceedings of the IS&T and SIDs Seventh Color Imaging Conference* (pp. 200-204).

Choobineh, J., Dhillon, G., Grimalla, M., & Rees, J. (2007). Management of information security: Challenges and research directions. *Communications of the AIS, 20*, 958–971.

Claburn, T. (2009, September 15). Government Embraces Cloud Computing, Launches App Store. *Information Week*. Retrieved October 1, 2010, from http://www.informationweek.com/news/government/cloud-saas/showArticle.jhtml?articleID=220000493

Clarkson, K. (n. d.). *A program for convex hulls*. Retrieved from http://www.netlib.org/voronoi/hull.html

CNN. (2005). *Rivers still strewn with bodies: Special Report – After the tsunami*. Retrieved from http://edition.cnn.com/2005/WORLD/asiapcf/01/03/otsc.aceh.chinoy/index.html

Cohen, N. (2009, June 20). *Twitter on the barricades: Six lessons learned*. Retrieved from http://www.nytimes.com/2009/06/21/weekinreview/21cohenweb.html

College Board. (2010). *College Search*. Retrieved from http://collegesearch.collegeboard.com/search

Comer, D. E. (2006). *Internetworking with TCP/IP - Principles, Protocols and Architecture*. Upper Saddle River, NJ: Pearson Prentice Hall.

Comes, T., Hiete, M., & Schultmann, F. (2010). A decision support system for multi-criteria decision problems under severe uncertainty in longer-term emergency management. In *Proceedings of the 25th Mini EURO Conference on Uncertainty and Robustness in Planning and Decision Making*, Coimbra, Portugal.

Comes, T., Hiete, M., Wijngaards, N., & Kempen, M. (2009). Integrating scenario-based reasoning into multi-criteria decision analysis. In *Proceedings of the 6th International Conference on Information Systems for Crisis Response and Management*, Gothenburg, Sweden.

Comes, T., Hiete, M., Wijngaards, N., & Schultmann, F. (2010). Enhancing robustness in multi-criteria decision-making: A scenario-based approach. In *Proceedings of the 2nd International Conference on Intelligent Networking and Collaborative Systems*, Thessaloniki, Greece.

Comes, T., Hiete, M., Wijngaards, N., & Schultmann, F. (in press). Decision maps: A framework for multi-criteria decision support under severe uncertainty. *Decision Support Systems*.

Comfort, L. K. (1999). *Shared risk: Complex seismic response*. New York, NY: Pergamon.

Comfort, L. K. (2005). Risk, security, and disaster management. *Annual Review of Political Science, 8*, 335–356. doi:10.1146/annurev.polisci.8.081404.075608

Comfort, L. K. (2007). Crisis management in hindsight: Cognition, communication, coordination, and control. *Public Administration Review, 67*, 189–197. doi:10.1111/j.1540-6210.2007.00827.x

Comfort, L. K., & Haase, T. W. (2006). Communication, coherence, and collective action: The impact of hurricane Katrina on communications infrastructure. *Public Works Management Policy, 10*(4), 328–343. doi:10.1177/1087724X06289052

Comfort, L. K., Sungu, Y., Johnson, D., & Dunn, M. (2001). Complex systems in crisis: Anticipation and resilience in dynamic environments. *Journal of Contingencies and Crisis Management, 9*(3), 144–158. doi:10.1111/1468-5973.00164

Comprehensive Crisis Management. (2008). *Human security in peacebuilding*. Kuopio, Finland: Crealab Oy.

Cordella, A., & Shaikh, M. (2003). Actor network theory and after: What's new for IS research? In *Proceedings of the 11th European Conference on Information Systems*.

Cordis. (2008). *Self-optimisation and self-configuration in wireless networks (SOCRATES). The network of the future* (ICT-2007.1.1.) Retrieved from http://cordis.europa.eu/

Craggs, S. (2009). *Cloud Computing Without the Hype; an Executive Guide*. Lustratus Research Limited.

Crandall, D. J., Backstrom, L., Huttenlocher, D. P., & Kleinberg, J. M. (2009). Mapping the world's photos. In *Proceedings of the 18th International World Wide Web Conference*.

Crichton, M. T., Flin, R., & Rattray, W. A. R. (2000). Training decision makers - Tactical decision games. *Journal of Contingencies and Crisis Management, 8*(4). doi:10.1111/1468-5973.00141

Cruz, A. M., Steinberg, L. J., & Vetere-Arellano, A. L. (2006). Emerging issues for Natech disaster risk management in Europe. *Journal of Risk Research, 9*(5), 483–501. doi:10.1080/13669870600717657

Cutter, S. L. (2003). GI science, disasters, and emergency management. *Transactions in GIS, 7*(4), 439–446. doi:10.1111/1467-9671.00157

D'addeiro, L. (2008). The performativity of routines: Theorising the influense of artefacts and distributed agencies on routine dynamics. *Research Policy, 37*, 769–789. doi:10.1016/j.respol.2007.12.012

Dantas, A., & Seville, E. (2006). Organisational issues in implementing an information sharing framework: Lessons from the Matata flooding events in New Zealand. *Journal of Contingencies and Crisis Management, 14*(1). doi:10.1111/j.1468-5973.2006.00479.x

Dausey, D. J., Buehler, J. W., & Lurie, N. (2007). Designing and Conducting Tabletop Exercises to Assess Public Health Preparedness for Manmade and Naturally Occurring Biological Threats. *BioMed Central Public Health, 7*, 92. Retrieved May 15, 2010, from http://search.ebscohost.com.proxy.longwood.edu

Davis, R., & Smith, R. G. (1983). Negotiation as a metaphor for distributed problem solving. *Artificial Intelligence, 20*(1), 63-109.

Dawes, S., Cresswell, A., & Cahan, B. (2004). Learning from Crisis: Lessons in human and Information Infrastructure from the World Trade Center Response. *Social Science Computer Review*, 52–66. doi:10.1177/0894439303259887

de Leoni, M., De Giacomo, G., Lespèrance, Y., & Mecella, M. (2009). On-line adaptation of sequential mobile processes running concurrently. In *Proceedings of the 2009 ACM Symposium on Applied Computing* (pp. 1345-1352).

de Leoni, M., De Rosa, F., Marrella, A., Mecella, M., Poggi, A., Krek, A., & Manti, F. (2007). Emergency Management: from User Requirements to a Flexible P2P Architecture. In *Proceedings of the 4th International Conference on Information Systems for Crisis Response and Management* (pp. 271-279).

de Leoni, M., Marrella, A., & Russo, A. (2010). *Process-Aware Information Systems for Emergency Management*. Paper presented at the International Workshop on Emergency Management through Service Oriented Architectures.

de Leoni, M., Mecella, M., & De Giacomo, G. (2007). Highly Dynamic Adaptation in Process Management Systems Through Execution Monitoring. In *Proceedings of the 5th International Conference on Business Process Management* (pp. 182-197).

De Silva, C., De Silva, R., Careem, M., Raschid, L., & Weerawarana, S. (2006). Sahana: Overview of a disaster management system. In *Proceedings of the IEEE International Conference on Information and Automation* (pp. 361-366).

de Souza, C. (2004). *The semiotic engineering of human-computer interaction*. Cambridge, MA: MIT Press.

Deloitte Consulting, L. L. P. (2004). *United Nations organizational integrity survey*. Retrieved from http://whistleblower.org/storage/documents/UN_Integrity_Survey.pdf

Dempsey, J. S., & Forst, L. S. (2010). *An Introduction to Policing* (5th ed.). Clifton Park, NY: Cengage.

Dhanjani, N. (2008, December 18). How terrorists may abuse micro-blogging channels like Twitter. *Nitesh Dhanjani Blog*. Retrieved October 1, 2010, from http://www.dhanjani.com/blog/2008 /12/how-terrorists-may-abuse-microblogging-channels-like-twitter.html

Dhillon, G. (2007). *Principles of information security: Text and cases*. Hoboken, NJ: John Wiley & Sons.

Dhillon, G., & Backhouse, J. (2001). Current directions in IS security research: Towards socio-organizational perspectives. *Information Systems Journal, 11*(2), 127–153. doi:10.1046/j.1365-2575.2001.00099.x

Dhillon, G., & Torkzadeh, G. (2006). Value-focused assessment of information system security in organizations. *Information Systems Journal, 16*(3), 293–314. doi:10.1111/j.1365-2575.2006.00219.x

Diehl, M., & Haimes, Y. (2004). Influence diagrams with multiple objectives and tradeoff analysis. *IEEE Transactions on Systems, Man, and Cybernetics. Part A, Systems and Humans, 34*(3), 293–304. doi:10.1109/TSMCA.2003.822967

Dillman, D. A. (1978). *Mail and telephone surveys: The total design method*. New York, NY: Wiley-Interscience.

Dillman, D. A. (1999). *Mail and Internet surveys: The tailored design method* (2nd ed.). New York, NY: John Wiley & Sons.

Diniz, V. B., Borges, M. R. S., Gomes, J. O., & Canós, J. H. (2008). Decision making support in emergency response. In Adam, F., & Humphreys, P. (Eds.), *Encyclopedia of decision making and decision support technologies* (1st ed., pp. 184–191). London, UK: Information Science Reference. doi:10.4018/978-1-59904-843-7.ch021

Disaster Accountability Project. (2010). *Report on the transparency of relief organizations responding to the 2010 Haiti earthquake*. Hartford, CT: Disaster Accountability Project.

Disasters Emergency Committee. (2001). *Independent evaluation: The DEC response to the earthquake in Gujarat (Vol. 1)*. London, UK: Disasters Emergency Committee.

Disastersrus. (2006). *Hurricane Katrina Reports*. Retrieved October 3, 2010, from http://www.disastersrus.org/katrina/

Dix, A. J., Finlay, J. E., Abowd, G. D., & Beale, R. (2004). *Human-computer interaction* (3rd ed.). London, UK: Prentice Hall.

Djurcilov, S., & Pang, A. (1999). Visualizing gridded datasets with large number of missing values. In *Proceedings of the Visualization Conference* (pp. 405-408).

Doherty, N. F., & Fulford, H. (2006). Aligning the information security policy with the strategic information systems plan. *Computers & Security, 23*(1), 55–63. doi:10.1016/j.cose.2005.09.009

Dugdale, J. (1996). A cooperative problem-solver for investment management. *International Journal of Information Management, 16*(2), 133–147. doi:10.1016/0268-4012(95)00074-7

Durbach, I., & Stewart, T. (2003). Integrating scenario planning and goal programming. *Journal of Multi-Criteria Decision Analysis, 124-125*, 261–271. doi:10.1002/mcda.362

Duvall, M. (2010). *CTOs Meet with Fed Officials to Discuss Cloud*. Retrieved October 3, 2010, from http://www.information-management.com/news/government-cloud-computing-push-BSA-10018810-1.html?msite=cloudcomputing

Dynes, R. R. (1994). Community emergency planning: False assumptions and inappropriate analogies. *International Journal of Mass Emergencies and Disasters*, *12*, 141–158.

Edelsbrunner, H., & Mucke, E. P. (1994). Three-dimensional alpha shapes. *ACM Transactions on Graphics*, *13*(1), 43–72. doi:10.1145/174462.156635

Ejigu, D., Scuturici, M., & Brunie, L. (2008). Hybrid approach to collaborative context-aware service platform for pervasive computing. *Journal of Computers, 16*(1), 40-50.

ELANSO. (2009). *Earthquake rescue – learning from disaster*. Retrieved from http://www.elanso.com/ArticleModule/HlVwUKSETDLcPzIsUfVcPAIi.html

Endsley, M. R. (2000). Theoretical underpinnings of situation awareness: A critical review. In Endsley, M. R., & Garland, D. J. (Eds.), *Situation awareness analysis and measurement*. Mahwah, NJ: Lawrence Erlbaum.

Endsley, M. R. (2004). Situation awareness: Progress and directions. In Banbury, S., & Tremblay, S. (Eds.), *A cognitive approach to situation awareness: Theory, measurement and application* (pp. 317–341). Aldershot, UK: Ashgate Publishing.

Energy Ideas. (n.d.). *Centralized versus Distributed - Computer Analogy*. Retrieved October 1, 2010, from http://energyideas.ca/Central_vs_Distributed_Computers.php

Engelbart, D. C. (1963). A conceptual framework for the augmentation of man's intellect. *Vistas in Information Handling*, *1*, 1–29.

Engelmann, H., & Fiedrich, F. (2009). DMT-EOC – A combined system for the decision support and training of EOC members. In *Proceedings of the 6th International Conference on Information Systems for Crisis Response and Management*, Gothenburg, Sweden.

ESRI. (2009). *ArcGIS*. Retrieved from http://www.esri.com/products/indexb.html

Fahland, D., & Woith, H. (2009). Towards process models for disaster response. In *Proceedings of the Business Process Management Workshops* (pp. 254-265).

Federal Emergency Management Agency (FEMA). (1980). *Aboveground Home Shelter*. Washington, DC: Government Printing Office.

Federal Emergency Management Agency (FEMA). (1983). *In Time of Emergency: A Citizen's Handbook*. Washington, DC: Government Printing Office.

Federal Emergency Management Agency (FEMA). (1985). *Protection in the Nuclear Age*. Washington, DC: Government Printing Office.

Federal Emergency Management Agency (FEMA). (2001). *State and Local Mitigation Planning, How-to Guide: Understanding Your Risks, Identifying Hazards and Estimating Losses* (Tech. Rep. No. 386-2). Washington, DC: Government Printing Office.

Federal Emergency Management Agency (FEMA). (2003). *National Urban Search and Rescue Response System: Field Operations Guide*. Retrieved from http://www.fema.gov/usr/index.shtm

Federal Emergency Management Agency (FEMA). (2009). *Developing and Maintaining State, Territorial, Tribal, and Local Government Emergency Plan: Comprehensive Preparedness Guide 101*. Retrieved February 2, 2010, from http://www.fema.gov

Feldman, M. S., & Pentland, B. (2005). Organizational routines and the macro-actor. In B. A. H. Czarniawska, T (Ed.), *Actor-network theory and organizing* (pp. 91-111). Copenhagen, Denmark: Copenhagen Business School Press.

Feldman, M. S., & Rafaeli, A. (2002). Organizational routines as sources of connections and understandings. *Journal of Management Studies*, *39*(3), 309–331. doi:10.1111/1467-6486.00294

FEMA. (2009). *Use of social media tools at FEMA*. Retrieved from http://tinyurl.com/yhsv9xn

Ferrante, P. (2010). Risk & Crisis Communication. *Professional Safety*, *55*(6). Retrieved January 20, 2011, from http://search.ebscohost.com.proxy.longwood.edu

Fields, J. W. (2009). 10 Steps to Creating a Campus Security Master Plan. *Campus Safety Magazine*. Retrieved January 19, 2010, from http://www.campussafetymagazine.com/Channel/Security-Technology/Articles/2009/03/10-Steps-to-Creating-A-Campus-Security-Master-Plan.aspx

Fischer, C., Jackson, R., Stueve, C., Gerson, K., McCallister, L., & Baldassare, M. (1977). *Networks and places: Social relations in the urban setting*. New York, NY: Free Press.

Fischer, H. W. (1998). The role of the new information technologies in emergency mitigation, planning, response, and recovery. *Disaster Prevention and Management*, 7(1), 28–37. doi:10.1108/09653569810206262

Fisher, B. S. (1995). Crime and Fear on Campus. *Annals of the American Academy of Political and Social Science, 539*, 85-101. Retrieved February 2, 2010, from http://www.jstor.org/stable/1048398

Foley, J. (2009, July). How Government's Driving Cloud Computing Ahead. *Information Week*. Retrieved September 5, 2010, from http://www.informationweek.com/story/showArticle.jhtml?articleID=218400025

Forgy, C. (1982). Rete: A fast algorithm for the many pattern/many object pattern match problem. *Artificial Intelligence, 19*, 17–37. doi:10.1016/0004-3702(82)90020-0

Fox, S., Zickuhr, K., & Smith, A. (2009). *Twitter and status updating*. Retrieved from http://tinyurl.com/yfmgg25

French, S., Maule, J., & Papamichail, N. (2009). *Decision behaviour, analysis and support*. Cambridge, UK: Cambridge University Press. doi:10.1017/CBO9780511609947

Gahegan, M., & Pike, W. (2006). A situated knowledge representation of geographical information. *Transactions in GIS, 10*, 727–749. doi:10.1111/j.1467-9671.2006.01025.x

Gaines, L. H., & Miller, R. L. (2009). *Criminal Justice in Action* (5th ed.). Belmont, CA: Thomson-Wadsworth.

Gamma, E., Helm, R., Johnson, R., & Vlissides, J. (1995). *Design patterns: Elements of reusable object-oriented software*. Reading, MA: Addison-Wesley.

Gasiorek-Nelson, S. (2002). *DAU hosts 9/11 first responder: challenges and logistics of responding to Pentagon terrorist attack - Logistics Preparedness*. Retrieved from http://findarticles.com/p/articles/mi_m0KAA/is_5_31/ai_94771290/pg_3?tag=content;col1

Geldermann, J., Bertsch, V., Treitz, M., French, S., Papamichail, K. N., & Hämäläinen, R. P. (2009). Multicriteria decision support and evaluation of strategies for nuclear remediation management. *Omega, 37*(1), 238–251. doi:10.1016/j.omega.2006.11.006

Gendreau, M., Laporte, G., & Semet, F. (1997). Solving an ambulance location model by tabu search. *Location Science, 5*(2), 75–88. doi:10.1016/S0966-8349(97)00015-6

Gentes, S. (2006). Rescue Operations and Demolition Works: Automating the Pneumatic Removal of Small Pieces of Rubble and Combination of Suction Plants with Demolition Machines. *Bulletin of Earthquake Engineering, 4*, 193–205. doi:10.1007/s10518-006-9006-1

Gheorghe, A. V., & Vamanu, D. V. (2001). Adapting to new challenges: IDSS for emergency preparedness and management. *International Journal of Risk Assessment and Management, 2*(3-4), 211–223. doi:10.1504/IJRAM.2001.001506

Giasson, J. (2010). *About the Exercise*. Retrieved September 19, 2010, from https://sites.google.com/a/inrelief.org/24/

Giasson, J. (2010). *Exercise24* [Lecture notes]. San Diego, CA: San Diego State University.

Giasson, J. (2010). *Injects and Updates*. Retrieved October 1, 2010, from https://sites.google.com/a/inrelief.org/24/injects

Gill, G. (2007). *Will a Twenty-first Century Logistics Management System Improve Federal Emergency Management Agency's Capability to Deliver Supplies to Critical Areas, During Future Catastrophic Disaster Relief Operations?* Unpublished master's thesis, University of Phoenix, AZ.

Gleicher, N. J., & Hwang, J. S. (2010). Geospatial information services: Balancing privacy and innovation. In *Proceedings of the AAAI Spring Symposium Series* (pp. 75-81).

Godet, M. (1990). Integration of scenarios and strategic management: Using relevant, consistent and likely scenarios. *Futures*, *22*(7), 730–739. doi:10.1016/0016-3287(90)90029-H

GoGrid. (2010). *The Cloud Computing Pyramid*. Retrieved September 1, 2010, from http://pyramid.gogrid. com/?ref=servepath

Golden, B. (2008). *Cloud Virtualization*. Retrieved October 6, 2010, from http://advice.cio.com/bernard_golden/cloud_virtualization

Golder, S., & Huberman, B. A. (2006). The structure of collaborative tagging systems. *Journal of Information Science*, *32*(2), 198–208. doi:10.1177/0165551506062337

Goldkuhl, G. (2005). Socio-instrumental pragmatism: A theoretical synthesis for pragmatic conceptualisation in information systems. In *Proceedings of the Third International Conference on Action in Language, Organisations and Information Systems*, Limerick, Ireland.

Gomez, E. A., & Turoff, M. (2007). Community crisis response teams: Leveraging local resources through ICT e-readiness. In *Proceedings of the 40th Annual Hawaii International Conference on System Sciences* (p. 24).

Gonsalves, C. (2005, January 28). The deadly bureaucracy in the Andamans. *The Indian Express*, p. 8.

Goodchild, M. F., & Glennon, J. A. (2010). Crowdsourcing geographic information for disaster response: a research frontier. *International Journal of Digital Earth*, *3*(3), 231–241. doi:10.1080/17538941003759255

Goser, K., Jurisch, M., Acker, H., Kreher, U., Lauer, M., Rinderle, S., et al. (2007). Next-generation Process Management with ADEPT2. In *Proceedings of the BPM Demonstration Program at the 5th International Conference on Business Process Management*.

Graham, R. L. (1972). An efficient algorithm for determining the convex hull of a finite planar set. *Information Processing Letters*, *1*, 132–133. doi:10.1016/0020-0190(72)90045-2

Granovetter, M. (1973). The strength of weak ties. *American Journal of Sociology*, *78*(6), 1360–1380. doi:10.1086/225469

Gray, R. (2009). Columbine 10 Years Later: The State of School Safety Today. *Campus Safety Magazine*. Retrieved January 8, 2010, from http://www.campussafetymagazine.com/Channel/School-Safety/Articles/2009/03/Columbine-10-Years-Later-The-State-of-School-Safety-Today.aspx

Greenes, R. A. (2006). *Clinical decision support: The road ahead*. Orlando, FL: Academic Press.

Grzenda, M., & Niemczak, M. (2004). Requirements and solutions for web-based expert system. In L. Rutkowski, J. H. Siekmann, R. Tadeusiewicz, & L. A. Zadeh (Eds.), *Proceedings of the 7th International Conference on Artificial Intelligence and Soft Computing* (LNCS 3070, pp. 866-871).

Gunes, A. E., & Kovel, J. P. (2000). Using GIS in emergency management operations. *Journal of Urban Planning and Development*, *126*(3), 136–149. doi:10.1061/(ASCE)0733-9488(2000)126:3(136)

Hagar, C. (2007). The information and social needs of farmers and use of ICT. In Nerlich, B., & Doring, M. (Eds.), *From mayhem to meaning: Assessing the social and cultural impact of the 2001 foot and mouth outbreak in the UK*. Manchester, UK: Manchester University Press.

Hagebölling, D., & de Leoni, M. (2009). Supporting Emergency Management through Process-Aware Information Systems. In *Proceedings of the Business Process Management Workshops* (pp. 298-302).

Harland, D., & Wahlstrom, M. (2001). *The role of OCHA in emergency United Nations operations following the earthquake in Gujarat, India - 26 January 2001*. Geneva, Switzerland: United Nations.

Harrald, J. R. (2006). Agility and discipline: Critical success factors for disaster response. *The Annals of the American Academy of Political and Social Science*, *604*(1), 256–272. doi:10.1177/0002716205285404

Harrald, J. R. (2009). Achieving agility in disaster management. *International Journal of Information Systems for Crisis Response and Management*, *1*(1), 1–11. doi:10.4018/jiscrm.2009010101

Harries, C. (2003). Correspondence to what? Coherence to what? What is good scenario-based decision making? *Technological Forecasting and Social Change*, *70*(8), 797–817. doi:10.1016/S0040-1625(03)00023-4

Heugens, P., & van Oosterhout, J. (2001). To boldly go where no man has gone before: Integrating cognitive and physical features in scenario studies. *Futures*, *33*(10), 861–872. doi:10.1016/S0016-3287(01)00023-4

Hevner, A. R., March, S. T., Park, J., & Ram, S. (2004). Design science in information systems research. *Management Information Systems Quarterly*, *28*(1).

Heymann, P., & Garcia-Molina, H. (2006). *Collaborative creation of communal hierarchical taxonomies in social tagging systems* (Tech. Rep. No. 2006-10). Stanford, CA: Stanford University.

Hiltz, S. R., & Turoff, M. (1985). Structuring computer-mediated communication systems to avoid information overload. *Communications of the ACM*, *28*(7), 680–689. doi:10.1145/3894.3895

Holguín-Veras, J., Perez, N., Ukkusuri, S., Wachtendorf, T., & Brown, B. (2007). Emergency Logistics Issues Affecting the Response to Katrina: A Synthesis and Preliminary Suggestions for Improvement. *Transportation Research Record: Journal of the Transportation Research Board*, 76-82.

Holmström, J., & Robey, D. (2005). Inscribing organizational change with information technology. In Czarniawska, B., & Hernes, T. (Eds.), *Actor-network theory and organising*. Copenhagen, Denmark: Copenhagen Business School Press.

Hoover, J. N. (2010, May 26). Gov 2.0: Google Readies Government Cloud. *Information Week*. Retrieved October 1, 2010, from http://www.information-week.com/news/government/cloud-saas/showArticle.jhtml?articleID=225200270

Hopcroft, J., & Tarjan, R. (1973). Algorithm 447: Efficient algorithms for graph manipulation. *Communications of the ACM*, *16*, 372–378. doi:10.1145/362248.362272

Horan, T. A., & Schooley, B. (2007). Time-critical information services. *Communications of the ACM*, *50*(3). doi:10.1145/1226736.1226738

Horrigan, J. (2007). *A typology of information and communication technology users*. Washington, DC: Pew Internet & American Life Project.

Huang, C.-M., Chan, E., & Hyder, A. A. (2010). Web 2.0 and Internet Social Networking: A New tool for Disaster Management? -- Lessons from Taiwan. (from http://search.ebscohost.com.proxy.longwood.edu). *BMC Medical Informatics and Decision Making*, *10*, Retrieved January 20, 2011. doi:10.1186/1472-6947-10-57

Hughes, A., & Palen, L. (2009). Twitter adoption and use in mass convergence and emergency events. In *Proceedings of the International Conference on Information Systems for Crisis Response and Management*, Gothenburg, Sweden

Hughes, A., Palen, L., Sutton, J., Liu, S., & Vieweg, S. (2008, May). "Site-seeing" in disaster: An examination of on-line social convergence. In *Proceedings of the 5th International Conference of the Information Systems for Crisis Response and Management*, Washington, DC.

Hutchins, E. (1996). *Cognition in the wild*. Cambridge, MA: MIT Press.

IBM. (n.d.). *Get the Facts on IBM vs the Competition - The Facts about IBM System Z 'Mainframe'*. Retrieved October 3, 2010, from http://www-03.ibm.com/systems/migratetoibm/getthefacts/mainframe.html

Idle, N., & Nunns, A. (Eds.). (2011). *Tweets from Tahrir* [Kindle DX version]. Retrieved from http://www.amazon.com

Immonen, A., Bali, R. K., Naguib, R. N. G., & Ilvonen, K. (2009). Towards a knowledge-based conceptual model for post-crisis public health scenarios. In *Proceedings of the IEEE International Conference on Humanoid, Nano-technology, Information Technology, Communication and Control, Environment, and Management*, Manila, Philippines (pp. 185-189).

Intagorn, S., Plangprasopchok, A., & Lerman, K. (2010). Harvesting geospatial knowledge from social metadata. In *Proceedings of the 7th International Conference on Information Systems for Crisis Response and Management*.

Inter-Agency Standing Committee. (2010). *Response to the humanitarian crisis in Haiti: Achievements, challenges, and lessons to be learned*. Geneva, Switzerland: Inter-Agency Standing Committee.

International Search and Rescue Response (INSAR). (2005). *International Search and Response: Guidelines*. Retrieved from http://www.reliefweb.int/undac/documents/insarag/guidelines/Id-mark.html

Jakobson, G., Kokar, M. M., Lewis, L., Buford, J., & Matheus, C. J. (2005). Overview of situation management at SIMA 2005. In *Proceedings of the Military Communications Workshop on Situation Management* (pp. 17-20).

James, A., & Rashed, T. (2006). In their own words: Utilizing weblogs in quick response research. In Leifeld, J. A. (Ed.), *Learning from catastrophe quick response research in the wake of Hurricane Katrina* (pp. 71–96). Boulder, CO: University of Colorado.

Jennex, M. E. (2004). Emergency response systems: The utility Y2K experience. *Journal of Information Technology Theory and Application, 6*(3), 85–102.

Jennex, M. E. (2007). Modeling emergency response systems. In *Proceedings of the 40th Hawaii International Conference on System Sciences*.

Jennex, M. E. (2010). Implementing social media in crisis response using knowledge management. *International Journal of Information Systems for Crisis Response and Management, 2*(4), 20–32.

Jennex, M. E., & Raman, M. (2009). Knowledge management in support of crisis response. *International Journal of Information Systems for Crisis Response and Management, 1*(3), 69–83. doi:10.4018/jiscrm.2009070104

Jevans, D. (2008). Cloud Security: The Need for Two-Factor Authentication in Cloud Computing. *Cloud Computing Journal*. Retrieved October 8, 2010, from http://cloudcomputing.sys-con.com/node/644838

Juhra, C., Ückert, F., Weber, T., Hentsch, S., Hartensuer, R., Vordemvenne, T., & Raschke, M. J. (2010). Improving communication in acute trauma. *ElectronicHealthcare, 8*(3), 3–8.

Jul, S. (2007). Who's Really on First? A Domain-Level User and Context Analysis for Response Technology. In *Proceedings of the 4th International Conference on Information Systems for Crisis Response and Management* (pp. 139-148).

Juszczyk, L., Psaier, H., Manzoor, A., & Dustdar, S. (2009). Adaptive Query Routing on Distributed Context - The COSINE Framework. In *Proceedings of the International Workshop on the Role of Services, Ontologies, and Context in Mobile Environments* (pp. 588-593).

Kahn, R., & Wilensky, R. (1995). *A framework for distributed digital object services*. Retrieved from http://www.cnri.reston.va.us/k-w.html

Kavouras, M., Kokla, M., & Tomai, E. (2006). Semantically-aware systems: Extraction of geosemantics, ontology engineering, and ontology integration. In Stefanakis, E., Peterson, M. P., Armenakis, C., & Delis, V. (Eds.), *Geographic hypermedia* (Vol. 3, pp. 257–273). Berlin, Germany: Springer-Verlag. doi:10.1007/978-3-540-34238-0_14

Keating, T., & Montoya, A. (2005). *Folksonomy extends geospatial taxonomy*. Directions Magazine.

Keeney, R., & Raiffa, H. (1976). *Decisions with multiple objectives*. New York, NY: John Wiley & Sons.

Kellerer, W., Schollmeier, R., & Wehrle, K. (2005). Peer-to-Peer in Mobile Environments. In Steinmetz, R., & Wehrle, K. (Eds.), *Peer-to-Peer Systems and Applications* (pp. 401–417). doi:10.1007/11530657_24

Kendra, J. M., & Wachtendorf, T. (2003). Elements of Resilience after the World Trade Center Disaster: Reconstituting New York City's Emergency Operations Centre. *Disasters, 27*(1), 37–53. doi:10.1111/1467-7717.00218

Kendra, J., & Wachtendorf, T. (2003). *Beyond September 11th: An account of post-disaster research* (pp. 121–146). Boulder, CO: University of Colorado, Natural Hazards Research and Applications Information Center.

Kennedy, K., Aghababian, R., Gans, L., & Lewis, C. (1996). Triage: Techniques and applications in decision-making. *Annals of Emergency Medicine, 28*(2), 136–144. doi:10.1016/S0196-0644(96)70053-7

Kevany, M. J. (2005). Geo-Information for Disaster Management: Lessons from 9/11. In van Oosterom, P., Zlatanova, S., & Fendel, E. (Eds.), *Geo-Information for Disaster Management* (pp. 443–464). Berlin, Germany: Springer. doi:10.1007/3-540-27468-5_32

Kiechle, G., Doerner, K. F., Gendreau, M., & Hartl, R. F. (2009). Waiting strategies for regular and emergency patient transportation. *Operations Research Proceedings, 2008*, 271-276.

Kirsch-Pinheiro, M., Villanova-Oliver, M., Gensel, J., & Martin, H. (2006, June). A personalized and context-aware adaptation process for web-based groupware systems. In *Proceedings of the 4th International Workshop on Ubiquitous Mobile Information and Collaboration Systems* (pp. 884-898).

Kleban, J., Moxley, M., Xu, J., & Manjunath, B. S. (2009). Global annotation on georeferenced photographs. In *Proceedings of the Conference on Image and Video Retrieval* (p. 12).

Klein, G. A. (1998). *Sources of power: How people make decisions*. Cambridge, MA: MIT Press.

Kotulic, A. G., & Clark, J. G. (2004). Why there aren't more information security research studies. *Information & Management, 41*(5), 597–607. doi:10.1016/j.im.2003.08.001

Kristensen, M., Kyng, M., & Palen, L. (2006). Participatory design in emergency medical service: Designing for future practice. In *Proceedings of the CHI Conference on Participatory Design*.

Kwon, O., Yoo, K., & Suh, E. (2005). UbiDSS: A proactive intelligent decision support system as an expert system deploying ubiquitous computing technologies. *Expert System Applications, 28*(1), 149-161.

La Rosa, M., & Mendling, J. (2009). Domain-Driven Process Adaptation in Emergency Scenarios. In *Proceedings of the Business Process Management Workshops* (pp. 290-297).

Lanz, A., Kreher, U., Reichert, M., & Dadam, P. (2010). Enabling Process Support for Advanced Applications with the AristaFlow BPM Suite. In *Proceedings of the Business Process Management Conference Demonstration Track*.

Latour, B. (1986). The power of association. In Law, J. (Ed.), *Power, action and belief: A new sociology of knowledge?* (pp. 264–280). London, UK: Routledge & Kegan Paul.

Latour, B. (1987). *Science in action*. Milton Keynes, UK: Open University Press.

Latour, B. (2005). *Reassembling the social: An introduction to actor-network theory*. Oxford, UK: Oxford University Press.

Lau, R. Y. K. (2007). Towards a web services and intelligent agents-based negotiation system for B2B eCommerce. *Electronic Commerce Research and Applications, 6*(3), 260-273.

Lawrence, T., Dyck, B., Maitlis, S., & Mauws, M. (2006). The underlying structure of continuous change. *MIT Sloan Management Review, 47*(4).

Le, B., Rondeau, R., Maldonado, D., Scaperoth, D., & Bostian, C. W. (2006, November 13-16). Signal recognition for cognitive radios. In *Proceedings of the Software Defined Radio Technical Conference*, Orlando, FL.

Leslie, A. J., & Stephenson, T. J. (1998). Transporting sick newborn babies. *Current Paediatrics, 8*(2), 98–102. doi:10.1016/S0957-5839(98)80127-9

Leuf, B., & Cunningham, W. (2001). *The Wiki way: Quick collaboration on the Web*. Reading, MA: Addison-Wesley.

Li, H., Srihari, R., Niu, C., & Li, W. (2003). InfoXtract location normalization: A hybrid approach to geographic references in information extraction. In *Proceedings of the Workshop on the Analysis of Geographic References NAACL-HLT*.

Liao, T., & Hu, T. (2002). A CORBA-based GIS-T for ambulance assignment. In *Proceedings of the IEEE International Conference on Application-Specific Systems, Architectures, and Processors* (pp. 371-380).

Liu, S. B., Iacucci, A. A., & Meier, P. (2010, August). Ushahidi in Haiti and Chile: Next generation crisis mapping. In *Proceedings of the American Congress on Surveying and Mapping*.

Liu, S. B., & Palen, L. (2009). Spatiotemporal mashups: A survey of current tools to inform next generation crisis support. In *Proceedings of the 6th International Information Systems for Crisis Response and Management Conference*, Gothenburg, Sweden.

Lorincz, K., Malan, D. J., Fulford-Jones, T. R., Nawoj, A., Clavel, A., & Shnayder, V. (2004). Sensor networks for emergency response: Challenges and opportunities. *IEEE Pervasive Computing / IEEE Computer Society [and] IEEE Communications Society, 3*, 16–23. doi:10.1109/MPRV.2004.18

Lotan, G. (2009). *A visual exploration of Twitter conversation threads in the days following the Iranian Elections of June 2009.* Retrieved from http://giladlotan.org/viz/iranelection/about.html

Lurie, N., Wasserman, J., Soto, M., Myers, S., Namkung, P., Fielding, J., & Valdez, R. B. (2004). Local Variation in Public Health Preparedness: Lessons from California. (from http://search.ebscohost.com.proxy.longwood.edu/). *Health Affairs, 23*, w341–w353. Retrieved May 15, 2010.

Lutrettung, D. R. F. (2011). *Besetzung der hubschrauber und ambulanzflugzeuge.* Retrieved from http://www.drf-luftrettung.de/rescuetrack.html

MacEachren, A. M., Gahegan, M., Pike, W., Brewer, I., Cai, G., Lengerich, E., & Hardisty, F. (2004). Geovisualization for knowledge construction and decision support. *IEEE Computer Graphics and Applications, 24*(1), 13–17. doi:10.1109/MCG.2004.1255801

Mahmoud, M., Liu, Y., Hartmann, H., Stewart, S., Wagener, T., & Semmens, D. (2009). A formal framework for scenario development in support of environmental decision-making. *Environmental Modelling & Software, 24*(7), 798–808. doi:10.1016/j.envsoft.2008.11.010

Maiyo, L., Kerle, N., & Köbben, B. (2010). Collaborative post-disaster mapping via geo web services. In Konecny, M., Ziatanova, S., & Bandrova, T. L. (Eds.), *Geographic information and cartography for risk and crisis management* (pp. 221–232). New York, NY: Springer. doi:10.1007/978-3-642-03442-8_15

Majchrzak, A., Jarvenpaa, S. L., & Hollingshead, A. B. (2007). Coordinating expertise among emergent groups responding to disasters. *Organization Science, 18*(1), 147–161. doi:10.1287/orsc.1060.0228

Majchrzak, T. A., Noack, O., Kuchen, H., Neuhaus, P., & Ückert, F. (2010). Towards a decision support system for the allocation of traumatized patients. In *Proceedings of the 7th International Conference on Information Systems for Crisis Response and Management.*

Manoj, B. S., & Hubenko Baker, A. (2007). Communication challenges in emergency response. *Communications of the ACM, 50*(3). doi:10.1145/1226736.1226765

March, S. T., & Smith, G. F. (1995). Design and natural science research on information technology. *Decision Support Systems, 15*(4), 251–266. doi:10.1016/0167-9236(94)00041-2

Marsden, P., & Lin, N. (Eds.). (1982). *Social structure and network analysis.* Thousand Oaks, CA: Sage.

Marshall, C. C., & Shipman, F. M. (1995). Spatial hypertext: Designing for change. *Communications of the ACM, 38*(8), 88–97. doi:10.1145/208344.208350

Martin, T. E. (1990). The Ramstein airshow disaster. *Journal of the Royal Army Medical Corps, 136*(1), 19–26.

Mathes, A. (2004). Folksonomies: Cooperative classification and communication through shared metadata. [University of Illinois Urbana-Champaign.]. *Urbana (Caracas, Venezuela), IL.*

May, A. (2006). *First informers in the disaster zone: The lessons of Katrina.* Washington, DC: The Aspen Institute.

Mayer-Schoenberger, V. (2002). *Emergency communications: The quest for interoperability in the United States and Europe.* Boston, MA: John F. Kennedy School of Government Harvard University.

McCarter, M. (2008). A New Standard for Campus Security. *Homeland Security Today.* Retrieved February 25, 2010, from http://hstoday.us

McGuigan, D. M. (2002). *Urban Search and Rescue and the Role of the Engineer.* Unpublished master's thesis, University of Canterbury, New Zealand.

Mecella, M. (2008). Adaptive Process Management. Issues and (Some) Solutions. In *Proceedings of the 2008 IEEE 17th Workshop on Enabling Technologies: Infrastructure for Collaborative Enterprises* (pp. 227-228). Washington, DC: IEEE Computer Society.

Mehrotra, S. (2007, May 20-23). Information technologies for improved situational awareness. In *Proceedings of the Digital Government Research Conference*, Philadelphia, PA (p. 333).

Mendonca, D., Jefferson, T., & Harrald, J. (2007). Collaborative adhocracies and mix-and-match technologies in emergency management. *Communications of the ACM, 50*(3), 44–49. doi:10.1145/1226736.1226764

Mendonça, D., & Wallace, W. A. (2004). Studying organizationally-situated improvisation in response to extreme events. *International Journal of Mass Emergencies and Disasters, 22*(2), 5–29.

Mendonça, D., & Wallace, W. A. (2007). A cognitive model of improvisation in emergency management. *IEEE Transactions on Systems, Man, and Cybernetics: Part A.*

Microsoft. (2010). *Microsoft Unveils New Government Cloud Offerings at Eighth Annual Public Sector CIO Summit.* Retrieved September 24, 2010, from http://www.microsoft.com/presspass/press/2010/feb10/02-24ciosummitpr.mspx

Mika, P. (2007). Ontologies are us: A unified model of social networks and semantics. *Web Semantics*, *5*(1), 5–15. doi:10.1016/j.websem.2006.11.002

Milanovic, N., Malek, M., Davidson, A., & Milutinovic, V. (2004). Routing and security in mobile ad hoc networks. *Computer*, *37*(2), 61–65. doi:10.1109/MC.2004.1266297

Miles, M. B., & Huberman, M. A. (1994). *Qualitative data analysis.* Thousand Oaks, CA: Sage.

Mileti, D. S., & Darlington, J. D. (1997). The role of searching in shaping reactions to earthquake risk information. *Social Problems*, *44*(1), 89–103. doi:10.1525/sp.1997.44.1.03x0214f

Milis, K., & van de Walle, B. (2007). IT for corporate crisis management: Findings from a survey in 6 different industries on management attention, intention and actual use. In *Proceedings of the 40th Annual Hawaii International Conference on System Sciences* (p. 24).

Mizuno, Y. (2001). *Collaborative Environments for Disaster Relief.* Unpublished master's thesis, Department of Civil & Environmental Engineering, Massachusetts Institute of Technology, Cambridge, MA.

Montibeller, G., & Belton, V. (2006). Causal maps and the evaluation of decision options - A review. *The Journal of the Operational Research Society*, *57*, 779–791. doi:10.1057/palgrave.jors.2602214

Montibeller, G., Gummer, H., & Tumidei, D. (2006). Combining scenario planning and multi-criteria decision analysis in practice. *Journal of Multi-Criteria Decision Analysis*, *14*(1-3), 5–20. doi:10.1002/mcda.403

Moradian, E. (2010). Integrating web services and intelligent agents in supply chain for securing sensitive messages. In I. Lovrek, R. J. Howlett, & L. C. Jain (Eds.), *Proceedings of the 12th International Conference on Knowledge-based Intelligent Information and Engineering Systems* (LNCS 5179, pp. 771-778).

Moxley, E., Kleban, J., & Manjunath, B. S. (2008). Spirittagger: A geo-aware tag suggestion tool mined from flickr. In *Proceedings of the 1st ACM International Conference on Multimedia Information Retrieval* (pp. 24-30).

Muller, R., Greiner, U., & Rahm, E. (2004). AGENT-WORK: a workflow system supporting rule-based workflow adaptation. *Data & Knowledge Engineering*, *51*, 223–256. doi:10.1016/j.datak.2004.03.010

Murhen, W., Van Den Eede, G., & Van de Walle, B. (2008). Sensemaking as a methodology for ISCRAM research: Information processing in an ongoing crisis. In *Proceedings of the 5th International ISCRAM Conference*, Washington, DC.

Murphy, T. (2010). *Transparencygate!* Retrieved from http://www.huffingtonpost.com/tom-murphy/transparencygate_b_695382.html

Murphy, T., & Jennex, M. E. (2006). Knowledge management, emergency response, and Hurricane Katrina. *International Journal of Intelligent Control and Systems*, *11*(4), 199–208.

Murphy, V. (2005, October 4). Fixing New Orleans' Thin Grey Line. *BBC News.* Retrieved October 3, 2010, from http://news.bbc.co.uk/2/hi/4307972.stm

Murthy, S., Maier, D., Delcambre, L., & Bowers, S. (2004). Putting integrated information into context: Superimposing conceptual models with SPARCE. In *Proceedings of the First Asia-Pacific Conference of Conceptual Modeling*, Denedin, New Zealand (pp. 71-80).

Nathan, M. L. (2004). How past becomes prologue: A sensemaking interpretation of the hindsight-foresight relationship given the circumstances of crisis. *Futures*, *36*, 181–199. doi:10.1016/S0016-3287(03)00149-6

National Commission on Terrorist Attacks Upon the United States. (2004). *9/11 Commission Report.* Retrieved from http://govinfo.library.unt.edu/911/report/911Report.pdf

National Research Council. (2007). *Improving disaster management: The role of IT in mitigation, preparedness, response and recovery.* Washington, DC: National Academies Press.

National Security Resources Board (NSRB). (1950). *Medical Aspects of Atomic Weapons.* Washington, DC: U.S. Government Printing Office.

National Security Resources Board (NSRB). (1950). *United States Civil Defense*. Washington, DC: U.S. Government Printing Office.

Neuhaus, P., Noack, O., Majchrzak, T., & Ückert, F. (2010). Using a business rule management system to improve disposition of traumatized patients. *Studies in Health Technology and Informatics, 160*(1), 759–763.

New York Magazine. (2002, September 11). *9/11 by the Numbers*. Retrieved September 20, 2010, from http://nymag.com/news/articles/wtc/1year/numbers.htm

News, B. B. C. (2001). *UK offers help to quake victims*. Retrieved from http://news.bbc.co.uk/1/hi/uk/1139081.stm

News, B. B. C. (2004, December 31). *UN Urges 'Special' Wave Response*. Retrieved September 28, 2010, from http://news.bbc.co.uk/2/hi/asia-pacific/4136153.stm

News, B. B. C. (2005, January 27). *Tsunami Aid: Who's Giving What*. Retrieved September 27, 2010, from http://news.bbc.co.uk/2/hi/asia-pacific/4145259.stm

News, B. B. C. (2006). *Tsunami disaster*. Retrieved from http://news.bbc.co.uk/1/hi/in_depth/world/2004/asia_quake_disaster/default.stm

News, B. B. C. (2010). *Haiti earthquake*. Retrieved from http://news.bbc.co.uk/1/hi/in_depth/americas/2010/haiti_earthquake/default.stm

Newsam, S., & Yang, Y. (2008). Integrating gazeteers and remote sensed imagery. *GIS, 26*.

Nonaka, I., & Takeuchi, H. (1995). *The knowledge-creating company: How Japanese companies create the dynamics of innovation*. Oxford, UK: Oxford University Press.

Nsiah-Kumi, P. A. (2008). Communicating Effectively With Vulnerable Populations During Water Contamination Events. (from http://search.ebscohost.com.proxy.longwood.edu). *Journal of Water and Health, 6*, 63–75. Retrieved January 20, 2011. doi:10.2166/wh.2008.041

Oestern, H. J., Huels, B., Quirini, W., & Pohlemann, T. (2000). Facts about the disaster at Eschede. *Journal of Orthopaedic Trauma, 14*(4), 287–290. doi:10.1097/00005131-200005000-00011

Office of the Director of National Intelligence. (2008). *Information Sharing Strategy*. Retrieved September 10, 2010, from http://www.dni.gov/reports/IC_Information_Sharing_Strategy.pdf

Olara, S. (2009). *Critique: Who will police the United Nations?* Retrieved from http://www.blackstarnews.com/?c=135&a=5536

Orlikowski, W. J. (2007). Sociomaterial practices: Exploring technology at work. *Organization Studies, 28*(9), 1435–1448. doi:10.1177/0170840607081138

Owen, J. (2005, July 11). London bombing pictures mark new role for camera phones. *National Geographic News*.

Pacific North West Economic Region. (2006). *BLUE CASCADES III: Critical infrastructure interdependencies exercise: Managing extreme disasters*. Bellevue, WA: Pacific North West Economic Region.

Paddock, R. C., & Magnier, R. C. (2004, December 30). Tsunami relief efforts mired in chaos. *Los Angeles Times*.

Palen, L., Anderson, K. M., Mark, G., Martin, J., Sicker, D., Palmer, M., & Grunwald, D. (2010, April 13-16). A vision for technology-mediated support for public participation & assistance in mass emergencies & disasters. In *Proceedings of the ACM-BCS Conference on Visions of Computer Science*, Edinburgh, UK (pp. 1-12). Retrieved September 11, 2010, from http://www.cs.colorado.edu/~palen/computingvisionspaper.pdf

Palen, L., & Liu, S. (2007). Citizen communications in disaster: Anticipating a future of ICT-supported public participation. In *Proceedings of the SIGCHI Conference on Human Factors in Computing Systems* (pp. 727-736).

Palen, L., & Vieweg, (2008). The emergence of online widescale interaction in unexpected events: Assistance, alliance & retreat. In *Proceedings of the ACM Conference on Computer Supported Cooperative Work* (pp. 117-126).

Palen, L., Vieweg, S., Liu, S., & Hughes, A. (2009). Crisis in a networked world: Features of computer-mediated communication in the April 16, 200 Virginia Tech event. *Social Science Computer Review, 27*(5), 1–14.

Panigrahy, R. (2008). An improved algorithm finding nearest neighbor using kd-trees. In *Proceedings of the 8th Latin American Symposium on Theoretical Informatics* (pp. 387-398).

Papamichail, K. N., & French, S. (2005). Design and evaluation of an intelligent decision support system for nuclear emergencies. *Decision Support Systems, 41*(1), 84–111. doi:10.1016/j.dss.2004.04.014

Patricelli, F., Beakley, J. E., Carnevale, A., Tarabochia, M., & von Lubitz, D. (2009). Disaster Management and Mitigation: The Telecommunications Infrastructure. [from http://search.ebscohost.com.proxy.longwood.edu]. *Disasters, 33*(1), 23–37. Retrieved January 20, 2011. doi:10.1111/j.1467-7717.2008.01060.x

Patton, D., & Flin, R. (1999). Disaster stress: An emergency management perspective. *Disaster Prevention and Management, 8*(4), 261–267. doi:10.1108/09653569910283897

Pavlin, G., Kamermans, M., & Scafes, M. (2010). Dynamic process integration framework: Toward efficient information processing in complex distributed systems. *Informatica, 34*, 477–490.

Payne, T., Singh, R., & Sycara, K. (2002) Communicating agents in open multi-agent systems. In *Proceedings of the First GSFC/JPL Workshop on Radical Agent Concepts*.

Pea, R. D. (1993). Practices of distributed intelligence and designs for education. In Salomon, G. (Ed.), *Distributed cognition: Psychological and educational considerations*. Cambridge, UK: Cambridge University Press.

Peña, A., Sossa, H., & Gutiérrez, A. (2008). Causal knowledge and reasoning by cognitive maps: Pursuing a holistic approach. *Expert Systems with Applications, 35*(1-2), 2–18. doi:10.1016/j.eswa.2007.06.016

Peña-Mora, F., Chen, A. Y., Aziz, Z., Soibelman, L., Liu, L. Y., & El-Rayes, K. (2010). A Mobile Ad hoc Network Enabled Collaborative Framework Supporting Civil Engineering Emergency Response Operations. *Journal of Computing in Civil Engineering, 24*(3), 302–312. doi:10.1061/(ASCE)CP.1943-5487.0000033

Peña-Mora, P., Aziz, Z., Chen, A. Y., Plans, A., & Foltz, S. (2008). Building Assessment during Disaster Response and Recovery. *Urban Design and Planning, 161*(4), 183–195. doi:10.1680/udap.2008.161.4.183

Pentland, B., & Feldman, M. (2007). Narrative networks: Patterns of technology and organization. *Organization Science, 18*(5), 781–795. doi:10.1287/orsc.1070.0283

Perry, R. W., & Lindell, M. K. (2003). Preparedness for Emergency Response: Guidelines for the Emergency Planning Process. [from http://search.ebscohost.com.proxy.longwood.edu]. *Disasters, 27*(4), 336–350. Retrieved February 21, 2010. doi:10.1111/j.0361-3666.2003.00237.x

Peskin, D., & Nachison, A. (2005). *We Media 2.0: Landfall synapse*. Retrieved from http://www.mediacenter.org/synapse/wemedia20_synapse_screen.pdf

Phelps, T., & Wilensky, R. (2000). Multivalent documents. *Communications of the ACM, 43*(6), 83–90. doi:10.1145/336460.336480

Pickrell, J. (2005). *Facts and figures: Asian tsunami disaster*. Retrieved from http://environment.newscientist.com/channel/earth/tsunami/dn9931-facts-and-figures-asian-tsunami-disaster-html

Pierre, R. E., & Gerhart, A. (2005, October 5). News of Pandemonium May Have Slowed Aid. *Washington Post*. Retrieved October 1, 2010, from http://www.washingtonpost.com/wp-dyn/content/article/2005/10/04/AR2005100401525.html

Plangprasopchok, A., & Lerman, K. (2009). Constructing folksonomies from user-specified relations on flickr. In *Proceedings of the International World Wide Web Conference*.

Pohl-Meuthen, U., Koch, B., & Kuschinsky, B. (1999). Rettungsdienst in der Europäischen Union. *Notfall & Rettungsmedizin, 2*(7), 442–450. doi:10.1007/s100490050175

Pryss, R., Tiedeken, J., & Reichert, M. (2010). Managing Processes on Mobile Devices: The MARPLE Approach. In *Proceedings of the CAiSE'10 FORUM*.

Quarantelli, E. L. (1997). Ten Criteria for Evaluating the Management of Community Disasters. [from http://search.ebscohost.com.proxy.longwood.edu]. *Disasters, 21*(1), 39–56. Retrieved February 21, 2010. doi:10.1111/1467-7717.00043

Ram, C., Montibeller, G., & Morton, A. (2010). Extending the use of scenario planning and MCDA for the evaluation of strategic options. *The Journal of the Operational Research Society, 62*, 817–829. doi:10.1057/jors.2010.90

Raman, M., & Jennex, M. E. (2010). Knowledge management systems for emergency preparedness: The way forward. *Journal of Information Technology Case and Application Research, 12*(3), 1–11.

Raman, M., Ryan, T., Jennex, M. E., & Olfman, L. (2010). Wiki technology and emergency response: An action research study. *International Journal of Information Systems for Crisis Response and Management, 2*(1), 49–69. doi:10.4018/jiscrm.2010120405

Rappaport, E. (1993). *Preliminary Report for Hurricane Andrew*. Miami, FL: National Hurricane Center. Retrieved October 1, 2010, from http://www.nhc.noaa.gov/1992andrew.html

Rashmi, S. (2005). *A cognitive analysis of tagging*. Retrieved from http://rashmisinha.com/2005/09/27/a-cognitive-analysis-of-tagging/

Rattenbury, T., & Naaman, M. (2009). Methods for extracting place semantics from Flickr tags. *ACM Transactions on the Web, 3*(1), 1–30. doi:10.1145/1462148.1462149

Rayport, J., & Heyward, A. (2009). Envisioning the cloud: The next computing paradigm. *Marketspace*. Retrieved September 20, 2010, from http://www. marketspaceadvisory.com/cloud/ Envisioning-the-Cloud.pdf

Redlener, I., & Berman, D. A. (2006). National Preparedness Planning: The Historical Context and Current State of the U.S. Public's Readiness, 1940-2005. [from http://search.ebscohost.com.proxy.longwood.edu]. *Journal of International Affairs, 59*(2), 81–103. Retrieved February 12, 2010.

Redlener, I., Grant, R., Abramson, D., & Johnson, D. (2008). *Annual Survey of the American Public by the National Center for Disaster Preparedness*. New York, NY: National Center for Disaster Preparedness, Columbia University Mailman School of Public Health and The Children's Health Fund. Retrieved May 23, 2010, from http://www.ncdp.mailman.columbia.edu/files/white_paper_9_08.pdf

Reese, G. (2000). Distributed Application Architecture. In *Database Programming with JDBC and Java* (2nd ed.). Sebastopol, CA: O'Reilly Media. Retrieved October 1, 2010, from http://java.sun.com/developer/Books/jdbc/ch07.pdf

Regan, H. M., Ben-Haim, Y., Langford, B., Wilson, W. G., Lundberg, P., & Andelman, S. J. (2005). Robust decision-making under severe uncertainty for conservation management. *Ecological Applications, 15*(4), 1471–1477. doi:10.1890/03-5419

Reynolds, B., & Seeger, M. (2005). Crisis and emergency risk communication as an integrative model. *Journal of Health Communication, 10*(1), 43–55. doi:10.1080/10810730590904571

Rheingold, H. (2002). *Smart mobs: The next social revolution transforming cultures and communities in the age of instant access*. Cambridge, MA: Basic Books.

Rhome, R., Brown, J., & Brown, D. P. (2005). *Tropical Cyclone Report: Hurricane Katrina: 23–30 August 2005*. Miami, FL: National Hurricane Center. Retrieved September 20, 2010, from http://w5jgv.com/downloads/Katrina/TCR-AL122005_Katrina.pdf

Ross, R. G. (2003). *Principles of the business rule approach*. Reading, MA: Addison-Wesley.

Russell, S., & Norvig, P. (2003). *Artificial intelligence*. Upper Saddle River, NJ: Prentice Hall.

Sanderson, M., & Croft, B. (1999). Deriving concept hierarchies from text. In *Proceedings of the 22nd Annual International ACM SIGIR Conference on Research and Development in Information Retrieval* (pp. 206-213).

Satyanarayanan, M. (1996). Fundamental challenges in mobile computing. In *Proceedings of the 15th Annual ACM Symposium on Principles of Distributed Computing* (pp. 1-7).

Scaffidi, C., Myers, B., & Shaw, M. (2007). Trial by water: Creating hurricane Katrina "person locator" web sites. In Weisband, S. (Ed.), *Leadership at a distance: Research in technologically-supported work* (pp. 209–222). Mahwah, NJ: Lawrence Erlbaum.

Schmitz, P. (2006) Inducing ontology from flickr tags. In *Proceedings of the Collaborative Web Tagging Workshop*.

Schnaars, S. (1987). How to develop and use scenarios. *Long Range Planning, 20*(1), 105–114. doi:10.1016/0024-6301(87)90038-0

Schneider, S., & Foot, K. (2004). Crisis communication & new media: The Web after September 11. In Howard, P., & Jones, S. (Eds.), *Society online: The Internet in context* (pp. 137–154). Thousand Oaks, CA: Sage.

Schoemaker, P. (1993). Multiple scenario development: Its conceptual and behavioral foundation. *Strategic Management Journal, 14*(3), 193–213. doi:10.1002/smj.4250140304

Scholz, R., & Tietje, O. (Eds.). (2002). *Embedded case study methods. Integrating quantitative and qualitative knowledge*. Thousand Oaks, CA: Sage.

Schooley, B., Marich, M., & Horan, T. (2007, May 20-23). Devising an architecture for time-critical information services: Inter-organizational performance data components for emergency medical services. In *Proceedings of the Digital Government Conference*, Philadelphia, PA (pp. 164-172).

Schwabe, D., & Rossi, G. (1995). The object-oriented hypermedia design model. *Communications of the ACM*, *38*(8), 45–46. doi:10.1145/208344.208354

Scott, P., & Rogova, G. (2004, June 28-July 1) Crisis management in a data fusion synthetic task environment. In *Proceedings of the 7th Conference on Multisource Information Fusion*, Stockholm, Sweden (pp. 330-337).

Shah, A. (2002). *Relief, rehabilitation and development: The case of Gujarat*. Retrieved from http://www.jha.ac/articles/a097.pdf

Shah, A. (2005). *Asian Earthquake and Tsunami Disaster*. Retrieved September 27, 2010, from http://www.global-issues.org/article/523/asian-earthquake-and-tsunami-disaster#Greatesteverpeacetimereliefoperationunderway

Sharifzadeh, M., Shahabi, C., & Knoblock, C. A. (2003). Approximate thematic maps from labeled geospatial data. In *Proceedings of International Workshop on Next Generation Geospatial Information*.

Shaw, J., Mulligan, M., Nadarajah, Y., Mercer, D., & Ahmed, I. (2010). *Lessons from tsunami recovery in Sri Lanka and India*. Melbourne, Australia: Monash University.

Sheetz, S., Kavanaugh, A., Quek, F., Kim, B. J., & Lu, S.-C. (2009, May). The expectation of connectedness and cell phone use in crises. In *Proceedings of the 6th International Conference on Information Systems for Crisis Response and Management*, Gothenburg, Sweden.

Shklovski, I., Burke, M., Kiesler, S., & Kraut, R. (2010). Technology adoption and use in the aftermath of hurricane Katrina in New Orleans. *The American Behavioral Scientist*, *53*(8), 1228–1246. doi:10.1177/0002764209356252

Shklovski, I., Palen, L., & Sutton, J. (2008). Finding community through information and communication technology in disaster events. In *Proceedings of the ACM Conference on Computer Supported Cooperative Work* (pp. 127-136).

Simon, H. A. (1996). *The sciences of the artificial* (3rd ed.). Cambridge, MA: MIT Press.

Siponen, M. T. (2005). Analysis of modern IS security development approaches: Towards the next generation of social and adaptable ISS methods. *Information and Organization*, *15*(4), 339–375. doi:10.1016/j.infoandorg.2004.11.001

Siponen, M. T., & Oinas-Kukkonen, H. (2007). A review of information security issues and respective research contributions. *The Data Base for Advances in Information Systems*, *38*(1), 60–80.

Smirnov, A., Kashevnik, A., Levashova, T. Pashkin, M., & Shilov, N. (2007). Situation modeling in decision support systems. In *Proceedings of the International Conference on Integration of Knowledge Intensive Multi-Agent Systems: Modeling, Evolution and Engineering* (pp. 34-39).

Smirnov, A., Levashova, T., Krizhanovsky, A., Shilov, N., & Kashevnik, A. (2009). Self-organizing resource network for traffic accident response. In *Proceedings of the International Conference on Information Systems for Crisis Response and Management*.

Smirnov, A., Pashkin, M., Chilov, N., Levashova, T., Krizhanovsky, A., & Kashevnik, A. (2005). Ontology-based users and requests clustering in customer service management system. In V. Gorodetsky, J. Liu, & V. Skormin (Eds.), *Proceedings of the International Workshop on Autonomous Intelligent Systems: Agents and Data Mining* (LNCS 3505, pp. 231-246).

Solis, C., & Ali, N. (2008). ShyWiki-A spatial hypertext Wiki. In *Proceedings of the International Symposium on Wikis*, Porto, Portugal (p. 10).

Solis, C., Canós, J. H., Penadés, M. C., & Llavador, M. (2006). A model-driven hypermedia development method. In *Proceedings of the World Wide Web Internet Conference* (pp. 321-328).

Solomon, J., & Higgins, A. (2005, January 24). Indonesia reviews claims of Graft in tsunami relief. *Wall Street Journal*, p. A16.

Sood, R., Stockdale, G., & Rogers, E. M. (1987). How the news media operate in natural disasters. *The Journal of Communication, 37*(3), 27–41. doi:10.1111/j.1460-2466.1987.tb00992.x

Sorenson, J., & Sorenson, B. (2006). Community processes: Warning and evacuation. In Rodríguez, H., Quarantelli, E. L., & Dynes, R. (Eds.), *Handbook of disaster research* (pp. 183–199). New York, NY: Springer.

Spitzer, M., Verst, H., Juhra, C., & Ückert, F. (2009). Trauma Network North-West - Improving holistic care for trauma patients by means of internet and mobile technologies. In *Proceedings of the European Congress for Medical Informatics* (pp. 371-375).

Stahl, B. C., Shaw, M., & Doherty, N. F. (2008). *Information systems security management: A critical research agenda.* Paper presented at the Association of Information Systems SIGSEC Workshop on Information Security and Privacy.

Starbird, K., & Palen, L. (2010). Pass it on? Retweeting in mass emergency. In *Proceedings of the International Conference on Information Systems for Crisis Response and Management*, Seattle, WA.

Starbird, K., & Palen, L. (2011). "Voluntweeters:" Self-organizing by digital volunteers in times of crisis. In *Proceedings of the ACM Conference on Computer Human Interaction*, Vancouver, BC, Canada.

Starbird, K., Palen, L., Hughes, A., & Vieweg, S. (2010). Chatter on *The Red*: What hazards threat reveals about the social life of microblogged information. In *Proceedings of the ACM Conference on Computer Supported Cooperative Work*, Savannah, GA (pp. 241-250).

Steinberg, L. (2006, May 19-22). E-Government and the preparation of citizens for natural disasters (National Science Foundation digital government grant # 429240). In *Proceedings of the Digital Government Conference*, San Diego, CA.

Sullum, J., Bailey, R., Taylor, J., Walker, J., Howley, K., & Kopel, D. B. (2005). *After the Storm Hurricane Katrina and the failure of public policy.* Retrieved from http://www.reason.com/news/show/36334.html

Sutton, J. (2009). Social media monitoring and the democratic national convention: New tasks and emergent processes. *Journal of Homeland Security and Emergency Management, 6*(1). doi:10.2202/1547-7355.1601

Sutton, J., Palen, L., & Shklovski, I. (2008). Backchannels on the front lines: Emergent use of social media in the 2007 southern California fires. In *Proceedings of the International Conference on Information Systems for Crisis Response and Management*, Washington DC.

Tabor, D. (2008, October 20). *LAFD's one-man geek squad brings Web 2.0 to firefighting.* Retrieved from http://www.wired.com/entertainment/theweb/magazine/16-11/st_firefight

Takeda, M., & Helms, M. (2006). Bureaucracy, meet catastrophe: Analysis of the tsunami disaster relief efforts and their implications for global emergency governance. *International Journal of Public Sector Management, 19*(2), 631–656. doi:10.1108/09513550610650446

TelaScience. (n.d.). *Haiti Earthquake.* Retrieved October 3, 2010, from http://hyperquad.telascience.org/haiti3/?zoom=13&lat=18.57027&lon=-72.32214&layers=BT

Templeton, D. E., Ellerman, G., & Branscome, T. (2009). Case Study: Radford University Overcomes Emergency Management Hurdles. *Campus Safety Magazine.* Retrieved February 21, 2010, from http://www.campussafetymagazine.com

Thomas, J. J., & Cook, K. A. (Eds.). (2005). *Illuminating the path: The research and development agenda for visual analytics.* Richland, WA: National Visualization and Analytics Center.

Thomsen, J., Vanrompay, Y., & Berbers, Y. (2009). Evolution of context-aware user profiles. In *Proceedings of the International Conference on Ultra Modern Telecommunications & Workshops* (pp. 1-6).

Torrey, C., Burke, M., Lee, M., Dey, A., Fussell, S., & Kiesler, S. (2007). Connected giving: Ordinary people coordinating disaster relief on the Internet. In *Proceedings of the 40th Annual Hawaii International Conference on System Sciences* (p. 179a).

Traumanetzwerk-Nordwest. (2011). *Startseite.* Retrieved from http://www.traumanetzwerk-nordwest.de/

Turban, E., Aronson, J. E., Liang, T., & Sharda, R. (2006). *Decision support and business intelligence systems* (8th ed.). Upper Saddle River, NJ: Prentice Hall.

Turban, E., & Watkins, P. (1986). Integrating expert systems and decision support systems. *Management Information Systems Quarterly, 10*(2), 121–136. doi:10.2307/249031

Turoff, M. (2002). On site: Past and future emergency response information systems. *Communications of the ACM, 45*(4), 29–32. doi:10.1145/505248.505265

Turoff, M., Chumer, M., Van de Walle, B., & Yao, X. (2004). The design of a dynamic emergency response management information system (Dermis). *Journal of Information Technology Theory and Application, 5*(4), 1–36.

Turoff, M., & Hiltz, S. R. (2009). The future of professional communities of practice. In C. Weinhardt, S. Luckner, & J. Stößer (Eds.), *Proceedings of the 7th Workshop on Designing E-business Systems, Markets, Services, and Networks* (LNBI 22, pp. 144-158).

U.S. Department of Education. (2009). *Action Guide for Emergency Management at Institutions of Higher Education*. Retrieved February 2, 2010, from http://www.ed.gov/emergencyplan

United Nations. (UN). (2005). *National post tsunami lessons learned and best practices workshop*. Retrieved from http://www.un.or.th/pdf/6months-govidn-idn-01jun.pdf

United Nations. (UN). (2010). *The United Nations at a glance*. Retrieved from http://www.un.org/en/aboutun/index.shtml

United Nations Development Program (UNDP). (1994). *Human development report*. New York, NY: Oxford University Press.

United Nations Development Program (UNDP). (2005). *The millennium development goals report 2005*. New York, NY: United Nations Development Program.

US Army Corps of Engineers. (2008). *Urban Search & Rescue Structures Specialist Field Operations Guide* (4th ed.). Washington, DC: Author.

van de Ven, J., van Rijk, R., Essens, P., & Frinking, E. (2008). Network centric operations in crisis management. In *Proceedings of the International Conference on Information Systems for Crisis Response and Management*, Washington, DC.

Van de Walle, B., & Turoff, M. (2007). Emergency response information systems: Emerging trends and technologies. *Communications of the ACM, 50*, 3.

van der Aalst, W. M. P. (1998). The Application of Petri Nets to Workflow Management. *Journal of Circuits. Systems and Computers, 8*(1), 21–66.

van der Aalst, W. M. P., & ter Hofstede, A. H. M. (2005). YAWL: Yet Another Workflow Language. *Information Systems, 30*(4), 245–275. doi:10.1016/j.is.2004.02.002

van der Pas, J., Walker, W., Marchau, V., Van Wee, G., & Agusdinata, D. (2010). Exploratory MCDA for handling deep uncertainties: the case of intelligent speed adaptation implementation. *Journal of Multi-Criteria Decision Analysis, 17*(1-2), 1–23. doi:10.1002/mcda.450

van Kamp, I., van der Velden, P. G., Stellato, R. K., Roorda, J., van Loon, J., & Kleber, R. J. (2006). Physical and mental health shortly after a disaster: First results from the Enschede firework disaster study. *European Journal of Public Health, 16*(3), 252–258. doi:10.1093/eurpub/cki188

van Notten, P. W. F., Rotmans, J., van Asselt, M. B. A., & Rothman, D. S. (2003). An updated scenario typology. *Futures, 35*(5), 423–443. doi:10.1016/S0016-3287(02)00090-3

Vetere, G., Faraotti, A., Poggi, A., & Salvatore, B. (2009). Information Management for Crisis Response in WORKPAD. In *Proceedings of the 6th International Conference on Information Systems for Crisis Response and Management.*

Vieweg, S., Palen, L., Liu, S., Hughers, A., & Sutton, J. (2008, May). Collective intelligence in disaster: Examination of the phenomenon in the aftermath of the 2007 Virginia Tech shooting. In *Proceedings of the 5th International Conference on Information Systems for Crisis Response and Management*, Washington DC.

Virginia State Crime Commission. (2006). *HJR 122 Final Report: Study on Campus Safety*. Retrieved December 15, 2009, from http://leg2.state.va.us/DLS

Vossen, G., & Hagemann, S. (2007). *Unleashing Web 2.0: From concepts to creativity*. San Francisco, CA: Morgan Kaufmann.

Walsham, G. (1997). Actor-network theory and IS research: Current status and future prospects. In Lee, A., Liebenau, J., & DeGross, J. (Eds.), *Information systems and qualitative research* (pp. 466–480). London, UK: Chapman Hall.

Wang, X., & Azman, L. (2008). IEEE 802.11s Wireless Mesh Networks: Framework and challenges. *Ad Hoc Networks, 6*(6), 970–984. doi:10.1016/j.adhoc.2007.09.003

Waugh, W. L., & Streib, G. (2006). Collaboration and leadership for effective emergency management. *Public Administration Review, 66*, 131–140. doi:10.1111/j.1540-6210.2006.00673.x

Wax, E. (2008, December 3). Gunmen used technology as a tactical tool. *Washington Post*. Retrieved October 10, 2010, from http://www.washingtonpost.com/wp-dyn/content/article /2008/12/02/AR2008120203519.html

Weaver, C. (2004, October). Building highly-coordinated visualizations in improvise. In *Proceedings of the IEEE Symposium on Information Visualization*, Austin, TX (pp. 159-166).

Weaver, C., Fyfe, D., Robinson, A., Holdsworth, D., Peuquet, D. J., & MacEachren, A. M. (2007). Visual exploration and analysis of historic hotel visits. *Information Visualization, 6*(1), 89–103.

Weiser, P. (2006). *Clearing the air: Convergence and the safety enterprise*. Washington, DC: The Aspen Institute.

Weiss, G. (2000). Introduction. In G. Weiss (Ed.), *Multiagent systems: A modern approach to distributed artificial intelligence*. Cambridge, MA: MIT Press.

Wellman, B. (1992). Which ties provide what kinds of support? *Advances in Group Processes, 9*, 207–235.

Wellnomics. (n.d.). *The Wellnomics® System Architecture*. Retrieved September 24, 2010, from http://wellanomics.com/pages/it_architecture.aspx

White, C., Plotnick, L., Kushma, J., Hiltz, S. R., & Turoff, M. (2009, May). An online social network for emergency management. In *Proceedings of the 6th International Conference on Information Systems for Crisis Response and Management*, Gothenburg, Sweden.

White, C. (2011). *Social media, crisis communications and emergency management: Leveraging Web 2.0 technology*. Boca Raton, FL: CRC Press.

Wickramasinghe, N., & Bali, R. (2008). Controlling chaos through the application of smart technologies and intelligent techniques. *International Journal of Risk Assessment and Management, 10*(1-2), 172–182. doi:10.1504/IJRAM.2008.021061

WindowSecurity. (2002). *Security: Secure Internet Data Transmission*. Retrieved October 1, 2010, from http://www.windowsecurity.com/whitepapers/security_secure_internet_data_transmission.html#WhatIsTransmissionSecurity

Winerman, L. (2009). Social Networking: Crisis communication. [from http://search.ebscohost.com.proxy.longwood.edu]. *Nature, 457*, 376–378. Retrieved January 20, 2011. doi:10.1038/457376a

Wireless and Mobile News. (2010). *WMN Exclusive: College Students Text Often & During All Sorts of Activities*. Retrieved January 20, 2011, from http://www.wirelessandmobilenews.com/2010/12/text-use-in-college-students.html

Wischgoll, T., & Meyer, J. (2007). *Earthquake visualization using large-scale ground motion and structural response simulations*. Retrieved from http://imaging.eng.uci.edu/~jmeyer/ PAPERS/c-25.pdf

Woltering, H. P., & Schneider, B. M. (2002). Das Unglück von Enschede am 13.05.2000. *Der Unfallchirurg, 105*(11), 961–967. doi:10.1007/s00113-002-0526-0

Word, I. Q. (2010). *Mainframe Computer*. Retrieved October 2, 2010, from http://www.wordiq.com/definition/Mainframe_computer

Wright, G., & Goodwin, P. (2009). Decision making and planning under low levels of predictability: Enhancing the scenario method. *International Journal of Forecasting, 25*(4), 813–825. doi:10.1016/j.ijforecast.2009.05.019

Yin, R. (1994). *Case study research: Design and methods* (2nd ed.). Thousand Oaks, CA: Sage.

Yun, J.-H. J., Park, S., & Avvari, M. V. (2011). Development and social diffusion of technological innovation. *Science, Technology & Society, 16*(2), 215–234. doi:10.1177/097172181001600205

About the Contributors

Murray E. Jennex is an Associate Professor at San Diego State University, Editor-in-Chief of the *International Journal of Knowledge Management*, Editor-in-Chief of IGI Global book series, Co-Editor in Chief of the *International Journal of Information Systems for Crisis Response and Management*, and President of the Foundation for Knowledge Management (LLC). Dr. Jennex specializes in knowledge management, system analysis and design, IS security, e-commerce, and organizational effectiveness. Dr. Jennex serves as the Knowledge Management Systems Track Co-Chair at the Hawaii International Conference on System Sciences. He is the author of over 100 journal articles, book chapters, and conference proceedings on knowledge management, end user computing, international information systems, organizational memory systems, e-commerce, security, and software outsourcing. He holds a BA in Chemistry and Physics from William Jewell College, a MBA and a MS in Software Engineering from National University, and a MS in Telecommunications Management and a PhD in Information Systems from the Claremont Graduate University. Dr. Jennex is also a registered professional mechanical engineer in the state of California and a Certified Information Systems Security Professional (CISSP).

* * *

Ignacio Aedo is a Full Professor in the computer science department at Universidad Carlos III de Madrid. He holds a Ph.D. from Universidad Politecnica de Madrid and has been working in the area of interactive systems since 1990. His main research interests are hypermedia, interactive systems in education, and Web systems and information systems for crisis situations. Since 2000 he collaborates with the Civil Protection Department of the Spanish Ministry of Interior in the development of emergency management information systems supporting interagency collaboration.

Rajeev K. Bali is a Reader in Healthcare Knowledge Management at Coventry University. His main research interests lie in clinical and healthcare knowledge management (from both technical and organisational perspectives). He founded and leads the Knowledge Management for Healthcare (KARMAH) research subgroup (working under BIOCORE). He is well published in peer-reviewed journals and conferences and has been invited internationally to deliver presentations and speeches. He serves on various editorial boards and conference committees and is the Associate Editor for the International Journal of Networking and Virtual Organisations as well as the International Journal of Biomedical Engineering and Technology.

John W. Barbrey, Ph.D. is an Assistant Professor of Criminal Justice Studies and is Director of Criminal Justice and Homeland Security Programs at Longwood University. His recent research involves the evolution of U.S. sentencing policies and the spatial analysis of crime. He is a member of the Southern Criminal Justice Association and the Academy of Criminal Justice Sciences. He is an academic advisor/member for two student organizations at Longwood: Alpha Phi Sigma (the Criminal Justice Honor Society) and Lambda Alpha Epsilon (a Criminal Justice Fraternity sponsored by the American Criminal Justice Association). Dr. Barbrey has a Master of Public Administration (MPA) from a Clemson University/University of S. Carolina joint program and a Ph.D. in Political Science (with subfields in policy process and judicial behavior) from the University of Tennessee, Knoxville.

Vikram Baskaran is an Assistant Professor in the Ted Rogers School of Information Technology Management at Ryerson University. He is an engineer by profession with an interest in biomedical computing. His research interest is on finding a viable application of the KM paradigm in healthcare. His special interest in developing HL7 messaging and healthcare informatics has provided opportunities to excel in these fields. His current activities overlap KM, e-health, AI and healthcare informatics. He has been previously engaged in a wide area of industrial projects (software and engineering at the senior level).

Marcos R. S. Borges is a Professor of Computer Science at the Federal University of Rio de Janeiro, Brazil. He earned his doctorate in Computer Science from the University of East Anglia, UK in 1986. From 1994 to 1996, he was a visiting research scholar and a member of the Object Technology Laboratory at Santa Clara University, California, USA. Dr. Borges has also served as Visiting Professor at University of Paris VI (2001) and the Polytechnic University of Valencia, Spain (2004–2005). He has published over a hundred research papers in international conferences and journals, including Decision Support Systems, Computers in Industry, Information Sciences and Expert Systems with Applications. His research interests include CSCW, Group Decision Support Systems, Resilience Engineering and Collective Knowledge. Since 2004 he has been working in the emergency management domain and has published the latest results on this topic in journals such as Journal of Decision Systems, Reliability Engineering and System Safety, Group Decision and Negotiation, and Journal of Loss Prevention in the Process Industries.

Manfred Bortenschlager is engineering manager at the Samsung Electronics Research Institute in Staines, UK. Before he has been working in applied research of mobile and location based services at Salzburg Research for over 7 years. His main research was in the domain of pervasive computing where he focused on coordination in pervasive environments. He was leading several national and international research projects. Bortenschlager holds a PhD in Computer Science and a MSc in Telecommunications Engineering.

Brian Brauer provides extensive expertise and guidance in Fire Fighting and emergency hands on experience to the research group. As Assistant Director of the IFSI, his current research embraces decision-making under stress, and adaptive teaching models. He is also coordinating the Illinois Homeland Security Research Center activities at IFSI.

George Bressler is responsible for the Leadership Agility Program at the U.S. Department of Homeland Security's Customs and Border Protection (CBP) Global Borders College. He is also an adjunct professor, and director of operational exercises for the homeland security master's program at San Diego State University. Dr. Bressler is the recipient of numerous awards to include the Federal 100 Award for successfully mobilizing representatives of 150 state, local, federal, tribal, academic, industry, and volunteer organizations for the anti-terrorism training exercise Operation Golden Phoenix in San Diego in July 2008. The mission of Operation Golden Phoenix was to demonstrate technologies and coordinated team response to threats as participating agencies together practiced command, communication and logistical skills for handling various disasters. Dr. Bressler is Director of X24, which successfully brought together over 12,772 people from 79 nations during X24 SoCal, and over 49,000 people from 92 nations during X24 Europe. The purpose of X24 is to develop collaborative resilient solutions to humanitarian assistance and disaster relief challenges. Dr. Bressler won his second Federal 100 Award for X24 SoCal. Dr. Bressler's present activities include working with the incredible team at the San Diego State Visualization Center, key leaders in Mexico, CBP Global Borders College, and US Northern Command on the next X24 event, X24 Mexico.

Bruce Campbell is faculty in the Web Design & Development and Young Artists programs at the Rhode Island School of Design and project director of the Watersheds Project with the Ocean Foundation in Washington, DC. While earning a Ph.D. in Systems Engineering from the University of Washington-Seattle, he spent twelve years as a research scientist with the Human Interface Technology Laboratory (HIT Lab), Center for Environmental Visualization (CEV), and PAcific Regional Visual Analytics Center (PARVAC) on campus. His research focuses on visualizing simulated natural and man-made processes in order to support training and planning for individual roles within team-based activities.

José H. Canós is an associate professor at the Department of Computer Science (DSIC) of the Universitat Politècnica de València, Spain, where he leads the Software Engineering and Information Systems Research Group. He holds a degree in Physics from the University of Valencia (1984) and a Ph.D. in Computer Science from the Universitat Politècnica de Valencia (1996). His current research interests are Digital Libraries, Document Engineering and Emergency Management Information Systems. He has participated in national, European and Iberoamerican research projects.

Tiziana Catarci received her PhD in Computer Science from the Sapienza-Università di Roma, where she is currently a full professor. She has published over 150 papers and 20 books in a variety of subjects comprising Database Access, Information Visualization, User Interfaces, Digital Libraries, Usability and Accessibility, Data Quality, Cooperative Information Systems, Data Integration, Web Access.

Albert Y. Chen is a Ph.D. Candidate in the Civil and Environmental Engineering Department at the University of Illinois at Urbana-Champaign. He has been working as a research assistant since August 2006 focusing on Computer-Aided Engineering with Geographic Information Systems, Artificial Intelligence, and Computer Networks for resource distribution in disaster response operations. Chen is interested in improving systems dynamics in disaster response to facilitate lifesaving operations.

Tina Comes studied mathematics, literature and philosophy at the Universität Trier, Germany, Université Lille I, France and the the Friedrich-Alexander Universität Erlangen-Nürnberg, Germany. She holds a diploma in mathematics from the Friedrich-Alexander Universität Erlangen-Nürnberg. At present, she is head of the interdisciplinary research unit on 'risk management' at the Institute for Industrial Production (IIP) at Karlsruhe Institute of Technology (KIT). Her major research areas are risk management, multi-criteria decision analysis, scenario analysis and intelligent distributed reasoning systems.

Claudine Conrado obtained her PhD in 1993 in theoretical non-linear physics from the Niels Bohr Institute in Copenhagen, Denmark. Subsequently, she worked as a research associate at the Imperial College, London, UK, on the modelling of multi-component and multiphase systems and the assessment of their environmental impact. In 1999, she joined Philips Research in Eindhoven, The Netherlands, to work on adaptive algorithms for home systems and later, on privacy enhancing technologies for digital content distribution, electronic identification systems and healthcare applications. Currently she works at Thales Research and Technology in Delft, The Netherlands, which she joined in 2009. Her areas of interest include decision making and problem solving in complex systems and real world applications as well as distributed reasoning systems involving the collaboration between humans and machines.

Massimiliano de Leoni is a Postdoc at the Faculty of Mathematics and Computer Science at Technical University of Eindhoven. Previously, he was a Research Associate at the Department of Computer Engineering at SAPIENZA-Università di Roma, where he also earned a PhD degree in Computer Science. He was visitor of Business Process Management Group at Queensland University of Technology, Brisbane; he made also a short visit at the Intelligent Agents group at RMIT University in Melbourne. His research interests range from business process management with special focus on process visualization, flexibility and mining, to mobile information systems, wireless networking in service-oriented architectures and logic programming for agents. Many of the research outcomes have been specifically applied to the domain of emergency management.

Paloma Díaz is Full Professor in the Computer Science Department of Universidad Carlos III de Madrid and head of the Interactive Systems Laboratory (DEI). Her research interests include web engineering methods, interaction design, e-learning and emergency management systems. Since 2000 she collaborates with the Civil Protection Department of the Spanish Ministry of Interior in the development of emergency management information systems supporting interagency collaboration.

Stuart Foltz is an US&R Structures Specialist (StS) with 12 years of experience. During 9/11, he was deployed to the Pentagon and to the WTC. His training includes basics US&R, specialized StS training and participation in a NATO disaster response exercise in Uzbekistan. He brings a connection to the US&R community (Corps of Engineers and USACE US&R cadre) critical to maintaining the relevance of the research.

Eric Frost directs the SDSU Viz Center and co-directs the Homeland Security Master's Program, which includes about 130 Homeland Security practioners from many Federal agencies, state and local government, NGOs, industry, and governments. Much of the program is based on actual interaction with real operational training in the US-Mexico border region, as well as many other international Homeland

Security groups such as in Mexico, Central Asia, India, Africa, and Indonesia. Dr. Frost and his colleagues use many new technologies and protocols that are enhanced during exercises such as Strong Angel III (http://www.strongangel3.net/) for situational awareness for many challenges including H1N1 using tools such as http://www.geoplayer.com/gateways for Banda Aceh, Katrina, Indonesia, and Haiti disasters (http://hypercube.telascience.org/haiti). Frost and co-workers work with sensor networks, wireless and optical communication, data fusion, visualization, and decision support for first responders and humanitarian groups, especially crossing the civilian-military boundary, especially in unusual coalition areas such as Somalia, Afghanistan, India. Haiti, and Mexico including using Cloud Computing (http://www.inrelief.org) with Navy to impact Humanitarian assistance like Haiti and Mexico earthquakes.

Chuck Goehring is a lecturer in the School of Communication at San Diego State University. He received a BA from the University of California at San Diego (1993), a MA from San Diego State University (2003), and a PhD in Communication Studies, with an emphasis in Rhetoric and Public Advocacy, from the University of Iowa (2008). His interest areas include rhetorical criticism, the rhetoric of social movements, feminist theory, and visual rhetoric. His current research examines the discourse surrounding the Abu Ghraib photographs and its effect on political policy.

Dan Harnesk is Assistant Professor in division of Information System Science at Lulea University of Technology (LTU), Sweden. Dr. Harnesk main research interests are information security and crisis management. He has particular interest in socio-technical issues of information systems/security design. Dr. Harnesk is a member of the editorial board of the Journal of Information Systems Security. He obtained his doctoral degree in computer and systems science at LTU, introducing a concept for social, transaction and IT alignment in SME interfirm relationships. His teaching duties at the University currently consist of Master level courses in Strategic Management of Information Security and Scientific Methods.

Heidi Hartikainen started her PhD work when joining the Nordic Safety and Security (NSS) project in 2009. NSS is funded by the European Commission and has its focus on emergency responders and emergency management. In NSS, Heidi studies how information security is maintained between different emergency actuators and between command leaders from different emergency organizations. As a final result of Heidi's studies, an IT enabled training concept will be introduced to emergency response practitioners.

Brianna Terese Hertzler is an adjunct faculty member for the Homeland Security Master's Program at San Diego State University. She received her BA in Communication, with an emphasis in International Relations, from San Diego State University and a MS in Homeland Security from San Diego State University. Her current research investigates how catastrophic disasters create global opportunity for the expansion of human trafficking.

Michael Hiete studied environmental sciences and received a PhD in natural sciences at Technical University of Braunschweig. He is an assistant lecturer at the Institute for Industrial Production and French-German Institute for Environmental Research of Karlsruhe Institute of Technology (KIT) where he was head of the research groups 'technique assessment and risk management' and 'Sustainable Construction'. His major research areas include multi-criteria decision support, sustainable built environment, industrial ecology and industrial risk management.

Austin W. Howe is a graduate of San Diego State University with a MS in Homeland Security with emphasis in emergency management, irregular warfare, and public health. His research interests are in designing effective collaboration, exercise analysis and design, and cloud computing solutions. He holds a BA in global studies and a BS in political science from Arizona State University. His knowledge of emergency management stems from post Hurricane Katrina planning with FEMA, counterterrorism training with the Israeli military, tracking of cholera in Haiti, and disaster relief from the 7.2 magnitude earthquake in Mexicali, Mexico. He is working to become a certified emergency manager with the International Association of Emergency Managers and has an interest in obtaining a PhD in public health.

Aapo Immonen's work aims to formulate a knowledge-based conceptual model which aims to enhance healthcare professionals' experience and knowledge in post-crisis (and crisis prevention) situations from the public health point of view. Mr Immonen works as a researcher at the Crisis Management Centre, Finland as well as at the Finnish Emergency Service College. Before this he worked as a researcher at the University of Kuopio (Finland) where his research areas were privacy, confidentiality and data security issues as well as wireless data communication technologies supporting public health. He has extensive experience in the areas of project and development management after working for several years in these areas in the public and private sectors as well as in numerous EU-funded projects. He is also a practising paramedic.

Suradej Intagorn is a PhD student in Computer Science at the University of Southern California. He received his B.S. in Information Engineering at King Mongkut's Institute of Technology Ladkrabang, Thailand, where He worked in Artificial Intelligence for Robocup research in Robotic Research Group. He has received scholarship from Ministry of Science, Thailand in 2006. Currently, he is working at the Information Sciences Institute under the supervision of Professor Lerman in the Information Integration Research Group. His research project is about using machine learning to extract geo spatial Folksonomy from social network data. His main research interests are Geo Spatial, Data Mining, Machine Learning, Social Networks and Databases.

Alexey Kashevnik, PhD, received his M.Eng. and M.Econ. at St. Petersburg State Technical University, Russia, in 2004 and his PhD in computer science at SPIIRAS in 2008. He is a senior researcher at the Computer Aided Integrated Systems Laboratory in SPIIRAS. He has been involved in various European research programs and Russian projects in areas of knowledge management, decision support systems, decision support systems, Web-based systems, and smart environments. His current research is in areas of profiling, Web-based systems, decision support systems, intelligent logistics, pervasive computing, and smart environments. A. Kashevnik has published more than 70 research works in reviewed journals and proceedings of international conferences and books.

Andrea Kavanaugh is a Senior Research Scientist in the Department of Computer Science at Virginia Tech and the Associate Director of the interdisciplinary research center for Human-Computer Interaction (HCI). A Fulbright scholar and Cunningham Fellow, her research lies in the areas of social computing, communication behavior and effects, and development communication. For over a decade she has been leading research on the use and social impact of information and communication technology funded primarily by the National Science Foundation, including the Digital Government Program. She is the author or editor of three books; her research is also published in *American Behavioral Scientist, Interacting with Computers, Journal for Computer Mediated Communication, Computer Supported Cooperative Work,* and *The Information Society,* among others. Prior to joining the HCI Center in 2002, she served as Director of Research for the community computer network known as the Blacksburg Electronic Village (BEV) from its inception in 1993. She holds an MA from the Annenberg School for Communication, University of Pennsylvania, and a PhD in Environmental Design and Planning (with a focus on telecommunications) from Virginia Tech. She serves on the Board of the Digital Government Society of North America, and formerly served on the Board of the International Telecommunications Society (2002-08).

B. Joon Kim is an Assistant Professor in the Department of Public Policy at Indiana University-Purdue University Fort Wayne. His research has appeared in Administration & Society, International Journal of Technology, Knowledge and Society, Information, Communication and Society, and The Global Studies Journal. His current research focuses on the role of community groups in E-government and E-democracy; the social impact of information technology on citizen interaction and collaborative governance; and the multi-level analysis on civic and political participation. He holds an MA from Korea University, an MPA from California State University, Hayward (Now, East Bay), and a PhD in Public Administration and Public Affairs from Virginia Tech.

Mark Latonero is the research director at the Annenberg Center on Communication Leadership and Policy at the University of Southern California. Latonero also holds a faculty appointment at the California State University at Fullerton, was a post-doctoral research scholar at the London School of Economics, and received a PhD from USC Annenberg School for Communication. Dr. Latonero's research investigates emerging technology and social change – from the application of social media to crisis and emergency management to the intersection of ICTs and human rights. The Haynes Foundation awarded Latonero a Faculty Fellowship, which provided the funding for this study. He is also a member of the resiliency task force at the Center for National Policy.

Brian Lehaney is an Independent Consultant operating in the UK.

Kristina Lerman is a Project Leader at the Information Sciences Institute and holds a joint appointment as a Research Assistant Professor in the USC Viterbi School of Engineering's Computer Science Department. Her research focuses on applying network- and machine learning-based methods to problems in social computing.

Tatiana Levashova, PhD, received her M.Eng. at St. Petersburg State Electrical Engineering University in 1986 and her PhD in computer science at SPIIRAS in 2009. She is a senior researcher at the Computer Aided Integrated Systems Laboratory in SPIIRAS. She has been involved in various European research programs and Russian projects in areas of networks of small and medium size enterprises, decision support systems, knowledge management, ontology-based context management, and distributed systems. Her current research is in areas of context aware systems, smart environments, Web-based systems, and pervasive computing. T. Levashova has published more than 170 research works in journals and proceedings of international conferences and books.

Manuel Llavador is a Ph. D. Student of Software Engineering and Declarative Programming (Quality Mention, http://www.mec.es/univ/) at the Technical University of Valencia (Universidad Politécnica de Valencia, http://www.upv.es). He has held pre-doctoral fellowships founded by Microsoft Research Cambridge Ltd and Spanish Government since 2003. Currently he is member of the Software Enginnering and Information Systems Research Group (http://issi.dsic.upv.es) and Project Leader at the Technology Transfer Software Institute of Valencia (ITI, http://www.iti.es). The main objective of his Ph. D. work is the development of a framework to improve interoperability in heterogeneous systems combining syntax, structure and semantics of data exchanged. He also works in the specification and (semi)automatic generation of flexible Emergency Management Systems including resolution procedures, presentation and visualization of data, information management and retrieval, communications, collaboration between participants, intelligence, context, and other aspects of the management and resolution of emergencies.

Tim A. Majchrzak is a research associate at the Department of Information Systems of the University of Münster, Germany, and the European Research Center for Information Systems (ERCIS). He received BSc and MSc degrees in Information Systems and a PhD in economics (Dr. rer. pol.) from the University of Münster. His research comprises both technical and organizational aspects of software engineering. He has also published work on several other interdisciplinary IS topics, particularly on IT&Healthcare. Tim has presented his work on computer science symposia as well as on IS conferences. He is a member of the IEEE, the IEEE Computer Society, and the Gesellschaft für Informatik e.V.

Alessio Malizia is an Associate Professor at Universidad Carlos III de Madrid, Spain. He holds a degree in Computer Science and Ph.D. in Computer Science from University of Rome "La Sapienza", Italy. From 1999 to 2002 he worked for the Rome IBM Tivoli Laboratory, Silicon Graphics and the Reply group. During the Summer 2003, he has spent three months for an internship at the XEROX PARC (Palo Alto Research Center, Palo Alto, CA, USA) in the ISTL (Information Science and Technology) Lab working in Human Document Interaction and after he has hold a Visiting Researcher contract within the same group. Until February 2007, he was a Research Fellow at the Computer Science Department of University of Rome "La Sapienza". In the past his research activities focused on theory and algorithms for pattern recognition, machine learning and visualization. Today he is working on Human Computer Interaction, Pervasive Computing and Social Networking.

Russell Mann is a Graduate Associate with the Biomedical Computing and Engineering Technologies (BIOCORE) Applied Research Group at Coventry University, UK.

Andrea Marrella is a PhD student in the Department of Systems and Computer Science and Engineering at Sapienza–Universitá di Roma. His research interests include business process management, automatic adaptation in process management systems through non-classical planning techniques, software architectures, mobile systems, disaster/crisis response & management, healthcare, Human-Computer Interaction (HCI) and User-Centered Design (UCD) methodologies. Marrella has a master degree in Computer Science and Engineering from Sapienza–Universitá di Roma.

Ian M. Marshall is Deputy Vice-Chancellor (Academic) at Coventry University. Professor Marshall's applied research interests are focused on the use of games in education and training and on development effort estimation for multimedia and other interactive courseware. He has worked extensively as a computer and information systems consultant, flexible learning material developer and trainer. In addition, he has extensive experience working with small- to medium-sized enterprises as well as with city councils and regional development agencies.

Massimo Mecella is an assistant professor in the Department of Systems and Computer Science and Engineering at Sapienza–Universitá di Roma. His research interests include CISs (Cooperative Information Systems) and Internet-based ones, service oriented computing, process management, User-Centered Design (UCD), Human-Computer Interaction (HCI), software architectures, distributed technologies and middleware, security. The previous interests are challenged in the application domains of eGovernment and eBusiness, domotics, healthcare, disaster/crisis response & management. Mecella has a PhD in computer science and engineering from Sapienza–Universitá di Roma. He is a member of the ACM, IEEE and the IEEE Computer Society.

Saumil Mehta earned a master's degree in Information Technology from MIT and another master's degree in Construction Management at UIUC. His research focus is on ad hoc computer networks in disaster response.

Scott Nacheman's experience involves disaster response initiatives for local, state and federal agencies as well as the investigation and mitigation of building failures and collapses for Thornton Tomasetti, Inc. He has participated in the response operations at the World Trade Center following the attacks of September 11, 2001, Hurricane Katrina, Hurricane Rita and Hurricane Ike. Mr. Nacheman also currently chairs the National Council of Structural Engineers Association (NCSEA) Structural Engineers Emergency Response Committee (SEER Committee).

Raouf N. G. Naguib is Professor of Biomedical Computing and Head of BIOCORE. Prior to this appointment, he was a Lecturer at Newcastle University, UK. He has published over 240 journals and conference papers and reports in many aspects of biomedical and digital signal processing, image processing, AI and evolutionary computation in cancer research. He was awarded the Fulbright Cancer Fellowship in 1995–1996 when he carried out research at the University of Hawaii in Mãnoa, on the applications of artificial neural networks in breast cancer diagnosis and prognosis. He is a member of several national and international research committees and boards.

Philipp Neuhaus is a research associate at the Department of Medical Informatics of the University Hospital of Münster, Germany. After his studies in computer science and mathematics he got his diploma in computer science and started his PhD studies about GSM-based localization of buried people. For his work on this project he has won the Regional Price of North Rhine-Westphalia in the European Satellite Navigation Competition 2010. Besides this, Philipp developed a system to support the disposition of traumatized patients for fire brigades and emergency physicians to improve the communication between emergency physicians and available hospitals participating in the Trauma Network North-West.

Oliver Noack is currently working as a consultant for Saracus Consulting GmbH, Germany. He studied Information System at the University of Münster, Germany and at the University Polytécnica de Madrid, Spain. In 2009 he received his diploma degree. He specialized in economics and information technology. During his years of study Oliver worked for 3 years as a student research assistant at the Institute of Medical Informatics of the University Hospital of Münster and gained first experience in IT&Healthcare. He started to research possibilities for the assignment of emergencies to a corresponding transportation and to a medical facility; eventually, he wrote his diploma thesis in this field of research.

Teresa Onorati is a PhD candidate at the Computer Science Department of Universidad Carlos III de Madrid, Spain. Currently, she is working at the definition of a common knowledge base for Accessibility, Semantic and Communication in the domain of Emergency Management. In 2007, she obtained her degree in Computer Science from University of Rome "La Sapienza", Italy, and in 2008, a MSc in Computer Science and Technology at Universidad Carlos III de Madrid, Spain. Today, she is a member of the Interactive Systems Laboratory (DEI) in the same university. Her research interests are about Human-Computer Interaction, Semantic Web and Interaction Design.

Feniosky Peña-Mora is the Dean of the Fu Foundation School of Engineering and Applied Science, and Morris A. and Alma Schapiro Professor of Civil Engineering and Engineering Mechanics, Earth and Environmental Engineering, and Computer Science at Columbia University. He received his Sc.D. in Civil Engineering at MIT and his research interest includes collaborative disaster preparedness, response, and recovery, visualization of construction progress monitoring with 4D augmented reality, dynamic project management, collaborative environments, and dispute resolution. He has lead various external funded projects including (1) IT-Based Collaboration Framework for Disasters Preparedness, Response, and Recovery Involving Critical Physical Infrastructures (2) Dynamic Planning and Control Methodology for Large scale Concurrent Design and Construction Projects, and (3) Initiation of the Interaction Space Theory for Geographically Distributed Teams Involved in Large scale Engineering Projects and Disaster Relief.

M. Carmen Penadés is an associate professor at the Department of Computer Science (DSIC) of the Universitat Politècnica de València, Spain. She holds a degree in Computer Science (1994) and a Ph.D. in Computer Science from the Universitat Politècnica de València (2002). She is member of the Software Engineering and Information Systems Research Group. Her current research interests are Business Processes Management, Document Engineering and Emergency Management Information Systems. She has participated in national, European and Iberoamerican research projects.

Albert Plans is a program manager at Vistaprint and a PhD student and lecturer in industrial engineering at Universitat Politècnica de Catalunya, Spain. His research focus is on the design, development and test of new IT tools supporting civil engineers in disaster management and construction management.

John Puentes holds an Electronics Engineering degree, a Masters of Image Processing and Artificial Intelligence, a PhD in Signal Processing and Telecommunications, and the Habilitation to Supervise Research. He worked as engineer, consultant and project manager for biomedical and telecommunications companies, before moving to the Image and Information Processing Department at Telecom Bretagne,

Brest - France, where he is Associate Professor. He is an invited associate editor and reviewer for several international journals, conferences and organizations. His primary research interests are medical practice support systems and medical information processing.

Francis Quek is a Professor of Computer Science at Virginia Tech. He also directs the Vision Interfaces and Systems Laboratory. Francis received both his B.S.E. *summa cum laude* (1984) and M.S.E. (1984) in electrical engineering from the University of Michigan. He completed his Ph.D. C.S.E. at the same university in 1990. Francis is a member of the IEEE and ACM. He performs research in embodied interaction, multimodal verbal/non-verbal interaction, physical computing, interactive multimodal meeting analysis, vision-based interaction, multimedia databases, medical imaging, collaboration technology, human computer interaction, computer vision, and computer graphics. He leads several multiple-disciplinary research efforts to understand the communicative realities of multimodal interaction.

Alan C. Richards is a Senior Lecturer in Information Technology and Systems at Coventry University, UK. His main research areas of interest are business organisations, business strategy, knowledge management, crisis management, urban health and water resource management. He has twenty years experience with Severn Trent Water (UK). He is currently undertaking a PhD with BIOCORE at Coventry University, UK.

Alessandro Russo is a PhD student and research assistant at the Department of Systems and Computer Science and Engineering at Sapienza-Università di Roma. He received his master degree in Computer Science and Engineering from the same university in 2008. His research interests include process-aware information systems, service oriented software architectures, mobile information systems, User-Centered Design (UCD) and Human-Computer Interaction (HCI), mainly targeted to the disaster/crisis response & management and healthcare application domains.

Frank Schultmann is professor at the Karlsruhe Institute of Technology (KIT) and director of the Institute for Industrial Production (IIP) and the French-German Institute for Environmental Research (DFIU). In addition, he is Adjunct Professor at the University of Adelaide, Australia. He studied Business Engineering at the Universität Karlsruhe (TH) and received a PhD in economics from the faculty of economics and business engineering of the Universität Karlsruhe (TH). Previous to his present positions he was professor at the department of computer science at the University of Koblenz-Landau and holder of the chair of business administration, construction management and economics at the University of Siegen.

Steven D. Sheetz is Director of the Center for Global e-Commerce and Associate Professor at the Pamplin College of Business at Virginia Tech. He received his Ph.D. in Information Systems from the University of Colorado at Boulder. His research interests include the design and use of standards for information systems development, the diffusion of software measures, computer supported collaboration systems including group cognitive mapping software and social networking for crisis situations, and information systems to support response to and recovery from crises. He has published articles in *Decision Support Systems, International Journal of Human-Computer Studies, Journal of Management Information Systems, Journal of Systems and Software, and Object-Oriented Systems.* He also holds a MBA from the University of Northern Colorado and a B.S. in Computer Science from Texas Tech University. He has substantial industry experience in database design and OO systems development.

Nikolay Shilov, PhD, received his M.Econ. at St. Petersburg State Technical University, Russia, in 1998 and his Ph.D. in computer science at SPIIRAS in 2005. He is a senior researcher at the Computer Aided Integrated Systems Laboratory in SPIIRAS. He has been involved in various European research programs and Russian projects in areas of multi-agent systems, knowledge management, decision support systems, networked organization management, intelligent logistics, and smart environments. His current research is in areas of ontology-based context management, Web-based systems, decision support systems, pervasive computing, and smart environments. N. Shilov has published more than 180 research works in journals and proceedings of international conferences and books.

Irina Shklovski is assistant professor at the IT University, of Copenhagen. Her work concerns the way information technologies are used by emergency response professionals and affected public in response to disaster events and during recovery. She has collaborated broadly with colleagues from Cal State Fullerton, CU Boulder and Carnegie Mellon University, to analyze how use of information and communication technologies changes emergency management practices and public-side disaster response and recovery. Irina's work is highly interdisciplinary, oriented toward future design and development of relevant supporting technologies for disaster response activities. Previously, Irina was a postdoctoral researcher in Informatics at the University of California, Irvine. She earned her PhD from Carnegie Mellon University.

Alexander Smirnov, Prof., received his M.Eng., PhD and DSc. degrees at St. Petersburg, Russia, in 1979, 1984, and 1994 respectively. He is a deputy director for research and the head of Computer Aided Integrated Systems Laboratory at St. Petersburg Institute for Informatics and Automation of the Russian Academy of Sciences (SPIIRAS). He is a full professor at St. Petersburg State Electrical Engineering University. He has been involved in various European research programs and Russian projects in areas of knowledge management, distributed decision support systems, intelligent systems, and advanced business systems. His current research is in areas of context management, multi-agent systems, decision support systems, and pervasive computing. Prof. Smirnov has published more than 300 research works in journals and proceedings of international conferences, books, and manuals.

Carlos Solis received his PhD from Universidad Politecnica de Valencia, Spain in 2008. His research interest include hypermedia, computer supported collaborative work, and knowledge management. Since 2009, he is a research fellow in Lero, the Irish Software Engineering Research Centre.

Renate Steinmann is a researcher at Salzburg Research within the group mobile and web-based information systems. Her research interests are Geographic Information Systems with special focus on user-centered design and the performance of user tests especially in the domains emergency management and pedestrian navigation. Steinmann has a master degree from the University of Salzburg in Geography with special focus on Geoinformatics.

Frank Ückert, MD PhD, is working at the University Medical Center of the Johannes Gutenberg University Mainz as Professor for Medical Informatics and is head of the department for medical informatics. After studying medicine and mathematics at the Westphalian Wilhelms-University Muenster he worked as a physician at the clinic for children's oncology. In Muenster and at the University Hospital in

Erlangen he specialised in Medical Informatics early. Since 1995 he managed several medical IT-projects and established a spin-off IT-company in 2003. In June 2005 he was appointed as Juniorprofessor for Medical Informatics in Muenster. So far his work was awarded for example with the "Peter L. Reichertz Memorial Prize" 2002 of the European Federation for Medical Informatics and the "Gesundheitspreis" 2004 of the federal state North Rhine-Westphalia. In April 2009 he was elected as Board Member of the TMF e.V.

Adriana S. Vivacqua holds a DSc in Systems Engineering and Computer Science from the Federal University of Rio de Janeiro and Université de Technologie de Compiègne. She is currently a professor in the Department of Computer Science at the Federal University of Rio de Janeiro and a collaborator at the Active Design Documents research laboratory. Her work focuses on Human Computer Interaction, with an emphasis on computer-supported cooperative work, intelligent user interfaces and context in information retrieval. She has published a number of research papers on these topics.

Chris Weaver is an assistant professor in the School of Computer Science and associate director of the Center for Spatial Analysis at the University of Oklahoma. After earning a PhD in Computer Science from the University of Wisconsin-Madison, he spent three years as a post-doctoral research associate with the GeoVISTA Center in the Department of Geography at Penn State, where he helped to found the North-East Visualization and Analytics Center. His research focuses on forms of multidimensional interaction in information visualization, the design space of highly interactive tools for visual analysis, and applications of visual analysis as methodological infrastructure for scholarship in the digital humanities.

Nilmini Wickramasinghe received her PhD from Case Western Reserve University, USA and currently is the Epworth Chair in Health Information Management and a professor at RMIT University, Australia. She researches and teaches within the information systems domain with a special focus on IS/IT solutions to effect superior, patient centric healthcare delivery. She has collaborated with leading scholars at various premier healthcare organizations throughout US and Europe. She is well published with more than 200 referred scholarly articles, 10 books, numerous book chapters, an encyclopedia and a well established funded research track record.

Niek Wijngaards, received his PhD in 1999 on the topic of self-modifying agent systems using a re-design process. Since 1998 he worked at the University of Canada as a postdoctoral-fellow and at the VUA, where he was an assistant professor from 2000 to 2004 at the Intelligent Interactive Distributed Systems group. Since October 2004 he works for Thales Research and Technology Netherlands as senior researcher and program manager. He is fully employed at D-CIS Lab. Wijngaards is involved in research on actor-agent teams as well as their practical applications at e.g. the Dutch Police organisation and the Dutch Railroads.

Index